物联网与人工智能开发系列丛书

物联网移动软件开发

廖义奎　编著

北京航空航天大学出版社

内 容 简 介

本书以一个物联网移动软件系统(物联网智能应用软件系统)的开发为主线,根据该软件实际的开发过程,结合移动软件开发的知识结构,从最简单的欢迎界面开始,一步一步深入讲解移动软件开发的知识体系和相关技术,最终完成该物联网移动软件系统的开发工作。

全书共 13 章,包括物联网移动软件开发概要、欢迎界面设计、登录界面布局设计、界面切换设计、列表视图界面设计、导航栏及滑动界面设计、Wi-Fi 物联网移动软件设计、蓝牙物联网移动软件设计、数据库及动态界面设计、嵌入网页的控制界面设计、传感器应用及拍照更换界面图片设计、苹果手机移动软件设计、跨平台移动软件设计。本书配套资料包括所有章节的程序代码,读者可以从北京航空航天大学出版社(www.buaapress.com.cn)的"下载专区"免费下载。

本书适合于从事物联网应用开发、物联网软件开发、移动软件开发的工程开发人员作为参考资料使用,也可作为本、专科物联网应用、移动软件开发等相关课程的教材。

图书在版编目(CIP)数据

物联网移动软件开发 / 廖义奎编著. -- 北京 : 北京航空航天大学出版社,2019.4
 ISBN 978-7-5124-2978-9

Ⅰ. ①物… Ⅱ. ①廖… Ⅲ. ①移动终端—应用程序—程序设计 Ⅳ. ①TN929.53

中国版本图书馆 CIP 数据核字(2019)第 060469 号

版权所有,侵权必究。

物联网移动软件开发
廖义奎 编著
责任编辑 宋淑娟 曹春耀

*

北京航空航天大学出版社出版发行

北京市海淀区学院路 37 号(邮编 100191) http://www.buaapress.com.cn
发行部电话:(010)82317024 传真:(010)82328026
读者信箱: emsbook@buaacm.com.cn 邮购电话:(010)82316936
艺堂印刷(天津)有限公司印装 各地书店经销

*

开本:710×1 000 1/16 印张:25 字数:562 千字
2019 年 9 月第 1 版 2019 年 9 月第 1 次印刷 印数:3 000 册
ISBN 978-7-5124-2978-9 定价:79.00 元

若本书有倒页、脱页、缺页等印装质量问题,请与本社发行部联系调换。联系电话:(010)82317024

前　言

物联网移动软件主要是指运行于手机、平板电脑、智能电视机、物联网专用触摸屏（HMI，人机界面）等移动设备和智能设备上的物联网应用软件。物联网移动软件的开发目标，主要是设计可以运行于手机、平板电脑、智能电视机、专用物联网设备上的软件系统，包括界面设计和底层功能设计两部分。例如，物联网移动软件可以通过手机实现对智能家居设备的监测与控制，也可以实现对智慧农业、智慧办公室、智慧医疗等设备与系统的监测与控制。

物联网移动软件的工作原理是将手机、控制目标设备、服务器通过无线路由器连接在一起，组成一个局域网，手机就可以通过无线网络对控制目标设备进行检测和控制。服务器是物联网系统结构的中心，一般采用 TCP 服务器或者 Web 服务器。TCP 服务器包括本地电脑作为 TCP 服务器、片式电脑作为 TCP 服务器、TCP 云服务器等。

物联网应用软件种类繁多，包括物联网底层（感知层）软件、通信（网络层）软件、服务程序（服务层）软件、客户端（应用层）软件等，本书介绍的物联网移动软件设计，重点在于客户端（应用层）软件的开发。物联网客户端（应用层）软件的开发，主要是指运行于手机、平板电脑、智能电视机、物联网专用触摸屏（HMI，人机界面）、普通电脑上的面向物联网应用系统的软件开发。除了少部分面向普通电脑的应用外，绝大部分都属于移动设备的应用，所以本书的重点在于介绍物联网的移动软件设计。

物联网移动软件其中一个重要的应用平台是手机、平板等移动设备。随着智能移动设备在我国的快速普及，使得国内的移动软件开发市场得以崛起，大量的开发者纷纷投入到移动设备应用软件的开发行列，致使移动软件开发市场异常火爆。物联网移动软件开发的重点是面向手机的应用软件开发与服务，从目前手机移动软件开发市场的繁华景象来看，未来物联网移动软件开发的发展前途一片光明。

物联网移动软件另外一个重要的应用平台是智能设备。智能设备包括机器人、智能电视机、智能眼睛、智能手表、智能手环、智能音箱、智能机顶盒、智能路由器、物联网专用触摸屏（HMI，人机界面）等。在智能设备开发方面，物联网移动软件设计也具有非常广阔的应用前景。据知名市场研究机构 Gartne 给出的预测，到 2020 年全球物联网设备接入量将会达到 260 亿，物联网产品和服务提供商们的收益预计会达到 3 000 亿美元量级。对于智能终端硬件的开发，物联网软件设计是一个特别重要的组成部分，也可以预测到物联网软件设计也将会具有非常广阔的应用前景。

本书的特点

本书以一个物联网移动软件系统（物联网智能应用软件系统）的开发为主线，根据该软件实际的开发过程,结合移动软件开发的知识结构,从最简单的欢迎界面开始,一步一步深入讲解移动软件开发的知识体系和相关技术,最终完成该物联网移动软件系统的开发工作。本书每一章首先介绍设计目标,然后介绍要达成该设计目标所需要的基础知识,最后介绍设计目标的实现方法。

本书主要为物联网及其相关专业读者提供物联网应用系统软件开发的基础知识和开发技术,重点是物联网移动智能终端的软件开发。

本书的结构

全书共13章,包括物联网移动软件开发概要、欢迎界面设计、登录界面布局设计、界面切换设计、列表视图界面设计、导航栏及滑动界面设计、Wi-Fi物联网移动软件设计、蓝牙物联网移动软件设计、数据库及动态界面设计、嵌入网页的控制界面设计、传感器应用及拍照更换界面图片设计、苹果手机移动软件设计、跨平台移动软件设计。

读者对象

本书的读者需要具有一定的Java基础。本书适合于从事物联网应用开发、物联网软件开发、移动软件开发的工程开发人员作为参考资料使用,也可作为本、专科物联网应用、移动软件开发等相关课程的教材。

配套资料及互动方式

配套资料中包括了所有章节的程序源代码,读者可以直接下载下来使用,并仿照这些程序源代码去快速开发新的应用程序。对本书相关配套的STM32开发板、控制模块、传感器模块、通信模块等有兴趣的读者,以及对本书相关知识有兴趣的读者,可以加入QQ群AI_IoT(群号784735940)联系、讨论,以便共同学习、共同进步。

致　谢

在本书的编写过程中,李昶春做了大量的准备工作,并参与了前几章部分内容的资料收集与整理工作,在此表示衷心的感谢。感谢蒙良桥、宋因建、殷徐栋、陈妍、张小珍、覃雪原、官玉恒、韦政、林宝玲、苏小艳、苏金秀分别审阅了本书各章节的内容。

本书在编写过程中参考了大量的文献资料,一些资料来自互联网和一些非正式出版物,书后的参考文献无法一一列举,在此对原作者表示诚挚的谢意。

限于作者水平,并且编写时间比较仓促,书中难免存在错误和疏漏之处,敬请读者批评指正。

作　者
2019.1

目 录

第 1 章　物联网移动软件开发 … 1
1.1　物联网移动软件开发目标 … 1
1.2　物联网移动软件开发概要 … 6
1.2.1　物联网移动软件开发简介 … 6
1.2.2　物联网移动软件的应用 … 7
1.3　物联网智能硬件 App 设计 … 9
1.4　物联网移动软件开发的发展趋势 … 9
1.4.1　物联网移动软件的特点 … 9
1.4.2　物联网移动软件开发的发展 … 11
1.4.3　移动云计算 … 12

第 2 章　欢迎界面设计 … 14
2.1　欢迎界面设计目标 … 14
2.2　物联网移动软件设计基础 … 15
2.2.1　物联网移动软件开发的内容 … 15
2.2.2　安卓物联网移动软件设计基础 … 15
2.3　开发工具 … 17
2.3.1　开发工具的选择 … 17
2.3.2　Eclipse 安装与配置 … 17
2.3.3　Android Studio 安装与配置 … 23
2.3.4　Obtain_Studio 安装与配置 … 26
2.4　移动软件开发 Hello World 程序 … 27
2.4.1　如何启动 Obtain_Studio 集成开发环境 … 27
2.4.2　创建 Android 项目 … 29
2.4.3　编译和运行 … 33
2.4.4　Obtain_Studio 集成开发系统常用技巧 … 36
2.5　Android 项目 … 39
2.5.1　Android 项目结构 … 39
2.5.2　Android 项目文件 … 40
2.5.3　Android 项目编译与配置文件 … 43
2.5.4　Android 项目全局配置文件 … 44

2.5.5　Android 资源文件 ………………………………………… 47
2.6　欢迎界面的实现 ……………………………………………………… 48
　　2.6.1　创建项目和编辑文件 …………………………………… 48
　　2.6.2　运行欢迎界面 …………………………………………… 50

第3章　登录界面布局设计　54
3.1　登录界面布局设计目标 ……………………………………………… 54
3.2　安卓界面布局 ………………………………………………………… 55
　　3.2.1　界面布局文件 ……………………………………………… 55
　　3.2.2　线性布局 …………………………………………………… 57
　　3.2.3　相对布局 …………………………………………………… 61
　　3.2.4　帧布局 ……………………………………………………… 63
　　3.2.5　绝对布局 …………………………………………………… 64
　　3.2.6　表格布局 …………………………………………………… 66
3.3　Android 常用控件 …………………………………………………… 67
　　3.3.1　Button 控件 ………………………………………………… 69
　　3.3.2　CheckBox 控件 ……………………………………………… 70
　　3.3.3　EditText 控件 ……………………………………………… 71
　　3.3.4　ImageButton 控件 …………………………………………… 74
　　3.3.5　ImageView 控件 …………………………………………… 75
　　3.3.6　ListView 控件 ……………………………………………… 76
　　3.3.7　ProgressBar 控件 …………………………………………… 78
　　3.3.8　RadioButton 控件 …………………………………………… 79
　　3.3.9　SeekBar 控件 ……………………………………………… 80
　　3.3.10　Spinner 控件 ……………………………………………… 81
　　3.3.11　TabHost/TabWidget（切换卡） …………………………… 83
　　3.3.12　Gallery 与 ImageSwitcher ………………………………… 85
3.4　自定义按钮背景 ……………………………………………………… 85
　　3.4.1　Shape 介绍 ………………………………………………… 85
　　3.4.2　Shape 使用步骤 …………………………………………… 87
　　3.4.3　Shape 常用属性 …………………………………………… 88
　　3.4.4　常见 Shape 标签的种类 …………………………………… 89
　　3.4.5　自定义背景的按钮 ………………………………………… 107
3.5　Selector 的使用 ……………………………………………………… 111
3.6　Android 沉浸式状态栏及悬浮效果 ………………………………… 114
3.7　登录界面布局的实现 ………………………………………………… 116

第4章　界面切换设计　121
4.1　界面切换设计目标 …………………………………………………… 121

  4.2 安卓应用程序组件 ·············· 122
  4.3 Activity ·············· 123
    4.3.1 Activity 类 ·············· 123
    4.3.2 Android 事件侦听器 ·············· 126
  4.4 Intent ·············· 128
    4.4.1 Intent 简介 ·············· 128
    4.4.2 Intent 实现两个 Activity 之间切换 ·············· 130
    4.4.3 Intent 实现两个 Activity 之间传递数据 ·············· 132
  4.5 Service ·············· 133
    4.5.1 Service 介绍 ·············· 133
    4.5.2 Service 启动流程 ·············· 134
  4.6 消息提示框和对话框 ·············· 134
    4.6.1 Toast 消息提示框 ·············· 134
    4.6.2 对话框 ·············· 137
  4.7 Android 程序生命周期 ·············· 139
  4.8 广播接收器 ·············· 141
  4.9 界面切换的实现 ·············· 143

第 5 章 列表视图界面设计 ·············· 149
  5.1 列表视图界面设计目标 ·············· 149
  5.2 ListView 应用 ·············· 150
    5.2.1 ListView 列表视图的工作原理 ·············· 150
    5.2.2 SimpleCursorAdapter ·············· 151
    5.2.3 SimpleAdapter ·············· 153
    5.2.4 有按钮的 ListView ·············· 155
    5.2.5 getView 应用 ·············· 159
  5.3 GridView 应用 ·············· 161
  5.4 RecyclerView 应用 ·············· 166
  5.5 列表视图界面的实现 ·············· 170

第 6 章 导航栏及滑动界面设计 ·············· 176
  6.1 导航栏及滑动界面设计目标 ·············· 176
  6.2 滑动界面设计 ·············· 177
    6.2.1 ViewPager 介绍 ·············· 177
    6.2.2 滑动界面实例 ·············· 178
  6.3 导航栏设计 ·············· 182
    6.3.1 导航栏设计方法 ·············· 182
    6.3.2 BottomNavigationView 底部导航栏 ·············· 182
  6.4 Fragment ·············· 185

6.4.1　Fragment 简介 ……………………………………………………… 185
　　6.4.2　Fragment 和 View 的比较 …………………………………………… 186
　　6.4.3　Fragment 应用 ……………………………………………………… 187
6.5　SurfaceView 与 TextureView ………………………………………………… 189
　　6.5.1　SurfaceView …………………………………………………………… 189
　　6.5.2　TextureView …………………………………………………………… 192
6.6　导航栏及滑动界面设计实例 ………………………………………………… 196
6.7　导航栏及滑动界面的实现 …………………………………………………… 202

第 7 章　Wi-Fi 物联网移动软件设计 ………………………………………… 209
7.1　Wi-Fi 物联网移动软件设计目标 ……………………………………………… 209
7.2　安卓通信程序设计 …………………………………………………………… 211
　　7.2.1　物联网 App 安卓端网络编程基础 …………………………………… 211
　　7.2.2　安卓 Socket 通信基础 ………………………………………………… 212
7.3　Wi-Fi 通信概要 ……………………………………………………………… 216
　　7.3.1　WLAN 通信 …………………………………………………………… 216
　　7.3.2　Wi-Fi 通信 …………………………………………………………… 217
　　7.3.3　ESP8266 模块的应用 ………………………………………………… 218
　　7.3.4　Smartconfig …………………………………………………………… 219
7.4　安卓 TCP 客户端程序实例 …………………………………………………… 220
7.5　Wi-Fi 物联网移动软件的实现 ………………………………………………… 223

第 8 章　蓝牙物联网移动软件设计 …………………………………………… 238
8.1　蓝牙物联网移动软件设计目标 ……………………………………………… 238
8.2　蓝牙通信概要 ………………………………………………………………… 239
　　8.2.1　蓝牙通信介绍 ………………………………………………………… 239
　　8.2.2　低能耗蓝牙(BLE) …………………………………………………… 241
8.3　CC2541 BLE 蓝牙模块应用 ………………………………………………… 242
　　8.3.1　CC2541 BLE 蓝牙模块介绍 ………………………………………… 242
　　8.3.2　Android 蓝牙 BLE 编程 ……………………………………………… 243
8.4　蓝牙物联网移动软件的实现 ………………………………………………… 247
　　8.4.1　蓝牙物联网移动软件界面设计 ……………………………………… 247
　　8.4.2　蓝牙物联网移动软件界面程序设计 ………………………………… 251
　　8.4.3　STM32 的蓝牙通信程序设计 ………………………………………… 261

第 9 章　数据库及动态界面设计 ……………………………………………… 263
9.1　数据库及动态界面设计目标 ………………………………………………… 263
9.2　物联网 App 安卓端数据存储 ………………………………………………… 264
　　9.2.1　使用 Shared Preferences 存储数据 ………………………………… 264
　　9.2.2　使用文件存储数据 …………………………………………………… 265

9.3 安卓端 SQLite 数据库应用设计 ……………………………………………… 269
　　9.3.1 安卓端 SQLite 数据库简介 ……………………………………… 269
　　9.3.2 SQLiteDatabase 介绍 …………………………………………… 271
　　9.3.3 SQLite 数据库编程方法 ………………………………………… 273
　　9.3.4 SQLiteOpenHelper ……………………………………………… 278
9.4 数据库及动态界面设计目标 ………………………………………………… 279

第 10 章 嵌入网页的控制界面设计

10.1 嵌入网页的控制界面设计目标 …………………………………………… 289
10.2 Android Http ………………………………………………………………… 291
　　10.2.1 Android Http 通信 ……………………………………………… 291
　　10.2.2 Okhttp …………………………………………………………… 296
10.3 WebView 应用 ……………………………………………………………… 300
　　10.3.1 WebView 介绍 …………………………………………………… 300
　　10.3.2 WebView 应用 …………………………………………………… 301
　　10.3.3 Android 与 JS 通过 WebView 互相调用方法 ………………… 302
10.4 嵌入网页的控制界面的实现 ……………………………………………… 303

第 11 章 传感器应用及拍照更换界面图片设计

11.1 传感器应用及拍照更换界面图片设计目标 ……………………………… 309
11.2 物联网 App 安卓端传感器编程 …………………………………………… 310
　　11.2.1 安卓传感器（OnSensorChanged）使用介绍 ………………… 310
　　11.2.2 方向传感器应用编程 …………………………………………… 312
　　11.2.3 安卓坐标系的定义 ……………………………………………… 314
　　11.2.4 安卓传感器 values 变量的定义 ………………………………… 315
11.3 摄像头及拍照应用 ………………………………………………………… 318
　　11.3.1 Camera2 应用 …………………………………………………… 318
　　11.3.2 使用 TensorFlow API 构建视频物体识别系统 ……………… 322
11.4 Android 拍照和选择照片 ………………………………………………… 326
　　11.4.1 Android 媒体库 MediaStore …………………………………… 326
　　11.4.2 Android 拍照和返回照片 ……………………………………… 327
　　11.4.3 Android 拍照和保存图片 ……………………………………… 328
11.5 拍照更换界面图片的实现 ………………………………………………… 332

第 12 章 苹果手机移动软件设计

12.1 苹果手机移动软件设计目标 ……………………………………………… 337
12.2 iOS 开发环境搭建 ………………………………………………………… 338
12.3 iOS 入门实例 ……………………………………………………………… 339
　　12.3.1 创建 iOS 项目 …………………………………………………… 339
　　12.3.2 编辑 main.storyboard 文件 …………………………………… 341

12.3.3 程序代码分析 346
12.3.4 main 函数及程序启动过程 346
12.3.5 UIResponder 类 347
12.4 Objective-c 348
12.4.1 Objective-c 介绍 348
12.4.2 Objective-c 特点 349
12.4.3 Objective-c 和 C++/Java 比较 350
12.5 iOS 基本控件 353

第 13 章 跨平台移动软件设计 357
13.1 跨平台移动软件设计目标 357
13.2 物联网 App 跨平台程序基础 357
13.2.1 物联网 App 跨平台程序简介 357
13.2.2 常见移动 Web 开发框架 358
13.2.3 常见 Hybrid App 平台 360
13.3 HTML5 362
13.4 PhoneGap 概述 363
13.4.1 PhoneGap 介绍 363
13.4.2 PhoneGap 实例 364
13.4.3 用 PhoneGap 开发 iOS 应用程序 365
13.5 jQuery Mobile 概要 367
13.5.1 jQuery Mobile 介绍 367
13.5.2 jQuery Mobile 应用 368
13.5.3 jQuery Mobile 页面链接 369
13.5.4 jQuery Mobile 内容格式 370
13.5.5 jQuery Mobile 导航 372
13.5.6 jQuery Mobile 工具栏 373
13.5.7 jQuery Mobile 按钮 378
13.5.8 jQuery Mobile 列表视图 379
13.6 跨平台移动软件的实现 383

参考文献 387

第1章 物联网移动软件开发

1.1 物联网移动软件开发目标

1. 物联网移动软件开发界面设计

物联网移动软件开发目标,主要是设计可以运行于手机、平板电脑、智能电视机、专用物联网设备上的软件系统,包括界面设计和底层功能设计两部分。例如,一个物联网移动软件可以通过手机实现对智能家居的监测与控制,其示意图如图1-1所示。

图 1-1 物联网移动软件示意图

物联网移动软件界面主要包括引导类、导航类、监控类和设置类界面等。

(1) 引导类界面

初始界面主要包括欢迎界面、引导界面、介绍界面、登录界面等,如图1-2所示。

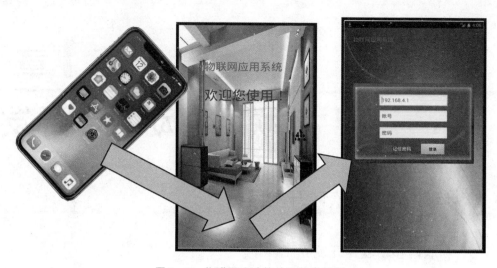

图 1-2 物联网移动软件引导类界面

（2）导航类界面

对于一个物联网移动软件，除了登录界面、主界面之外，其核心之一是导航类界面。导航类界面包括地点选择界面、设备选择界面等，如图 1-3 所示。

图 1-3 物联网移动软件导航类界面

（3）监控类界面

物联网移动软件另外一个核心部分是监控类界面，主要包括控制界面、监测界面等，如图 1-4 所示，监控类界面最能体现物联网应用特征。

第1章 物联网移动软件开发

图1-4 物联网移动软件监控类界面

(4) 设置类界面

设置类界面主要包括地点界面、设备设置界面、控制设置界面、用户设置界面、系统设置界面等,如图1-5所示。

图1-5 物联网移动软件设置类界面

2. 物联网移动软件工作原理

物联网移动软件的工作原理是将手机、控制目标设备、服务器通过无线路由器连接在一起,组成一个局域网,手机就可以通过无线网络对控制目标设备进行检测和控制。服务器是物联网系统结构的中心,一般采用 TCP 服务器或者 Web 服务器。TCP 服务器包括本地电脑作为 TCP 服务器、片式电脑作为 TCP 服务器、TCP 云服务器等。

(1) 本地电脑作为 TCP 服务器

采用本地电脑作为 TCP 服务器的物联网应用系统结构如图 1-6 所示。

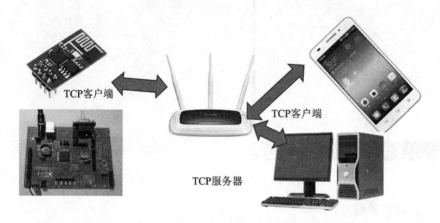

图 1-6 本地电脑作为 TCP 服务器的物联网应用系统结构

(2) 片式电脑作为 TCP 服务器

采用片式电脑(树莓派、香蕉派、香橙派等)作为 TCP 服务器的物联网应用系统结构如图 1-7 所示。

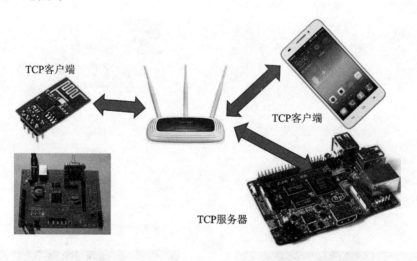

图 1-7 片式电脑作为 TCP 服务器的物联网应用系统结构

(3) TCP 云服务器

采用 TCP 云服务器的物联网应用系统结构如图 1-8 所示,可以实现远程的检测和控制。

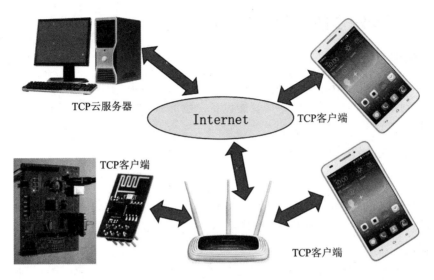

图 1-8　TCP 云服务器的物联网应用系统结构

上述几种物联网移动软件工作模式的感知层都是基于 Wi-Fi 通信模式。如果采用蓝牙模式,则可以用手机直接监测与控制物联网智能设备。如果采用 NB-IoT,也可以直接接入 Internet。如果采用 Zigbee 和 LoRa 等通信模式,则需要通过专用的网关和路由模块才能接入 Internet,结构如图 1-9 所示。

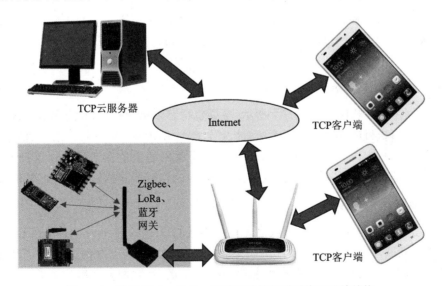

图 1-9　采用 Zigbee、LoRa、蓝牙网关的物联网应用系统结构

1.2 物联网移动软件开发概要

1.2.1 物联网移动软件开发简介

1. 物联网

物联网(IoT,Internet of Things)就是物物相连的互联网,或者更加准确地说是万物相连的网络,这里所说的物,是指智能设备。物联网的基本元素是智能设备,而智能设备的核心是单片机和嵌入式系统。

2. 物联网应用软件设计

物联网应用软件设计是指与物联网相关的应用软件设计。物联网应用软件种类繁多,包括物联网底层(感知层)软件、通信(网络层)软件、服务程序(服务层)软件、客户端(应用层)软件等,本书介绍的物联网移动软件设计,重点在于客户端(应用层)软件的开发。

物联网客户端(应用层)软件的开发,主要是指运行于手机、平板电脑、智能电视机、物联网专用触摸屏(HMI,人机界面)、普通电脑上的面向物联网应用系统的软件开发。除了少部分面向普通电脑的应用,绝大部分都属于移动设备的应用,所以本书的重点在于介绍物联网的移动软件设计。

3. 物联网移动软件设计

物联网移动软件开发是指面向手机、平板电脑、智能电视机等移动终端的物联网应用软件设计。物联网、云计算、人工智能、共享经济、AR、VR、3D打印等不同的互联网关键词显然是对互联网行业这一发展阶段最好的概括和呈现,它们无一不是在暗示着"移动化"的到来。

4. 物联网系统结构

物联网系统结构图如图 1-10 所示,主要包括感知层、网络层和应用层。物联网是以感知与应用为目的的物物互联系统,涉及传感器、RFID、安全、网络、通信、信息处理、服务技术、标识、定位、同步等众多技术领域。物联网的价值在于让物体也拥有了"智慧",从而实现人与物、物与物之间的沟通,物联网的特征在于感知、互联和智能的叠加。

5. 物联网移动设备

"移动化"并不是一个陌生的概念。据新华社北京 2017 年 8 月 4 日电(记者高亢朱基钗)中国互联网络信息中心 3 日在京发布的第 38 次《中国互联网络发展状况统计报告》显示,截至 2017 年 6 月,我国网民规模达 7.51 亿,互联网普及率达到 54.3%,其中手机网民占比达 96.3%,网民规模连续 10 年位居全球首位。

中国移动互联网用户规模已经超过 PC 互联网用户规模,中国网民上网的主流行

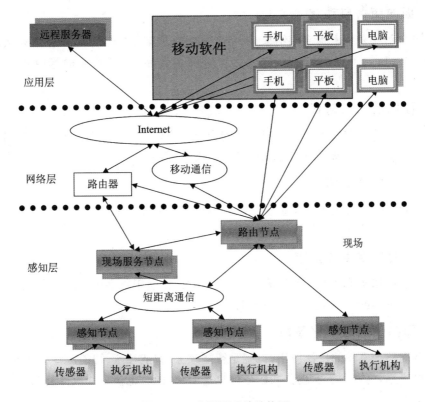

图 1-10 物联网系统结构图

为已偏向移动化。在未来的十年也许所有东西都是移动的,移动化成为自然而然的一件事情,移动化的趋势会更加深入,移动化的概念将越来越向外扩展,渗透到生活当中的所有事物中。

物联网移动设备已经成为日常生活中不可或缺的重要组成部分。物联网移动设备包括物联网移动电脑类设备、物联网便携移动设备、物联网多媒体移动设备三大类别:

- 物联网移动电脑类设备包括平板电脑、触摸笔记本电脑、触摸一体机电脑、触摸台式电脑等。
- 物联网便携移动设备包括智能手机、安卓 GPS 导航、智能手表、智能音乐盒、智能眼镜、智能相机等。
- 物联网多媒体移动设备包括智能电视、网络电视、安卓机顶盒、网络播放器、智能视频监控系统等。

1.2.2 物联网移动软件的应用

物联网移动软件的应用涵盖物联网系统的各个方面,其重点是物联网的移动软件开发,在今后的一段时间内,应该重点关注以下几个方面:

1. 面向商务自动化

商务自动化是技术带来的结果,消费者可能会把某些零售体验外包给算法已经智能化的设备,比如手机、机器人,这就意味着筛选商品、谈判和采购等内容都是智能化的。

2. 面向后人口时代

今后手机软件开发可能会出现跨性别的时装模特、为老人开设的企业家课程等,不断变化的社会以及消费者行为构成将开启后人口时代。

3. 面向虚拟实体

人工智能技术的出现让虚拟实体进行有意义的对话,所以虚拟语音软件可能会从助手变成自己的朋友。

4. 面向数字生活需求

消费者的期望值如今已经达到很高的需求,这得益于数字生活的开启,手机软件的功能将会通过不断完善来适应用户不断调整的需求。

5. 面向信息透明化革命

物联网移动软件的广泛应用,特别是智能政务、智能社区、智能医疗、智能养老等智能系统的应用,将有利于实现公共服务信息透明化,同时也让公共服务资源的使用更加公平公正。

6. 面向个性化用户体验

大多移动应用都将提供更加个性化的内容和服务,那些只在 Web 上可用的服务也将过渡到移动应用当中。新应用将允许用户去创建、修改、分享和购买个性化的产品和服务。新开发的技术能够有效地使用由移动分析工具提供的大数据来推动应用程序服务的个性化。

通过对用户移动设备上相关信息的收集,来提供与之相关的特定服务。这将使用户把更少的时间花费在挑选自己喜欢的事情上,因为他们看到的结果全是根据自身的喜好显示出来的内容。因此,个性化将成为移动应用领域最重要的一方面。

7. 面向移动支付

通过使用近距离无线通信技术(NFC)的移动应用的支付正在开展行动,因为最近 Apple 的升级引入了一个带有 NFC 的移动支付系统,并说服了大量的商家和企业去接受这个系统的支付。移动支付将有望随着安全移动应用的开发而实现快速增长,像 Apple Pay、Google Wallet 和 MCX 的 Current C 之类的移动支付解决方案也昭示着移动支付快速增长的趋势。为 Android、iOS 和 Windows Phone 设计的集成支付系统的应用,让用户通过智能手机就能够安全地购买产品和服务。

8. 面向企业应用

灵活性已经成为一个企业服务交付的重要方面,大多数的企业需要快速响应更新发布的变化,因为他们的企业应用有特定的时间约束。像 HTML 5 这样技术的发展将促使使用很少的开销就能开发出丰富的企业应用。

1.3 物联网智能硬件 App 设计

智能硬件是物联网系统的重要组成部分,是物联网的基础。智能硬件是指通过将硬件和软件相结合对传统设备进行智能化改造。而智能硬件移动应用则是软件通过应用连接智能硬件,操作简单,开发简便,各式应用层出不穷,也是企业获取用户的重要入口。

智能硬件是物联网的基本元素、基础单元,是物联网节点。随着智能硬件的快速发展,智能硬件已经从可穿戴设备延伸到智能电视、智能家居、智能汽车、医疗健康、智能玩具、机器人等领域,在这些繁杂的智能硬件组成的物联网中,沟通交流尤为重要。当然,这里说的沟通并不是最基本的人与人的沟通,而且人与智能硬件的沟通、智能硬件之间的沟通。其需要做的就是通过社交关系链,将人与人、人与设备、设备与设备连接在一起,打造"硬件社交"。

智能硬件是继智能手机之后的一个科技概念,通过软硬件结合的方式,对传统设备进行改造,进而让其拥有智能化的功能。智能化之后,硬件具备连接的能力,实现互联网服务的加载,形成"云＋端"的典型架构,具备了大数据等附加价值。

改造对象可能是电子设备,例如手表、电视和其他电器;也可能是以前没有电子化的设备,例如门锁、茶杯、汽车,甚至房子。

智能硬件已经从可穿戴设备延伸到家居、汽车、医疗、机器人等领域。典型的智能硬件包括 Google Glass、Samsung Gear、Fitbit、Apple Watch 等。

在 2018 年 1 月 8 日至 1 月 12 日的 CES 2018 大会上,百度公司携旗下多款科技新品亮相,其中 Raven H 内核搭载百度 DuerOS 2.0,目前已有 50 余项原生技能和 150 余项第三方技能,渡鸦还在 DuerOS 2.0 平台上为 raven H 深度定制了音乐、视频、生活资讯和智能家居等场景。

1.4 物联网移动软件开发的发展趋势

1.4.1 物联网移动软件的特点

物联网移动软件早已不是只能远程控制智能单品开关的初级形态,它正在经历超级化、H5 化、去 App 化和人性化等演进过程。

做物联网产品,物联网移动软件开发是一道绕不开的环节。虽说远程控制本身远非智能化的最终形态，但无论从灵活的客制化还是控制功能的完整性来看，App的入口作用在相当长远的未来都将是不可或缺的。

了解物联网移动软件开发的行业现状和未来发展趋势，对产品功能开发和提升市场竞争力具有重大意义。App可能是智能产品功能序列中最幸运，也是最倒霉的一个。说它幸运，是因为自从终端智能化的概念兴起，App就一直是各方关注的中心，自定义、场景化、多产品集成化、一键控制，甚至去App化，智能产品提供的各项功能和服务中，游离于物理产品之外的App所承载的用户期待居然是最多的；说它不幸，也是因为被关注得最多，被动刀的机会也最多，最后甚至会被改造到四分五裂，甚至踪迹难觅的程度。

(1) 超级化

受智能终端的泛在化和场景化发展影响，物联网产品App向集成了多终端控制能力的超级App演进是一个明显的趋势。一方面，移动互联网领域的微信、淘宝，以及与司南物联合作的思源集团的toon App等，都已经实现了平台化。另一方面，继承了多终端控制能力的物联网超级App，其天生就具备平台化的属性。

(2) H5化

若论及近年来能够影响软件开发格局的技术，继Java之后，就要数Html 5了。与生俱来的网络化任务能力，以及得到充分强化的本地资源调度能力，让H5编码可以胜任用以开发nativeApp的工作。虽然效率方面还比不上本地化语言，但是H5作为前端功能开发的明日之星的地位，显然已经是毋庸置疑的了。包括淘宝、微信和司南物联的智能App，都已经在不同程度上采用了H5页面技术，这种技术可以让前端页面功能脱离App本身，存储在云端；只需在云端进行统一更新，即可完成相应的功能升级，而不必对本地App做更改，从而完美地解决了传统App更新过程中会遇到的OS和App版本碎片化的难题。

(3) 去App化

从智能终端诞生伊始，就一直在不断地收到"有App控制就是智能化了吗""开个灯我也要打开App点一下我累不累"这种质疑。本着为用户不断减负的宗旨，行业内一直都在试图简化App操作，甚至想把App本身给简化掉。

但是无论App控制终端设备的功能被削减到怎样的程度，其作为用户端配置入口的作用将永远存在。因此，我们看到那些现象级App一个接一个地提供App嫁接功能，QQ有"我的设备"，微信有服务号，和司南物联有着深度合作的京东微联也是集成了第三方设备控制功能的超级App。正是因为有了这些"App嫁接和运行平台"的存在，加之前述H5技术的发展，为物联网终端的去App化提供了可能。

(4) 人性化

物联网移动软件开发的核心功能在于配置和控制。尽管去App化是物联网产品行业的必然趋势，但是其作为用户端配置入口的作用却无法被精简掉。在终端App发展演进的过程中，受相关产业链的利益驱使，其自身一定会持续优化，不断强化自身功能和对用户的贡献，以避免被精简掉的命运。

1.4.2 物联网移动软件开发的发展

1. 物联网移动软件开发的发展

这是个言必称物联网的时代。相关数据显示,中国仅智能家居市场 2018 年增长至 1 000 亿元人民币规模,55% 的中国消费者对于在家中使用一些智能设备或者是互联互通的设备表示非常感兴趣,这个比例相比其他国家要高得多。

可是仅仅连接在一起就够了吗?Marvell[1] 公司 Marvell 全球无线及 IoT 业务总经理 Philip Poulidis 日前在 Marvell 物联网方案巡展中国站上指出,如果不能搭建出一个完全智能的互联网络,或者称之为认知物联网(CIoT),消费者与产业界对智能家居、可穿戴设备、智能工业、智能汽车等 IoT 相关应用的兴趣就不可能被真正激发出来。

2. 最好的 GUI 就是没有 GUI

"只关心如何通过 App 或是云服务把物与物连接起来,这是非常初级的 IFTTT 阶段。"Philip 借编程语言中的 IF…Then… 句式,暗指现在的物联网其实是"被动的""非智能的"。而人们真正需要的,是将人工智能和认知技术与物联网结合起来,构建认知物联网,使之成为连接现实世界(物体、传感器)与虚拟世界(人类行为、社会语境)的桥梁,让所有相连的物体最终都具备自我学习、思考与理解的能力。要实现这样的功能,需要包括传感器融合算法、机器深度学习、语音、视觉和手势控制在内的更多新技术。

Philip 认为,要想和一个智能设备进行交互,用户根本不需要打开 App。一个真正意义上的 IoT 设备应该能够主动感知你的存在、位置、动作、手势和语音指令,或者至少应该能够根据过往记录和语境信息做出主动反应。另一方面,消费者需要的是能够自我配置、自我管理的设备。未来将有 500 亿台不同设备实现互联,单靠人的力量去对其进行逐一配置是不可能的,这就要求设备能够自动进行连接,并能根据某项特定指令进行功能整合。同时,相应的服务也能够根据用户行为反馈实现自适应功能。

Marvell 已经在嵌入式系统层面为认知物联网时代的到来做好了准备——这是 Philip 本人,也是整个 Marvell 公司通过路演想传达的理念,而增强版 EZ-Connect 系列无线微控制器 SoC、开源的 Kinoma IoT 开发套件,以及先进语音处理、情景感知、传感器融合、位置定位、机器视觉、手势识别等技术,将是实现这一理念的保证。

3. 物联网移动软件的未来是没有 App 的

2015 年 10 月 22 日,Ayla Networks[2] 携手亚马逊 AWS 主办的"赢在风口——2015 物联网 & 产品智能化研讨会"在上海举行。Ayla NetworksCTO、硅谷物联网大咖 Adrian Caceres 与亚马逊资深解决方案架构师及 Ayla 物联网生态系统合作伙伴

---

[1] Marvell(迈威科技集团有限公司,现更名为美满),成立于 1995 年,总部在硅谷,在中国上海设有研发中心,是一家提供全套宽带通信和存储解决方案的全球领先半导体厂商,是一个针对高速,高密度,数字资料存贮和宽频数字数据网络市场,从事混合信号和数字信号处理集成电路设计、开发和供货的厂商。

[2] Ayla Networks,总部设在美国圣克拉拉市的物联网公司。

USI、CleNET 瞬联软件等现场分享了物联网未来趋势、如何重新定义物、硬体商如何将传统产品与物联网健康结合并成功落地等课题。

Ayla CTO Adrian 认为"物联网的未来,是没有 App 的,不需要 App。现阶段许多的联网设备都在使用 App 控制,通过 App 实现人机交互,物联网的未来则是去 App 化,不论是微信这类的超级 App,还是企业自身的 App"。

在他看来,物联网精髓在于能够支持设备自学习迭代,通过不断升级部件,改善服务,提升用户体验。而作为领导团队研发出第四代 Kindle 并成功上市的硅谷技术牛,他认为,Kindle 能够占据市场多数份额的原因是利用了物联网最重要特性,使得设备能够持续更新,不断进化与学习,并能收集消费习惯等数据,最终为交易商城提供指导性参考意见,最终形成生态闭环。

Adrian 表示,Ayla IoT 平台在设计之初,就支持客户的产品迭代,客户可以不断完善其联网产品。未来,设备制造商可以无需更换硬件,只需通过在云端配置即可直接为设备追加新功能。此外,使用物联网技术后,制造商可以第一时间知道产品损耗程度,进行实时维护,为用户节约成本。

1.4.3 移动云计算

1. 移动云计算介绍

物联网移动软件开发的另外一个发展方向是面向移动云计算。云计算的发展并不局限于 PC,随着移动互联网的蓬勃发展,基于手机等移动终端的云计算服务已经出现。

基于云计算的定义,移动云计算是指通过移动网络以按需、易扩展的方式获得所需的基础设施、平台、软件(或应用)等的一种 IT 资源或(信息)服务的交付与使用模式。移动云计算是云计算技术在移动互联网中的应用。

2. 移动云计算优势

(1) 突破终端硬件限制;

(2) 便捷的数据存取;

(3) 智能均衡负载;

(4) 降低管理成本;

(5) 按需服务降低成本。

云计算技术在电信行业的应用必然会开创移动互联网的新时代,随着移动云计算的进一步发展,以及移动互联网相关设备的进一步成熟和完善,移动云计算业务必将会在世界范围内迅速发展,成为移动互联网服务的新热点,使得移动互联网站在云端之上。

3. 移动云计算成功实例

(1) 加拿大 RIM 公司面向众多商业用户提供的黑莓企业应用服务器方案,可以说是一种具有云计算特征的移动互联网应用。在这个方案中,黑莓的邮件服务器将企业应用、无线网络和移动终端连接在一起,让用户通过应用推送(Push)技术的黑莓终端远程接入服务器访问自己的邮件账户,从而可以轻松地远程同步邮件和日历,查看附件和地址本。除黑莓终端外,RIM 同时也授权其他移动设备平台接入黑莓服务器,享用黑莓服务。

(2) 苹果公司推出的"iCloud"云服务,该服务可以让所有苹果设备实现无缝对接。iCloud 服务提供英语、中文、俄语、西班牙语、葡萄牙语、德语和菲律宾语多种语言。

(3) 微软公司推出的"Onedrive"(以前 SkyDrive,Windows Live SkyDrive,Windows Live Folders)是一个文件托管服务,是微软在线服务套件的一部分。它允许用户将文件以及其他个人数据存储在云端、进行 Windows 设置或用 BitLocker 恢复密钥。文件可以同步到 PC 和 Web 浏览器或移动设备上进行访问,以及公开或与特定的人共享。

(4) 作为云计算的先行者,Google 公司积极开发面向移动环境的 Android 系统平台和终端,不断推出基于移动终端和云计算的新应用,包括整合移动搜索、语音搜索服务、定点搜索以及 Google 手机地图、Android 上的 Google 街景等。

第 2 章

欢迎界面设计

2.1 欢迎界面设计目标

在一个物联网移动软件之中,最基本也最简单的功能和界面是欢迎界面,欢迎界面差不多是所有移动 App 的第一个界面。欢迎界面结构如图 2-1 所示,包括背景图片、软件名称、欢迎词等。部分欢迎界面做成多个滑动界面的切换功能,用于介绍软件的功能或者简单介绍软件的使用方法。

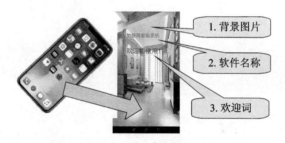

图 2-1 欢迎界面结构

本章的设计目标是设计一个最简单的欢迎界面,如图 2-1 所示,该界面包括一个背景图片以及两个静态文本,分别显示软件名称和欢迎词。该欢迎界面是本书介绍的第一个移动软件设计实例,也就是移动软件开发的 Hello World 程序。

欢迎界面设计过程如图 2-2 所示,包括创建项目、程序设计、编译、安装和运行四个过程,而程序设计过程包括 XML 布局设计、Java 导入布局、注册活动界面。

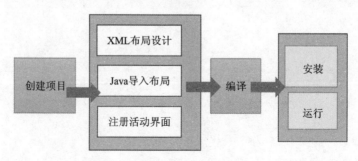

图 2-2 欢迎界面设计过程

2.2 物联网移动软件设计基础

2.2.1 物联网移动软件开发的内容

根据相关调查显示,在2019年将会有三分之二的消费者愿意选择物联网,到了2020年全球将有260亿商业和工业物联网设备,是2009年的三十倍。随着智能终端硬件的开发,物联网移动软件开发是一直都甩不开的话题,也可以预测到物联网移动软件开发将会具有非常广泛的应用前景。

物联网移动软件开发的四个层面:

(1)首先物联网终端设备比手机大得多,而且本身没有显示界面,通常只是能够通过特定网络协议回传数据的传感器(直接连入互联网或者通过网关设备),也就是说在物联网大数据汇聚的前端,数据的汇入是自动化进行的,App开发的重点是后端的汇聚层。

(2)分析师JefferyHammond认为,物联网App后端汇聚层需要有一个智能化软件系统(通常运行于数据中心)来管理物联网设备(包括固件升级等)、网络、处理海量数据,并提供给用户。

(3)在设备层、汇聚层之外,物联网App还需要一个分析层负责处理物联网设备产生的大数据。

(4)最后是最终用户层,负责将有用的数据分析结果以可视化的方式展示到用户的终端设备中,这个层面的开发,可以是移动Web网站,也可以是一个手机App。

2.2.2 安卓物联网移动软件设计基础

安卓正在成为物联网产品首选的操作系统。目前采用Android的设备众多,小到鞋子传感器大到喷气发动机监视器,无所不在。由于物联网设备在市场上的兴起,谷歌进一步在互联世界领先苹果和微软。Android之所以快速增长,部分原因是其开放性。谷歌允许设备制造商对安卓进行修改。谷歌本身则通过安卓手机和平板电脑上的广告和其他服务获得利润。

谷歌在2005年收购了Android公司。这家搜索巨头对这款基于Linux的操作系统进行了精简和优化。但是谷歌并不是唯一推出基于Linux的操作系统的高科技公司。英特尔开发的Linux版本名为Moblin,而诺基亚则开发Maemo,这两个版本也是Meego的前身。Palm的WebOS也是基于Linux核心。正如发生在二十世纪90年代的个人电脑操作系统大战,硝烟过后,只有一个赢家就是微软的Windows。而现在除了服务器或PC,其他硬件操作系统的赢家就是Android。

不同类型的屏幕、移动芯片和传感器都可以与Android完美配合,任何人都可以对Android进行优化,使它胜任各种工作。例如,美国宇航局艾姆斯研究中心的工程师利用Android系统开发了宇航飞船的大脑,硬件大小不过一颗棒球。卫星通常需要花费数百万美元进行建造和发射。而这款控制系统的造价不过一万五千美元。他们希望高中实验室和个人爱好者就可以负担这款系统进行空间科学实验。而其之所以造价如此低就是因为采用了开源 Android 系统。

同样,Xively 公司的物联网作品——一款基于 Android 的农业灌溉系统。这款系统利用一个小型防水芯片建立调节水量的网络。有了 Android,可以开发一些低功耗的产品,它很容易进行用户界面和触摸控制的开发,同样很容易处理数据传输。

1. 开放平台带给应用更多可能性

Android 在物联网的优势,在于其操作系统是完全免费开放的,任何厂商都可以不经过 Google 和开放手持设备联盟的授权使用 Android 操作系统,开发者亦可以通过在 Android NDK(Android Native 开发包)中使用 C 语言或者 C++语言,来作为编程语言开发应用程序。

此外,Google 还推出了 Google App Inventor 开发工具,该开发工具可以快速地构建应用程序,以方便初次开发 Android 的工程师在短时间内上手,而资深工程师则可大展身手,设计出不同的应用软件。而 Hardware independence 的概念,让开发 Android 应用软件的设计者只需要考虑到 Framework 层,不会再被硬件商绑死,如此一来这些应用软件开发商就无须再担忧更换硬件时会增加额外的成本,愿意投入更多资源在 Android 的应用上。

因为物联网的应用很广,包括零售、仓储、物流、医疗、车载等,而装置(Device)特性差异更大,从 PDA、Tablet、车机或是其它移动式计算机,Android 所提供的开放语言架构,不仅有很多开发者投入,相对终端使用者更愿意更换成 Android 的平台,是非常大的优势。

2. 更适合于物联网应用

不同于微软针对不同操作平台推出 Windows Embedded OS、Windows CE(现在已经改名为 Windows Compact),到智能型手机使用的 Windows Phone,针对不同产品的特性,提供精简但高速或是高效能,可以负责复杂运算的操作系统;Android 仅提供一个版本运行于智能型手机以及平板计算机,且仅提供娱乐用的装置并不讲求效能,Android 更加容易应用于物联网系统。

Android 的崛起是物联网相关行业者不容忽视,也需要跟进的操作系统,不过 Google 如何看待物联网这群相对于消费者为少数但可以长期经营的族群,将会是各家 OEM 以及品牌商关注的焦点。

全球工业计算机的领先者——威强工业计算机,在工业计算机垂直产业上的努力不遗余力,在 2010 年成立 IEIMobile 部门专注于产业用平板计算机和产业用 PDA,IEI Mobile 除了既有的 Wintel 的能力和合作伙伴关系外,也看中 Android 未来在物联网的发展性,并且基于持续提供客户创新平台的观念,与主力芯片业者合作,选择适当的 Android 版本,开发产业用 PDA 及产业用中小尺寸的平板计算机。

2.3 开发工具

2.3.1 开发工具的选择

1. Obtain_Studio

本书的程序主要在 Obtain_Studio 环境下开发，所有的项目都兼容 Android Studio 和 Eclipse，所以读者无须担心程序的兼容性问题。对于不熟悉 Android Studio 和 Eclipse 的读者，或者感觉 Android Studio 和 Eclipse 安装过程太过于复杂的初学者，可以采用本节所介绍的 Obtain_Studio，Obtain_Studio 解压之后无须安装可以直接运行，也无须配置即可以直接进行 Android 开发。对于喜欢使用 Android Studio 和 Eclipse 的读者，本书介绍的程序都可以直接在 Android Studio 打开和运行，也可以导入到 Eclipse 中运行。

本书选择 Obtain_Studio 的原因是因为 Obtain_Studio 无须安装和配置，方便初学者快速使用。同时 Obtain_Studio 还集成了本书所介绍的移动软件开发配套的 STM32 单片机程序开发、Windows 和 Linux TCP 服务器程序开发等，方便读者在 Obtain_Studio 搭建和开发整个物联网系统程序。

2. Android Studio

Android Studio 是一个 Android 开发环境，基于 IntelliJ IDEA。类似 EclipseADT，Android Studio 提供了集成的 Android 开发工具用于开发和调试。为了简化 Android 的开发力度，以重点建设 Android Studio 工具，Google 已宣布停止支持 Eclipse 等其他集成开发环境。而随着 Android Studio 正式版的推出和完善，Android 开发者们转向 Android Studio 开发平台也将是大势所趋。

2.3.2 Eclipse 安装与配置

1. 下载 JDK[①]

进入 Oracle 官网，下拉到如图 2-3 所示页面，根据操作系统选择合适的 JDK 版本。例如 64 位 Win10 操作系统，在同意下载协议后，选择 Windows x64，点击 jdk-8u181-windows-x64.exe 后下载。

2. 安装并配置 Java 环境

建议默认安装在 C 盘，如图 2-4 所示。在安装完 JDK 后，还会弹出一个安装框提示安装 Jre，如果是默认安装只需点下一步，若更改了安装目录，Jre 的安装路径也应修改（非必须）。

---

① 下载地址：http://www.oracle.com/technetwork/java/javase/downloads/jdk8-downloads-2133151.html。

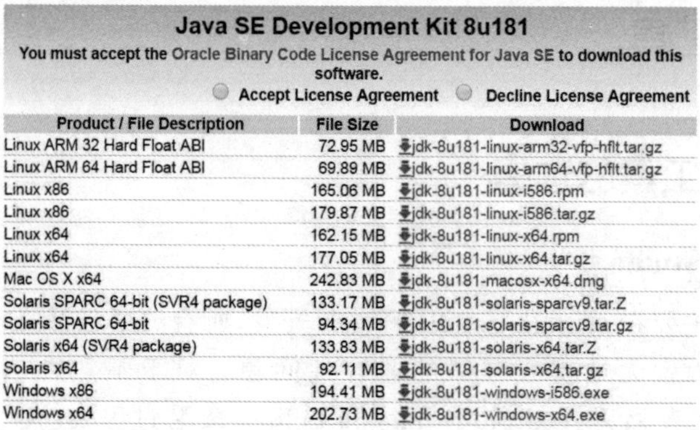

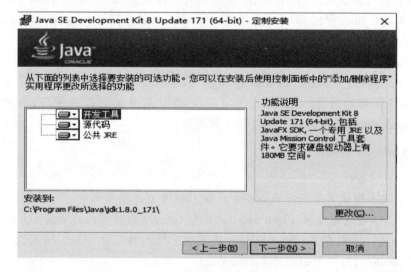

图 2-3　JDK 下载页面

图 2-4　Java 安装

至此,Java 的安装工作已经完成。下面配置环境变量。

(1) 右键单击"此电脑",选择属性,看到如图 2-5 所示窗口,选择"高级系统设置"。

(2) 点击"环境变量",在"系统变量"中,设置如表 2-1 所列的三个变量。前面四步设置环境变量对搭建 Android 开发环境不是必须的,可以跳过。

表 2-1　环境变量设置

变量	值
CLASSPATH	;%JAVA_HOME%lib;%JAVA_HOME%lib\tools.jar;
JAVA_HOME	C:\Program Files\Java\jdk1.7.0_79
Path	;C:\Program Files\Java\jdk1.7.0_79\bin;C:\Program Files\Java\jre7\bin

安装完成之后,可以再检查 JDK 是否安装成功。打开 cmd 窗口,输入 java –version,查看 JDK 的版本信息。如果正确显示出 JDK 版本信息则表示安装成功。

第 2 章 欢迎界面设计

图 2-5 电脑属性选择

3. Eclipse 安装

在如图 2-6 所示的页面下载 Eclipse[②]IDE for Java Developers(181M)的 Windows 32 bit 版,解压之后即可使用。

4. Android SDK 安装

在 Android Developers[③] 下载 android-sdk,下载完成后解压到任意路径。

- 运行 SDK Setup.exe,单击 Available Packages。如果没有出现可安装的包,单击 Settings,选中 Misc 中的"Force https://..."这项,再单击 Available Packages。
- 选择希望安装的 SDK 及其文档或者其他包,单击 Installation Selected、Accept All、Install Accepted,开始下载安装所选包。
- 在用户变量中新建 PATH 值为:Android SDK 中的 tools 绝对路径(本机为 D:\AndroidDevelop\android-sdk-windows\tools)。

确定重新启动计算机。进入 cmd 命令窗口,检查 SDK 是不是安装成功。运行 android -h 如果 CMD 窗口内显示 android 命令的使用参数提示消息,表明安装成功。

5. ADT 安装

打开 Eclipse IDE,进入菜单中的"Help"→"Install New Software",单击"Add..." 按钮,弹出对话框要求输入 Name 和 Location:Name 自己随便取,Location 输入

---

② Eclipse 下载地址:http://www.eclipse.org/downloads/eclipse-packages/。
③ 下载地址:http://developer.android.com/sdk/index.html。

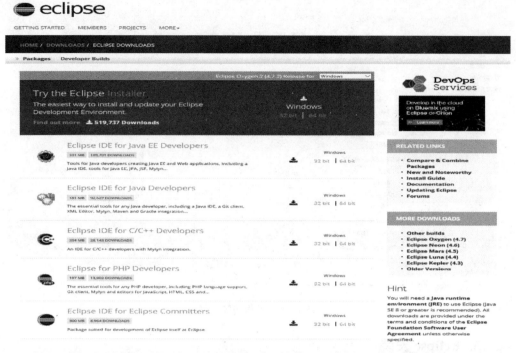

图 2-6 Eclipse 下载

http://dl-ssl.google.com/android/eclipse，如图 2-7 所示。

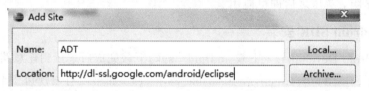

图 2-7 ADT 安装

确定返回后，在 Work with 后的下拉列表中选择刚才添加的 ADT，会看到下面出有 Developer Tools，展开它会有 Android DDMS 和 Android Development Tool，勾选它们，如图 2-8 所示。

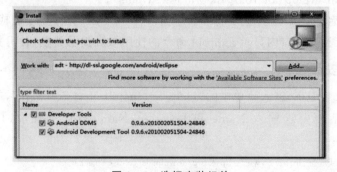

图 2-8 选择安装组件

然后就是按提示一步一步 Next。完成之后,选择 Window > Preferences。在左边的面板选择 Android,然后在右侧单击 Browse…并选中 SDK 路径,例如本机为:D:\AndroidDevelop\android-sdk-windows,单击 Apply、OK 按钮。配置完成。

6. 创建 AVD

为使 Android 应用程序可以在模拟器上运行,必须创建 AVD。方法如下:
- 在 Eclipse 中。选择 Windows > Android SDK and AVD Manager;
- 单击左侧面板的 Virtual Devices,再右侧单击 New;
- 填入 Name,选择 Target 的 API,SD Card 大小任意,Skin 随便选,Hardware 目前保持默认值;
- 单击 Create AVD 即可完成创建 AVD。

注意:如果单击左侧面板的 Virtual Devices,再右侧单击 New,而 target 下拉列表没有可选项时,单击左侧面板的 Available Packages,再右侧勾选 https://dl-ssl.google.com/android/repository/repository.xml,如图 2-9 所示。

图 2-9　Available Packages

然后单击 Install Selected 按钮,接下来就是按提示做就行了。要做这两步,原因是在 Android SDK 安装中没有安装一些必要的可用包(Available Packages)。

通过 File→New→Project 菜单,建立新项目"Android Project",然后填写必要的参数,如图 2-10 所示。

相关参数的说明:

(1) Project name:包含这个项目的文件夹的名称。

(2) Package name:包名,遵循 Java 规范,用包名来区分不同的类是很重要的,这里作用的是 helloworld.test。

(3) Activity name:这是项目的主类名,这个类将会是 Android 的 Activity 类的子类。一个 Activity 类是一个简单的启动程序和控制程序的类。它可以根据需要创建界面,但不是必须的。

(4) Application name:应用程序标题。

(5) 在"选择栏"的"Use default location"选项,允许选择一个已存在的项目。

单击 Finish 按钮后,单击 Eclipse 的 Run 菜单选择"Run Configurations…"。

选择"Android Application",单击在左上角(按钮像一张纸上有个"+"号)或者双击"Android Application",有个新的选项"New_configuration"(可以改为喜欢的名字)。

在右侧 Android 面板中单击"Browse…",选择 HelloWorld。在 Target 面板的

图 2-10　Eclipse 创建 Android 项目

Automatic 中勾选相应的 AVD，如果没有可用的 AVD 的话，需要单击右下角的"Manager…"按钮，然后新建相应的 AVD，如图 2-11 所示。

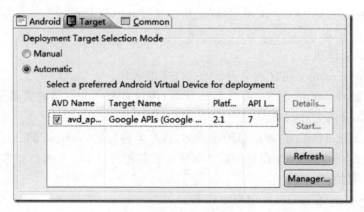

图 2-11　选择模拟器

然后单击 Run 按钮即可，运行成功的话会显示 Android 的模拟器界面。

2.3.3 Android Studio 安装与配置

1. 安装 JDK

需要 JDK1.8 以上,配置 JAVA_HOME=C:\Program Files\Java\jdk1.8.0
Path=%JAVA_HOME%\bin。

此处提醒几点:

(1) Android Studio 要求 JDK 版本为 JDK8 或更高版本。

(2) 确认操作系统是 32 位还是 64 位,下载对应的 JDK 版本:"Windows x86"——对应 Windows 32 位机器,"Windows x64"——对应 Windows 64 位机器。否则,安装好 Android Studio 后,由于与 JDK 不匹配,打开时会报错。

(3) JDK 的环境变量按链接中的要求支配好,即使用传统的 JAVA_HOME 环境变量名称,否则,打开 Android Studio 时会因为找不到 JDK 的路径同样报错。

2. Android Studio 安装

下载 Android Studio 安装包,可以从 http://www.android-studio.org/下载最新版本,这里采用 3.0 版本进行演示,对应安装包为 android-studio-ide-171.4408382-windows.exe,安装包大小 681 MB,安装包不带 SDK3. Android Studio 安装。

Android Studio 安装过程如下:

(1) 双击 Android Studio 的安装文件,进入安装界面 1,如图 2-12 所示。

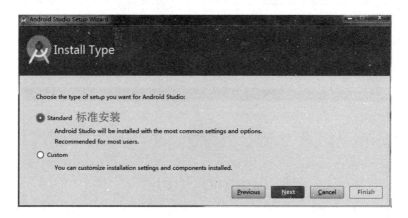

图 2-12 Android Studio 安装界面 1

第一个是 Android Studio 主程序,必选。第二个是 Android SDK,会安装 Android 5.0 版本的 SDK,也勾上。第三个和第四个是虚拟机和虚拟机的加速程序,如果要在电脑上使用虚拟机调试程序,就勾上。完成后单击 Next 按钮进入下一步。

(2) 选择 Android Studio 和 SDK 的安装目录。

AS:推荐使用默认安装地址,Android Studio 安装界面 2 如图 2-13 所示。

SDK:改变地址,不应该和 AS 安装路径相同,例如:可以修改为 D:\Android\sdk,

安装路径不要有中文和空格。

图 2-13　Android Studio 安装界面 2

（3）设置虚拟机硬件加速器可使用的最大内存。

如果电脑配置较高，默认设置 2GB 即可；如果配置较低，选择 1GB 即可，否则过大的话也会影响运行其他软件。

（4）下一步后，就进入自动安装模式了，其他的步骤采用默认即可。安装完成后，即可创建安卓项目，Android Studio 启动界面如图 2-14 所示。

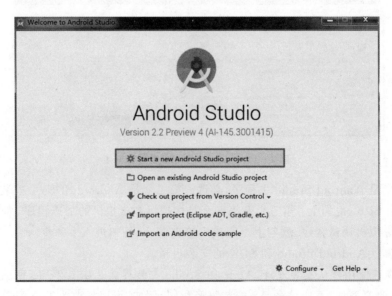

图 2-14　Android Studio 启动界面

第 2 章　欢迎界面设计

3. Android Studio 项目结构

Android Studio 项目主要目录如下：

（1）.gradle 和.idea：这两个目录下放置的都是 Android Studio 自动生成的一些文件，无须关心，也不要去手动编辑。

（2）app：项目中的代码、资源等内容几乎都是放置在这个目录下，后面的开发工作也基本都是在这个目录下进行的，后面还会对这个目录单独展开进行讲解。

（3）build：这个目录也不需要过多关心，它主要包含了一些在编译时自动生成的文件。

（4）gradle：这个目录下包含了 gradle wrapper 的配置文件，使用 gradle wrapper 的方式不需要提前将 gradle 下载好，而是会自动根据本地的缓存情况决定是否需要联网下载 gradle。Android Studio 默认没有启动 gradle wrapper 的方式，如果需要打开，可以选择 Android Studio 导航栏→File→Settings→Build，Execution，Deployment→Gradle，进行配置更改。

（5）.gitignore：这个文件是用来将指定的目录或文件排除在版本控制之外的。

（6）build.gradle：这是项目全局的 gradle 构建脚本，通常这个文件中的内容是不需要修改的。

（7）gradle.properties：这个文件是全局的 gradle 配置文件，在这里配置的属性将会影响到项目中所有的 gradle 编译脚本。

（8）gradlew 和 gradlew.bat：这两个文件是用来在命令行界面中执行 gradle 命令的，其中 gradlew 是在 Linux 或 Mac 系统中使用的，gradlew.bat 是在 Windows 系统中使用的。

（9）HelloWorld.iml：iml 文件是所有 IntelliJ IDEA 项目都会自动生成的一个文件（Android Studio 是基于 IntelliJ IDEA 开发的），用于标识这是一个 IntelliJ IDEA 项目，不需要修改这个文件中的任何内容。

（10）local.properties：这个文件用于指定本机中的 Android SDK 路径，通常内容都是自动生成的，并不需要修改。除非本机中的 Android SDK 位置发生了变化，那么就将这个文件中的路径改成新的位置即可。

（11）settings.gradle：这个文件用于指定项目中所有引入的模块。由于 HelloWorld 项目中就只有一个 app 模块，因此该文件中也就只引入了 app 这一个模块。通常情况下模块的引入都是自动完成的，需要手动去修改这个文件的场景可能比较少。

除了 app 目录之外，大多数的文件和目录都是自动生成的，并不需要进行修改，app 目录结构如下：

build：这个目录和外层的 build 目录类似，主要也是包含了一些在编译时自动生成的文件，不过它里面的内容会更多更杂，不需要过多关心。

libs：如果项目中使用到了第三方 Jar 包，就需要把这些 Jar 包都放在 libs 目录下，放在这个目录下的 Jar 包都会被自动添加到构建路径里去。

AndroidTest：此处是用来编写 Android Test 测试用例的，可以对项目进行一些自动化测试。

java：放置所有 Java 代码的地方，展开该目录，将看到刚才创建的 HelloWorld Activity 文件就在里面。

res：项目中使用到的所有图片、布局、字符串等资源都要存放在这个目录下。这个目录下还有很多子目录，图片放在 drawable 目录下，布局放在 layout 目录下，字符串放在 values 目录下。

AndroidManifest.xml：这是整个 Android 项目的配置文件，在程序中定义的所有四大组件都需要在这个文件里注册，另外还可以在这个文件中给应用程序添加权限声明。

test：此处是用来编写 Unit Test 测试用例的，是对项目进行自动化测试的另一种方式。

.gitignore：这个文件用于将 app 模块内的指定的目录或文件排除在版本控制之外，作用和外层的.gitignore 文件类似。

app.iml：IntelliJ IDEA 项目自动生成的文件，不需要关心或修改这个文件中的内容。

build.gradle：这是 app 模块的 gradle 构建脚本，这个文件中会指定很多项目构建相关的配置。

proguard-rules.pro：这个文件用于指定项目代码的混淆规则，当代码开发完成后打成安装包文件，如果不希望代码被破解，通常会将代码混淆，从而让破解者难以阅读。

2.3.4　Obtain_Studio 安装与配置

1. 关于 Obtain_Studio

对于不熟悉 Eclipse 的读者，或者感觉 Eclipse 安装过程太过于复杂的初学者，也可以采用本节所介绍的 Obtain_Studio 的 Android 版本。Obtain_Studio 解压之后无须安装可以直接运行，也无须配置即可以直接进行 Android 开发。

本书后面章节所介绍的例子，大部分在 Obtain_Studio 开发平台中开发，Obtain_Studio 的 Android 项目与 Eclipse 项目完全兼容。即 Obtain_Studio 的 Android 项目可以直接在 Eclipse 中打开和运行，Eclipse 项目也可以直接在 Obtain_Studio 中直接打开和运行。

本书使用 Obtain_Studio 的原因是因为 Obtain_Studio 在不需要安装的情况下就可以直接开发 Eclipse 项目，有利于初学 Android 开发的人员可以快速入门并且可以把精力集中在编程上，而不需要花费大量的时间去安装 Android 开发软件。入门之后，可以进一步学习 Eclipse 的使用方法。

Obtain_Studio 的本质，是一个记事本与编译批处理文件的合成。相当于把用记事本写程序以及用批处理文件编译程序的功能合成到了 Obtain_Studio 之中，方便用户使用命令方式来编写应用程序和编译程序。

Obtain_Studio 平台与微软的 Visual Studio 相类似，也可进行 Windows 可视化程

序设计,前者具有更好的嵌入式软件开发支持,例如支持 MS51、AVR 等单片机开发,支持 ARM7、ARM9 软件开发,SystemC 仿真与自动生成 Verilog HDL 代码,支持 Java、JSP、Struts、Spring 开发,支持 Windows 桌面操作系统和 Windows CE 嵌入式操作系统应用软件开发,支持 Smartwin＋＋、QT、计算机视觉开源库 OpenCV 和 3D 开源库 irrlicht 应用软件开发,还支持 Android 操作系统应用程序的开发。

2. 安装与配置

Obtain_Studio 并不需要安装与配置,这里所说的安装其实就是下载和解压 Obtain_Studio 软件包。

(1) Obtain_Studio 直接把压缩包解压即可用,无须安装。解压目录不能有中文的目录名或放桌面上。尽量放在"D:\ Obtain_Studio""E:\ Obtain_Studio""F:\ Obtain_Studio"等目录。

(2) 软件启动文件为 Obtain_Studio.exe,可以直接运行。该文件在 Obtain_Studio 解压目录下的 bin 子目录之中。

(3) 特别需要注意,Obtain_Studio 的应用程序都以项目的形式存在,项目一般都放在工作目录下。工作目录必须直接入在 Obtain_Studio 解压的目录下,默认为 Workie,例如"D:\Obtain_android\WorkDir"。Android 的工作目录也可以是 android work。

Obtain_Studio 新创建的项目,默认在 Obtain_Studio 目录下的 Workie 子目录中。

3. Obtain_Studio 的特点

Obtain_Studio 具有如下特点:

(1) Obtain_Studio 是永远性全免费集成开发环境,任何人都可以使用、拷贝、配置、修改,也可以免费用于商业用途。

(2) Obtain_Studio 是绿色软件,不用安装,也不用配置系统环境变量和全局环境变量,压缩包解压后双击 bin 目录下的 Obtain_Studio.exe 文件即可运行,不想用时直接删除 Obtain_Studio 所在目录即可。

(3) Obtain_Studio 可以与任何编译器配合使用,只要通过批处理文件进行临时环境变量配置和通过批处理文件进行项目编译即可。

(4) Obtain_Studio 集成开发环境中提供有许多应用模板,可以快速生成各种应用项目。

2.4 移动软件开发 Hello World 程序

2.4.1 如何启动 Obtain_Studio 集成开发环境

1. Obtain_Studio 的安装

Obtain_Studio 并不需要安装,只需要将压缩包解压到某个硬盘根目录下或一个简单的英文目录名下即可。具体方式如下：
- 将光盘中的 Obtain_Studio 集成开发环境拷贝或者解压缩到硬盘某个盘的根目录下（全部路径都使用英文路径），例如在"E:\"目录下，即"E:\Obtain_Studio"；
- Obtain_Studio 目录下的 Bin 子目录，有一个名为 Obtain_Studio.exe 的可执行文件，用鼠标双击该文件即可以启动 Obtain_Studio 集成开发环境。

由于 Obtain_Studio 中的 WinARM 编译环境是采用动态环境变量设置的方式（即绿色版方式），因此，运行 Obtain_Studio 集成开发环境不会影响到其他的 GCC 软件的运行。

大部分 IDE 都是采用设置固定环境变量的方式，如果计算机中已经安装有其他版本的 GCC 编译环境，很可能会影响到 Obtain_Studio 动态环境变量的自动获取。因此，如果发现 Obtain_Studio 不能正常编译，应该首先考虑卸载已经安装在计算机中的 GCC 编译环境，改用非安装版的（即动态环境变量设置的版本）。

2. Obtain_Studio 的运行

运行 X:\Obtain_Studio\bin 目录下的 Obtain_Studio.exe 文件，此时将出现如图 2-15 所示的软件主界面。

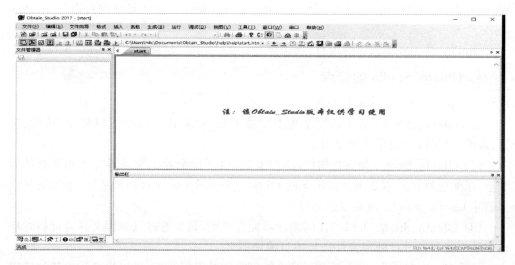

图 2-15　Obtain_Studio 主界面

3. 恢复 Obtain_Studio 原始界面状态

有时候不小心把 Obtain_Studio 的界面弄乱了，或者关闭了某些窗口而找不到打开的地方，可以采用如下方式恢复 Obtain_Studio 原始界面状态：
- 退出 Obtain_Studio；
- 选择电脑桌面"开始"→"运行"，输入 regedit；
- 在注册表里，找到 Obtain 一项，并显示"HKEY_CURRENT_USER\Software\Obtain"，删除 Obtain 这一项的所有内容；

● 重新启动 Obtain_Studio。

2.4.2 创建 Android 项目

1. 创建 Android 项目

在 Obtain_Studio 开发环境下,选择菜单"文件"→"新建项目",打开新建项目对话框,或在 Obtain_Studio 主界面左边的项目资源管理器单击鼠标右键新建项目菜单,也可进入新建项目对话框。项目类型下拉类别中选择"android 项目",在右边的模版名称框中选择"hello 模板"创建一个名为 Android_001 的项目,项目保存路径采用默认的路径"C:\Users\hilic\Documents\Obtain_Studio\WorkDir\",如图 2-16 所示。

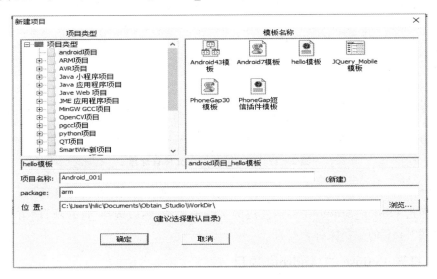

图 2-16 Obtain_Studio 创建新项目模板

在 Obtain_Studio 创建项目时,如果选择在 Obtain_Studio 里的 Android Studio 项目模板,则 Obtain_Studio 生成的项目与 Android Studio 相同和完全兼容。也就是说,可以直接拿到 Android Studio 里编译和运行。

需要特别注意的是,所有的 Obtain_Studio 项目,都建议放在 Obtain_Studio 所在的目录下的 WorkDir 子工作目录下,否则,某些类型的项目可能不能正确编译。主要是因为 Obtain_Studio 并没有配置全局变量,编译时头文件的查找一般都采用了相对于 WorkDir 工作子目录。

单击"确定"按钮之后,即从"Android 项目"→"hello 模板"中生成了一个新的 Android 项目,在所生成的项目中包含了 Key、Led 相关的一些初始化以及做演示的完整源代码,编译之后,即可下载到目标模板上面运行。项目文件结构从 Obtain_Studio 文件管理器中可以看出,如图 2-17 示。

2. 打开 Obtain_Studio 的 Android 项目

对于 Obtain_Studio 的 Android 项目,在 Obtain_Studio 主界面中,可以选择菜单

图 2-17 Obtain_Studio 文件管理器

"文件"→"打开项目"来选择扩展名为".prj"的项目文件,然后直接打开即可。也可以用鼠标单击 Obtain_Studio 主界面左边的文件管理器,然后选择"打开项目"菜单,选择扩展名为".prj"的项目文件来打开。

3. 打开 Eclipse 的 Android 项目

对于打开 Eclipse 的 Android 项目,在 Obtain_Studio 主界面中,可以选择菜单"文件"→"打开项目",在打开项目对话框中,在最下边的"文件类型"列表框中,选择"eclipse 项目文件(*.project)",如图 2-18 所示,然后选择扩展名为".project"的项目

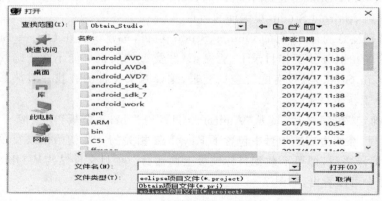

图 2-18 选择文件类型

文件打开即可。也可以用鼠标单击 Obtain_Studio 主界面左边的文件管理器,然后选择"打开项目"菜单,在打开项目对话框中,在最下边的"文件类型"列表框中,选择"eclipse 项目文件(＊.project)",最后选择扩展名为".project"的项目文件来打开。

4．打开 Android Studio 项目

在 Obtain_Studio 中,可以打开和创建 Android Studio 项目,这时 Obtain_Studio 与 Android Studio 项目完全兼容。打开 Android Studio 项目的方法与上述打开Eclipse 的 Android 项目完全相同。需要特别注意的是,Android Studio 项目与 Eclipse 的 Android 项目都必须拷贝到 Obtain_Studio 软件的工作目录之下才能正常编译和运行。

5．编辑 Java 程序

源程序放在\src 目录下,其中 hello.java 是主程序。Src\目录下放的是 Studio_GUI 库文件。在 Obtain_Studio 左边的项目文件资源管理器中双击相应的文件名,即可以打开文件,然后对文件进行编辑。

6．编辑 Android 界面布局文件

例如,main.xml 是上面 hello 项目的主界面布局文件。在 Obtain 左边的工具栏中,包括了常用的 Android 控件布局代码,选择想要添加的控件拖到右边的源程序编辑界面中,然后进一步修改布局参数即可,如图 2-19 所示。

图 2-19 拖动添加的控件

7．Android 中的 Java、XML 源文件编码问题

(1) UTF-8 编码

Android 项目的 Java、XML 源文件编码默认格式是 UTF-8 编码,在这种情况下,Obtain_Studio 可以直接打开 UTF-8 编码的 Java、XML 源文件。保存时,也将自动保存为 UTF-8 编码格式。

(2) 非 UTF-8 编码

对于非 UTF-8 编码的 Java、XML 源文件，在 Obtain_Studio 的 Android 项目中直接打开将出现乱码或打开的文件不完整，如图 2-20 所示。

图 2-20 非 UTF-8 编码

解决办法是在 Obtain_Studio 的 Android 项目左边的文件管理器中，选择准备打开的文件单击鼠标右键，在弹出菜单中选择"打开方式"，如图 2-21 所示。

在文件打开方式对话框中，选择"java 源文件(非 UTF-8 转 UTF-8)"，如图 2-21 所示。

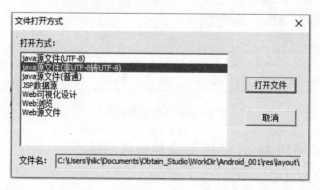

图 2-21 选择非 UTF-8 转 UTF-8 打开方式

然后单击"打开文件"按钮,可以正常打开和显示非 UTF-8 编码的文件,如图 2-22 所示。打开完成之后,再保存该文件,即可以按 UTF-8 编码格式保存文件。

图 2-22　非 UTF-8 转 UTF-8 编码

2.4.3　编译和运行

1. 编译并生成 APK 安装包

Obtain_Studio 开发环境下,编译项目有两种方法:

(1) 单击工具条上的编译按钮。Obtain_Studio 编译工具条如图 2-23 所示。只要单击工具条上的编译按键即可。

图 2-23　Obtain_Studio 编译工具条

(2) 选择主菜单"生成—保存并编译"。编译的结果可以从 Obtain_Studio 下面的输入栏看到。如果编译不成功,所有的错误和警告消息,都在输入栏显示。

如果编译时没有错误,则在 Obtain_Studio 主界面下边的输出栏中输出如下消息:

```
……(其他省略)
debug:
[propertyfile] Updating property file: F:\Obtain_Studio\WorkDir\android_001\bin\build.prop
    BUILD SUCCESSFUL
    Total time: 6 seconds
```

其中最后几行显示的内容包含"BUILD SUCCESSFUL",代表编译正常。

编译完成之后,生成 APK 安装包,该安装包位于项目的 BIN 目录下,扩展名为". apk"。例如 hello-debug. apk、hello-debug-unaligned. apk 等。这些安装包可以在模拟器上运行,也可以下载到安卓手机或安卓平板电脑上运行。

2. 启动 Android 模拟器

Android 模拟器,就是 Android 手机/平板电脑的虚拟机,是为了程序开发者开发应用而提供的一个开发工具。Android 模拟器 Android SDK 自带一个移动模拟器,是一个可以运行在电脑上的虚拟设备。Android 模拟器可以无须使用物理设备即可预览、开发和测试 Android 应用程序。

Android 模拟器能够模拟除了接听和拨打电话外的所有移动设备上的典型功能和行为。Android 模拟器提供了大量的导航和控制键,可以通过鼠标或键盘单击这些按键来为应用程序产生事件。同时,它还有一个屏幕用于显示 Android 自带应用程序和自己的应用程序。为了便于模拟和测试应用程序,Android 模拟器允许的应用程序通过 Android 平台服务调用其他程序、访问网络、播放音频和视频、保存和传输数据、通知用户、渲染图像过渡和场景。在 Obtain_Studio 中,启动 Android 模拟器的工具如图 2-24 所示。

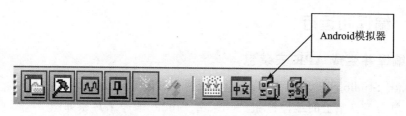

图 2-24　启动 Android 模拟器工具

启动 Android 模拟器的过程特别慢,请耐心等待。启动完成之后,可以保持模拟器的运行状态,不要关闭。Android 模拟器开锁之后,如图 2-25 所示。

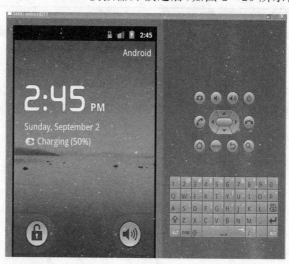

图 2-25　Android 模拟器

模拟器 90°旋转,快捷键:Ctrl+F11 或 Ctrl+F12。

在 Obtain_Studio 中,运行 Android 应用程序的工具如图 2-26 所示。

图 2-26　运行 Android 应用程序的工具

运行 Android 应用程序之后,出现如下消息,代表已经成功运行:

```
F:\Obtain_Studio\WorkDir\android_001\bin>adb uninstall
    com.android.hello
* daemon not running. starting it now on port 5037 *
* daemon started successfully *
Success
- waiting for device -
F:\Obtain_Studio\WorkDir\android_001\bin>adb - s emulator - 5554 install hello - debug.apk
    pkg: /data/local/tmp/hello - debug - unaligned.apk
Success
6 KB/s (6153 bytes in 1.000s)
F:\Obtain_Studio\WorkDir\android_001\bin>adb - s emulator - 5554 shell am start - n com.android.hello/.hello
Starting: Intent { cmp = com.android.hello/.hello }
```

在模拟器中,可以看到运行的效果,如图 2-27 所示。

图 2-27　程序运行效果

2.4.4 Obtain_Studio 集成开发系统常用技巧

1. 如何在项目中添加新文件

可以选择 Obtain_Studio 主菜单"文件—新建文件",或单击左上边工具条的新建文件按钮,进入新建文件对话框,在文件对话框中下边的"位置"框中选择文件将创建的目录,在文件对话框上边的"文件类型"框中选择所需要的文件类型,在对话框中间的"文件名"框中输入要创建的文件名,最后单击"确定"按钮即可创建一个新文件。

也可以在左边文件管理器栏中对着要创建新文件的目录单击鼠标右键,然后选择菜单"新建文件",这样文件对话框中的目录已经变为鼠标单击位置的目录,选择文件类型并录入文件名然后单击"确定"按钮即可以创建新文件。

新文件是从一些模板文件中直接拷贝过来的,如果用户想自己创建一些模板文件,可以把模板文件放到 Obtain_Studio 所在目录下的 bin\模板\新建立文件的样板\newfile 子目录中,即可变成公共的模板文件。

如果想创建某种项目类型专用的模板文件,可以到项目模板所在目录下创建一个名为 newfile 的新目录,然后把模板文件放到该目录下即可。例如,要创建 STM32F4_GCC 项目模板的模板文件,到 Obtain_Studio_mini 所在目录的 bin\模板\ARM 项目\STM32 项目\STM32F4_GCC 项目模板子目录下,创建一个 newfile 新目录,然后把模板文件放到该目录下即可。

2. 如何创建项目模板

用户可以自己创建项目模板,方法是:

(1) 把已经编辑好并且可以正常编译的项目拷贝到 Obtain_Studio 所在目录的\bin\模板目录的某种类型的项目子目录下,例如,把 STM32 的项目模板放到"\bin\模板\android 项目"目录下。如果不是这种类型,也可以新创建目录,目录名字以"项目"两个字结束。

(2) 把新拷贝过来的项目目录名修改为用户想要的名称,目录名字以"模板"两个字结束,例如"hello 模板"。

(3) 把该模板目录下扩展名为". prj"的项目文件名修改成"prj. prj",然后用记事本打开 prj. prj 文件,最关键是要修改 <typepath> 一项,该项目的内容要与当前模板目录完全相同,例如"\android 项目\hello 模板",另外, <title> 也应该修改成新的模板名称。修改完成之后的内容格式如下:

```
<projectname>hello 模板</projectname>
<projecttype>android</projecttype>
<title>android 项目_hello 模板</title>
<package>arm</package>
<typepath>\android 项目\hello 模板</typepath>
<projectpath>F:\Obtain_Studio\bin\模板\android 项目\hello 模板
```

```
</projectpath><envionment_variables_batfile>
    bin\config\android.bat
    </envionment_variables_batfile>
    <compile>
    ant debug
    </compile>
    <emulator>
    cd..
    cd..
    cd .\android_AVD\.android\avd
    emulator – avd android223
    </emulator>
    <run>
    cd bin
    adb uninstall com.android.hello
    adb – s emulator – 5554 install hello – debug – unaligned.apk
    adb – s emulator – 5554 shell am start – n com.android.hello/.hello
    </run>
```

prj.prj 文件采用 XML 的格式编写,格式为"<名称>内容</名称>",其中<名称>表示某项内容的开始,</名称>表示某项内容的结束,中间是该项目的内容。prj.prj 文件各项内容的意义如下:

- <projecttype>android</projecttype>代表项目类型的 android 类型。
- <projectname>一项的内容可以不写,项目名称。
- <title>一项是模式的名称,这一项一定要写,用户根据需要自己写一个想要的模板名字。
- <package>这一项可以不写,只在 Java 项目中用。
- <typepath>这一项一定要写,代表模板所在的目录,是一个相对目录,只能写出"Obtain_Studio\bin\模板\"目录之后的那一部分。
- <projectpath>这一项可以不写,是项目当前所在路径。
- <envionment_variables_batfile>这一项一定要写,是默认的用于编译时设置环境变量的批处理文件名以及相对目录,相对于 Obtain_Studio 所在的目录。例如,"bin\config\android.bat",android.bat 文件就是 android JDK 编译器的环境变量设置文件。
- <compile>这一项一定要写,是编译时执行编译的批处理文件名。

3. 用户自己配置新的编译器环境变量

例如,从网上下载一个新的 android JDK 编译器,如何配置环境变量让 Obtain_Studio 里的项目在编译时调用该新的编译器呢?

方法就是在 Obtain_Studio 所在目录下的 config 子目录中创建一个新的批处理文

件。下面是 ARM GCC 编译器的环境变量设置文件 android.bat 的内容,用户可以参考该文件自己配置新的编译器环境变量。

例如,下载下来的 android JDK 编译器解压(或安装)在 Obtain_Studio 所在目录的 \ARM\jdk 目录下,那么 android.bat 内容可以写成如下所示:

```
@rem ------------------------------------
@rem ---下行是设置 MINGW 安装根目录,把{{{}}}用安装目录代替即可,例如 MINGW_ROOT = F:\Obtain_Studio ---------------
@rem --这里是自动环境设置,请不要把{{{}}}删除,程序会自动用 F:\Obtain_Studio 代替 - ------------

@echo Setting environment for using ant and android
@set ANT_HOME = {{{}}}\ant
@set JAVA_HOME = {{{}}}\jdk
@set ANDROID_BIN = {{{}}}\android\platform-tools;{{{}}}\android\tools
@set CLASSPATH = {{{}}}\jdk\lib\tools.jar
@set PATH = %PATH%;%JAVA_HOME%\bin;%ANT_HOME%\bin;%ANDROID_BIN%;{{{}}}\jdk\jre\bin;{{{}}}\android
@set ANDROID_SDK_HOME = {{{}}}/android_avd
```

上面的配置使用了相对地址,相对于 Obtain_Studio 所在的目录。如果 android JDK 编译器解压(或安装)不在 Obtain_Studio 所在目录之下,那么就得配置一个绝对地址。例如,JDK 和 ant 安装在 D 盘的 mydir 目录下,那么批处理文件可以改写成如下形式:

```
@rem ------------------------------------
@rem ---下行是设置 MINGW 安装根目录,把 d:\mydir 用安装目录代替即可,例如 MINGW_ROOT = F:\Obtain_Studio ---------------
@rem --这里是自动环境设置,请不要把 d:\mydir 删除,程序会自动用 F:\Obtain_Studio 代替 -------------

@echo Setting environment for using ant and android
@set ANT_HOME = d:\mydir\ant
@set JAVA_HOME = d:\mydir\jdk
@set ANDROID_BIN = d:\mydir\android\platform-tools;
              d:\mydir\android\tools
@set CLASSPATH = d:\mydir\jdk\lib\tools.jar
@set PATH = %PATH%;%JAVA_HOME%\bin;%ANT_HOME%\bin;%ANDROID_BIN%;d:\mydir\jdk\jre\bin;d:\mydir\android
@set ANDROID_SDK_HOME = d:\mydir/android_avd
```

4. 为关键词自动提示功能添加新的关键词

在编译 C、C++、Java 等源程序时,只要输入关键词的前面部分字符,Obtain_Studio 会自动提示后面部分字符供选择。这些关键词都放在 Obtain_Studio 所在目录 config 子目录的 CFileModule.config 文件中,用记事本打开该文件,在其他已经定义好的关键词之后添加新的一行并且录入新的关键词即可。

Obtain_Studio 可以自动认识一些类、对象和变量然后产生自动录入提供功能,但提示的内容并不完整。对于一些常用的类,用户可以按\Obtain_Studio\bin\config\C_TXT 目录下扩展名为.txt 的那么文件的格式,自己创建提示文件内容。

编写规则是把包含有类定义的.h、.c、.cpp 扩展的文件拷贝到\Obtain_Studio\bin\config\C_TXT 目录下,并把扩展名改变.txt。然后用记事本打开这些文件,在类定义中的变量成员和成员函数名前插入一个"|"符号(在键盘 Enter 键上边),然后保存退出即可。这样 Obtain_Studio 会自动到这里查找提示的内容。

2.5 Android 项目

2.5.1 Android 项目结构

1. Android 项目目录

Android 项目的结构包含了如表 2-2 所列的文件和目录,本章将对该项目之中的文件进行详细的分析。Android 项目文件目录结构如表 2-2 所列,包括了项目根目录之下的应用程序声明文件 AndroidManifest.xml 以及相关子目录。

表 2-2 Android 项目结构

文件	说明
AndroidManifest.xml	应用程序声明文件
build.xml	Ant 的构建文件
default.properties project.properties	保存编译目标,由 Android 工具自动建立,不可手工修改
build.properties ant.properties	保存自定义的编译属性
local.properties	保存 Android SDK 的路径,仅供 Ant 使用
/<路径>/myActivity.java	Activity 文件
/bin	编译脚本输出目录
/gen	保存 Ant 自动生成文件的目录,例如 R.java
/libs	私有函数库目录
/res	资源目录
/src	源代码目录
/tests	测试目录
/assets	应用系统需要使用到的诸如 mp3、视频类的文件

Android ADT 14.0.0 改名,将 default.properties 改成了 project.properties, build.properties 改成了 ant.properties,然后是使用 ADT 自动生成项目,ADT 会自动对这些文件改名。

gen 文件夹下面有个 R.java 文件,R.java 是在建立项目时自动生成的,这个文件是只读模式的,不能更改。R.java 文件中定义了一个类——R,R 类中包含很多静态类,且静态类的名字都与 res 中的一个名字对应,即 R 类定义该项目所有资源的索引。

通过 R.java 可以很快地查找需要的资源,另外,编译器也会检查 R.java 列表中的资源是否被使用到,没有被使用到的资源不会编译到软件中,这样可以减少应用在手机占用的空间。

2. 使用命令行创建新项目

在开发 Android 应用程序过程中,除了使用现成的项目,以及使用集成开发环境(IDE)创建新项目之外,也可以使用命令行来建新项目。建立过程:开始→运行→CMD 启动 CMD 并进入 <Android SDK>/tools 目录,输入以下命令:

```
android create project - n HelloCommandline - k
    edu.hrbeu.HelloCommandline - a HelloCommandline - t 2 - p
e:\Android\workplace\HelloCommandline
```

或者使用以下命令:

```
android create project -- name HelloCommandline -- package
edu.hrbeu.HelloCommandline -- activity HelloCommandline
-- target 2 -- path e:\Android\workplace\HelloCommandline
```

新工程的名称为 HelloCommandline,包名称为 edu.hrbeu.Hello Commandline,Activity 名称是 HelloCommandline,编译目标的 ID 为 2,新工程的保存路径是 E:\Android\workplace\HelloCommandline。

2.5.2 Android 项目文件

1. 项目文件内容

Obtain 中的 Android 项目文件的扩展名为".prj",在项目的根目录下可以找到。典型的 Obtain 中的 Android 项目文件的内容如下:

```
<projectname>android_001</projectname>
<projecttype>android</projecttype>
<title>android 项目_hello 模板</title>
<package>arm</package>
<typepath>\android 项目\hello 模板</typepath>
<projectpath>F:\Obtain_Studio\WorkDir\android_001</projectpath>
```

```
<envionment_variables_batfile>
bin\config\android.bat
</envionment_variables_batfile>
<compile>
ant debug
</compile>

<emulator>
cd ..\..\android_AVD\.android\avd
emulator – avd android223
</emulator>
<run>
cd bin
adb uninstall com.android.hello
adb – s emulator – 5554 install hello – debug – unaligned.apk
adb – s emulator – 5554 shell am start – n com.android.hello/.hello
</run>
```

（1）envionment_variables_batfile 项，是环境变量的配置文件，使用 bin\config\android.bat；

（2）compile 项是编译配置编译命令，使用 ant debug 命令；

（3）emulator 项是启动 Android 模拟器命令；

（4）run 是运行 Android 应用程序命令。

2. Ant 项目编译

Android 项目采用的编译命令保存在项目文件的 <compile> 一项之中，默认命令是"ant debug"，这个是编译成调试版本。如果希望编译成发行版本，则把项目文件中 <compile> 一项内容修改为"ant release"。

Ant 编译 Android 项目时，可选择参数如下：
- debug:带调试用签名的构建 release:构建应用程序，生成的 apk 必须签名才可以发布；
- install:安装调试构建的包到运行着的模拟器或者设备；
- reinstall:重新安装；
- uninstall:卸载。

3. 启动 Android 模拟器

启动 Android 模拟器的命令保存在项目文件的 <emulator> 一项之中，默认命令如下：

cd ..\..\android_AVD\.android\avd

emulator —avd android223

前三行是转到Obtain_Studio自带的Android模拟器所在目录,然后调用emulator命令启动模拟器。

(1) 模拟器常用功能

模拟器常用功能如下:

1) 查看当前支持版本(在列出的版本中需要记住id值,这个值在第2步中使用):

android list target

2) 创建AVD:

android create avd -n magicyu -t 2

-n后面接需要创建avd的名字,-t后面接需要创建虚拟器的类型,2即为步骤1)中得到的类型id号。

3) 查看是否创建成功(如果成功会显示刚才创建的avd信息):

android list avd

4) 启动模拟器:

emulator -avd magicyu

其中-avd后接的是已经创建过的avd名字。

5) 选择启动的皮肤:

emulator -avd magicyu -skin QVGA

skin后面接所要启动皮肤的类型,所有的类型可以在/skins目录下找到,例如,HVGA、QVGA、WVGA800、WVGA854等。按Ctrl+F11,可以直接改变模拟器的横纵摆放。

(2) 为模拟器加上SD卡

为模拟器加上SD卡的命令如下:

emulator -sdcard <path>/sdcard.img

创建"sdcard.img"文件的命令如下:

mksdcard 1024M D:/sdcard.img

传文件到SDCard的命令如下:

adb push <目录/audio.mp3> </sdcard/audio.mp3>

4. Android项目的运行

Obtain_Studio中运行Android项目的运行命令保存在项目文件的<run>一项之中,默认命令是:

cd bin

adb -s emulator-5554 uninstall com.android.hello

adb -s emulator-5554 install hello-debug-unaligned.apk

adb -s emulator-5554 shell am start -n com.android.hello/.hello

其中emulator-5554是模拟器名称,模拟器上方显示有该名称。如果需要连接手机真机安装,则把"emulator-5554"修改为电脑连接手机时的手机名称即可。

ADB(Android Debug Bridge)是 Android 提供的一个通用的调试工具,借助这个工具,可以管理设备或手机模拟器的状态,还可以进行以下的操作:
- 快速更新设备或手机模拟器中的代码,如应用或 Android 系统升级;
- 在设备上运行 shell 命令;
- 管理设备或手机模拟器上的预定端口;
- 在设备或手机模拟器上复制或粘贴文件。

其他常用的 ADB 操作:
- 上传文件:adb push <PC 文件> </tmp/...>;
- 下载文件:adb pull </tmp/...> <PC 文件>;
- 安装程序:adb install <*.apk>;
- 卸载软件:adb shell rm /data/app/<*.apk>。

5. 动态环境变量配置

Obtain_Studio 采用了动态环境变量配置的方式,这样 Obtain_Studio 无须安装即可运行,同时也无须修改系统的环境变量,避免影响到其他软件的运行。Obtain_Studio 在编译、运行 Android 项目时,会首先配置动态环境变量,其配置命令在项目文件的 <envionment_variables_batfile> 一项之中定义,Android 项目默认的动态环境变量配置命令是:

bin\config\android.bat

android.bat 是一个批处理文件,内容如下:

```
@rem ------------------------------------
@rem ---下行是设置 MINGW 安装根目录,把{{{}}}用安装目录代替即可,例如 MINGW_ROOT = F:
\Obtain_Studio---------------
@rem ---这里是自动环境设置,请不要把{{{}}}删除,程序会自动用 F:\Obtain_Studio 代替-
------------

@echo Setting environment for using ant and android
@set ANT_HOME = {{{}}}\ant
@set JAVA_HOME = {{{}}}\jdk
@set ANDROID_BIN = {{{}}}\android\platform-tools;{{{}}}\android\tools
@set CLASSPATH = {{{}}}\jdk\lib\tools.jar
@set PATH = %PATH%;%JAVA_HOME%\bin;%ANT_HOME%\bin;%ANDROID_BIN%;{{{}}}\jdk\jre\bin;{{{}}}\android
@set ANDROID_SDK_HOME = {{{}}}/android_avd
```

2.5.3 Android 项目编译与配置文件

build.xml 是 Ant 的构建过程文件。当开始一个新的项目时,首先应该编写 Ant 构建文件。构建文件定义了构建过程,并被团队开发中每个人使用。Ant 构建文件默认命名为build.xml,也可以取其他的名字。只不过在运行的时候把这个命名当作参数传给 Ant。构建文件可以放在任何位置。一般做法是放在项目顶层目录中,这样可以保持项目的简洁和清晰。下面是一个典型的项目层次结构。

> src 存放文件;
> class 存放编译后的文件;
> lib 存放第三方 JAR 包;
> dist 存放打包,发布以后的代码。

Ant 构建文件是 XML 文件。每个构建文件定义一个唯一的项目(Project 元素)。每个项目下可以定义很多目标(target 元素),这些目标之间可以有依赖关系。当执行这类目标时,需要执行他们所依赖的目标。每个目标中可以定义多个任务,目标中还定义了所要执行的任务序列。Ant 在构建目标时必须调用所定义的任务。任务定义了 Ant 实际执行的命令。Ant 中的任务可以为三类:

> 核心任务:核心任务是 Ant 自带的任务;
> 可选任务:可选任务是来自第三方的任务,因此,需要一个附加的 JAR 文件;
> 用户自定义的任务:用户自定义的任务/用户自己开发的任务。

2.5.4　Android 项目全局配置文件

AndroidManifest.xml 是 Android 应用程序中最重要的文件之一。它是 Android 程序的全局配置文件,是每 Android 程序中必须的文件。它位于应用程序的根目录下,描述了 package 中的全局数据,包括 package 中暴露的组件(activities、services 等),以及它们各自的实现类,各种能被处理的数据和启动位置等重要信息。

AndroidManifest.xml 文件提供了 Android 系统所需要的关于该应用程序的必要信息,即在该应用程序的任何代码运行之前系统所必须拥有的信息。

一个典型的 AndroidManifest.xml 文件内容如下:

```
<? xml version = "1.0" encoding = "utf - 8"? >
<manifest xmlns:android = "http://schemas.android.com/apk/res/android"
    package = "com.android.hello"
    android:versionCode = "1"
    android:versionName = "1.0">
  <application android:label = "@string/app_name" >
      <activity android:name = "hello"
            android:label = "@string/app_name">
        <intent - filter >
            <action android:name = "android.intent.action.MAIN" />
            <category android:name = "android.intent.category.LAUNCHER" />
        </intent - filter >
      </activity>
  </application >
</manifest>
```

1. AndroidManifest.xml 主要功能

AndroidManifest.xml 主要包含以下功能：
- 说明应用的 Java 数据包，数据包名是 application 的唯一标识；
- 描述应用的元件(component)；
- 说明应用的元件运行在哪个过程(process)下；
- 声明应用所必须具备的权限，用以访问受保护的部分 API，以及与其他应用的交互；
- 声明应用其他的必备权限，用以元件之间的交互；
- 列举应用运行时需要的环境配置信息，这些声明信息只在程序开发和测试时存在，发布前将被删除；
- 声明应用所需要的 Android API 的最低版本级别，比如 1.0,1.1,1.5；
- 列举应用所需要链接的库。

2. AndroidManifest.xml 文件主要内容

AndroidManifest.xml 文件的结构、元素，以及元素的属性，可以在 Android SDK 文档中查看详细说明。而在看这些众多的元素以及元素的属性前，需要先了解一下这些元素在命名、结构等方面的规则。

(1) 元　素

在所有的元素中只有 < manifest > 和 < application > 是必需的，且只能出现一次。如果一个元素包含有其他子元素，必须通过子元素的属性来设置其值。处于同一层次的元素的说明是没有顺序的。

(2) 属　性

按照常理，所有的属性都是可选的，但是有些属性是必须设置的。那些真正可选的属性，即使不存在，其也有默认的数值项说明。除了根元素 < manifest > 的属性，所有其他元素属性的名字都是以 android:前缀的。

(3) 定义类名

所有的元素名都对应其在 SDK 中的类名，如果自己定义类名，必须包含类的数据包名，如果类与 application 处于同一数据包中，可以直接简写为"."。

(4) 多数值项

如果某个元素有超过一个数值，这个元素必须通过重复的方式来说明其某个属性具有多个数值项，且不能将多个数值项一次性说明在一个属性中。

(5) 资源项说明

当需要引用某个资源时，其采用如下格式：@[package:]type:name。例如：

    < activity android:icon＝"@drawable/ icon ">

(6) 字符串值

类似于其他语言，如果字符中包含有字符，则必须使用转义字符。

3. AndroidManifest.xml 文件的主要元素

下面将介绍 AndroidManifest.xml 文件之中包括的各个元素的功能。

(1) 包名(package)

指定本应用内 Java 主程序包的包名。当没有指定 apk 的文件名时,编译后产生程序包将以此命名。本包名应当在 Android 系统运行时唯一。

(2) 认证(certificate)

指定本应用程序所授予的信任级别,目前有的认证级别有 platform(system)、shared、media,以及应用自定义的认证。不同的认证可以享受不同的权限。

(3) 权限组(permission-group)

权限组的定义是为了描述一组具有共同特性的权限。

(4) 权限(permission)

权限用来描述是否拥有做某件事的权力。Android 系统中权限是分级的,分为普通级别(Normal)、危险级别(dangerous)、签名级别(signature)和系统/签名级别(signature or system)。系统中所有预定义的权限根据作用的不同,分别属于不同的级别。对于普通和危险级别的权限,称之为低级权限,应用申请即授予。其他两级权限,称之为高级权限或系统权限,应用拥有 platform 级别的认证才能申请。当应用试图在没有权限的情况下做受限操作,应用将被系统摧毁以警示。系统应用可以使用任何权限。权限的声明者可无条件使用该权限。

(5) 权限树(permission-tree)

权限树的设置是为了统一管理一组权限,声明于该树下的权限所有者归属该应用。系统提供了 API 函数 PackageManager.addPermission(),应用可以在运行时动态添加。

(6) 使用权限(uses-permission)

应用需要的权限应当在此处申请,所申请的权限应当被系统或某个应用所定义,否则视为无效申请。同时,使用权限的申请需要遵循权限授予条件,非 platform 认证的应用无法申请高级权限。

(7) SDK(uses-sdk)

标识本应用运行的 SDK 版本。高兼容性的应用可以忽略此项。

(8) application:application

application:application 是 Android 应用内最高级别(top level)的模块,每个应用内最多只能有一个 application,如果应用没有指定该模块,一个默认的 application 将被启用。application 将在应用启动时最先被加载,并存活在应用的整个运行生命周期。因此,一些初始化的工作适合在本模块完成。Application 元素有许多属性,其中:"persistent"表示本应用是否为常驻内存,"enable"表示本应用当前是否应当被加载。例如:

```
<application android:icon = "@drawable/icon" android:label = "@string/
            app_name">
    <activity android:name = ".HelloOPhone"
            android:label = "@string/app_name">
        <intent-filter>
            <action android:name = "android.intent.action.MAIN" />
            <category android:name = "android.intent.category.LAUNCHER"/>
        </intent-filter>
    </activity>
</application>
```

在 AndroidManifest.xml 文件中,运行时模块的定义都作为本模块的子元素。当运行时模块被调用,如果应用没有启动,将首先启动应用进行初始化,然后调用对应模块。

(9) Activity:Activity

Activity:Activity 是 application 模块的运行时子元素,标识了一个 UI。除了 Application,一个应用可以声明并实现零至多个其他运行时模块,Activity 也同样。Activity 也包含了许多定义它工作状态的属性,其中"name"是必须的,它指定了该 Activity 所在的文件名,如果该文件所属包不同于该应用的包名(即本描述文件的最开始处),那么名字前面需要加入所在包名。Activity 通过增加 intent-fliter 来标识哪些 Intent 可以被处理,同时 Intent 也是调用 Activity 的主要参数。

例如,intent-filter 如下:

      `<action android:name="android.intent.action.MAIN" />`

      `<category android:name="android.intent.category.LAUNCHER" />`

intent-filters 告诉 Android 系统哪个 Intent 它们可以处理,activities、services 和 broadcast receivers 必须设置一个或者多个 Intent 过滤器。每个过滤器描述了组件的一种能力,它过滤掉不想要的 Intent,留下想要的。显示意图则不用考虑这些。

一个过滤器中包含一个 Intent object 中的三个属性 action、data、catrgory。一个隐式意图必须要通过这三项测试才能传递到 包含该过滤器的组件中。

一个应用程序可以有多个 Activity,每个 Activity 是同级别的,那么在启动程序时,最先启动哪个 Activity 呢?有些程序可能需要显示在程序列表里,有些不需要,怎么定义呢?

android.intent.action.MAIN 决定应用程序最先启动的 Activity;

android.intent.category.LAUNCHER 决定应用程序是否显示在程序列表里。MAIN 和 LAUNCHER 同时设定才有意义。

2.5.5 Android 资源文件

Android 支持字符串,位图和许多其他类型的资源。每一种资源定义文件的语法和格式及保存的位置取决于其依赖的对象。通常,可以通过三种文件创建资源:XML

文件(除位图和原生文件外)、位图文件(作为图片)和原生文件(所有其他的类型,比如声音文件)。Android 资源类型如表 2-3 所列。

表 2-3 Android 资源类型

目 录	资源类型
res/anim	XML 文件编译为桢序列动画或者自动动画对象
res/drawable	.png,9.png,.jpg 文件被编译为 Drawable 资源子类型: ➢ 使用 Resources.getDrawable(id)可以获得资源类型 ➢ 位图文件 ➢ 9-patchs(可变位图文件)
res/layout	资源编译为屏幕布局器
res/values	XML 文件可以被编译为多种资源。不像其他 res 下的目录,这个目录可以包含多个资源描述文件。XML 文件元素类型控制着这些资源被 R 类放置在何处。这些文件可以自定义名称,这里有一些约定俗成的文件,例如: ➢ arrays.xml 定义数组 ➢ colors.xml 定义可绘制对象的颜色和字符串的颜色。使用 Resources.getDrawable() 和 Resources.getColor()都可以获得这些资源。 ➢ dimens.xml 定义尺度。使用 Resources.getDimension()可以获得这些资源 ➢ strings.xml 定义字符串(使用 Resources.getString()或者更适合的 Resources.getText()方法获得这些资源。Resources.getText()方法将保留所有用于描述用户界面样式的描述符,保持复杂文本的原貌。 ➢ styles.xml 定义样式对象
res/xml	自定义的 XML 文件。这些文件将在运行时编译进应用程序,并且使用 Resources.getXML()方法可以在运行时获取
res/raw	自定义的原生资源,将被直接拷贝到设备。这些文件将不被压缩进应用程序。使用带有 ID 参数的 Resources.getRawResource()方法可以获得这些资源,例如 R.raw.somefilename

资源被最终编译进 APK 文件。Android 创建包装类 R,可以用它找回资源。R 包含一些与资源所在目录同名的子类。

2.6 欢迎界面的实现

2.6.1 创建项目和编辑文件

1. 新建一个 Android 项目

在 Obtain_Studio 中,创建一个 Android 项目,选择与 Android Studio 项目相兼容的模板"\android 项目\AndroidStudio 项目\AndroidStudio43 模板"。项目名称为

"IoT_IS_App_001",代表物联网智能系统 App(Internet of things intelligent system App),如图 2-28 所示。

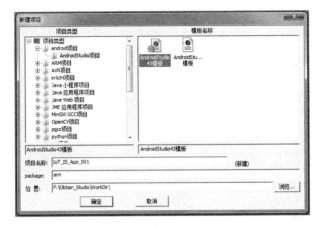
(a) 新建项目界面

(b) 项目目录结构

图 2-28　AndroidStudio43 模板

找到 Obtain_Studio 软件工作目录 WorkDir 子目录 IoT_IS_App_001\app\src\main\java\com\example\administrator\myapplication 的 Java 文件,把文件名"MainActivity.java"修改为"WelcomeActivity.java",打开 WelcomeActivity.java 文件,提取 Window 对象,然后隐藏了标题栏,程序代码如下:

```
package com.example.administrator.myapplication;
import android.app.Activity;
import android.os.Bundle;
import android.view.View;
import android.widget.Button;
import android.widget.EditText;
import android.widget.Toast;
import android.view.Gravity;
import android.view.Window;
import android.view.WindowManager;
public class WelcomeActivity extends Activity {
    @Override
    protected void onCreate(Bundle savedInstanceState) {
        super.onCreate(savedInstanceState);
        final Window window = getWindow();                    //获取当前的窗体对象
        window.setFlags(WindowManager.LayoutParams.FLAG_FULLSCREEN,
            WindowManager.LayoutParams.FLAG_FULLSCREEN);      //隐藏了状态栏
        requestWindowFeature(Window.FEATURE_NO_TITLE);        //隐藏了标题栏
        setContentView(R.layout.welcome);
    }
}
```

把\IoT_IS_App_001\app\src\main\res\layout 目录下的"activity_main.xml"文件修改为"welcome.xml",界面代码如下:

```xml
<?xml version = "1.0" encoding = "utf-8"?>
<LinearLayout xmlns:android = "http://schemas.android.com/apk/res/android"
    xmlns:tools = "http://schemas.android.com/tools"
    android:layout_width = "fill_parent"
    android:layout_height = "match_parent"
    android:background = "@drawable/welcome"
    android:orientation = "vertical" >
    <TextView
        android:layout_width = "fill_parent"
        android:layout_height = "wrap_content"
        android:layout_marginRight = "90dip"
        android:layout_marginLeft = "90dip"
        android:layout_marginTop = "130dip"
        android:textSize = "40sp"
        android:textColor = "#F83333"
        android:text = "物联网智能系统"
        />
    <TextView
        android:layout_width = "fill_parent"
        android:layout_height = "wrap_content"
        android:layout_marginRight = "90dip"
        android:layout_marginLeft = "90dip"
        android:layout_marginTop = "50dip"
        android:textSize = "50sp"
        android:textColor = "#3333F8"
        android:text = "欢迎您使用!"
        />
</LinearLayout>
```

修改 AndroidManifest.xml 之中的 activity 内容,把 MainActivity 修改为 WelcomeActivity。修改 IoT_IS_App_001.prj 文件,把启动的类名 MainActivity 修改为 WelcomeActivity。由于在界面中还用到了一个背景图片 welcome,因此,需要在\IoT_IS_App_001\app\src\main\res\drawable 目录下放置名为 welcome 的背景图片,例如放置 welcome.jpg 文件。

2.6.2 运行欢迎界面

1. 在模拟器上运行

完成上述初步之后,可以编译、启动模拟器和运行,运行效果如图 2-29 所示,完成了本章最开始所介绍的设计目标。

注意:启动模拟器需要的时间比较长,可以在打开项目之后首先启动模拟器;另外,启动模拟器之后不要关闭,可以一直用,就算重新启动项目或者启动其他项目,也可以共用该模拟器;模拟器上的数字5554就是TCP通信的端口,只要启动一个模拟器即可,如果启动多个模拟器,则上面的数字(端口号)会不同,希望连接到哪个模拟器,可以修改prj文件(本项目为"IoT_IS_App_001.prj")里Run命令中的端口号,例如,安装App命令adb-s emulator-5554 install app-debug.apk,5554就是对应安装到端口号为5554的模拟器上。

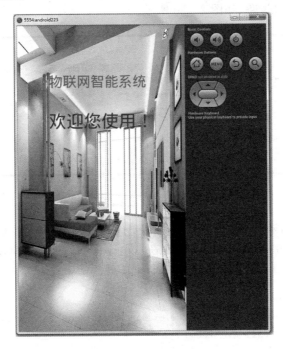

图2-29 欢迎界面App在模拟器上的运行效果图

Android SDK自带的ADB模拟器是通过TCP通信方式连接,由于部分电脑的防火墙、杀毒软件等原因,无法连接上ADB模拟器,那么也可以到网上自行下载一些安卓模拟器,网上有很多用于电脑玩安卓游戏的模拟器很好用,通过拖放APK软件包就可以安装。

2. 手机在线运行

1) 手机开启Use调试,点击手机Menu键(菜单键),在弹出的菜单中选择"设置"(Setting),或在应用程序中找到设置程序点击进入,进入设置界面的应用程序,即可打开USB调试模式。

2) 用数据线将手机与电脑相连,在手机弹出的对话框里选择"文件传输模式"。

3) 修改IoT_IS_App_001.prj里的运行选项,去掉"-s emulator-5554",例如:

```
<run>
cd app\build\outputs\apk\debug
adb    uninstall com.example.administrator.myapplication
adb    install app-debug.apk
adb    shell am start -n com.example.administrator.myapplication/.WelcomeActivity
</run>
```

保存修改，然后运行，即可在手机上安装 app-debug.apk 软件包并运行。也可以把"-s emulator-5554"修改成手机的序列号。首先，查看手机序列号，在电脑上磁盘列表之中会出现手机设备，如图 2-30 所示。

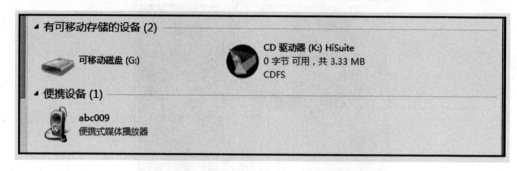

图 2-30　电脑上显示的手机设备

鼠标右键点击该设备，选择属性，可以查看到设备序列号。假设手机序列号为"APU0215C0500999"，则修改 IoT_IS_App_001.prj 里的运行选项，例如：

```
<run>
cd app\build\outputs\apk\debug
adb-s APU0215C0500999 uninstall com.example.administrator.myapplication
adb-s APU0215C0500999 app-debug.apk
adb-s APU0215C0500999 shell am start -n com.example.administrator.myapplication
                      /.WelcomeActivity
</run>
```

3. 把 App 拷贝到手机上安装和运行

IoT_IS_App_001 项目生成的 APK 安装包在 app\build\outputs\apk\debug 目录下，文件名为 app-debug.apk，如图 2-31 所示，如果 Obtain_Studio 软件所在的目录是"F:\Obtain_Studio"，那么 IoT_IS_App_001 项目生成的 APK 所在的目录是"F:\Obtain_Studio\WorkDir\IoT_IS_App_001\app\build\outputs\apk\debug"。

第 2 章 欢迎界面设计

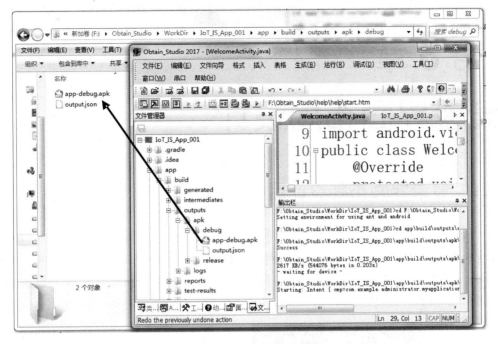

图 2-31 生成的 APK 所在的目录

第 3 章

登录界面布局设计

3.1 登录界面布局设计目标

用户登录界面也是物联网移动软件最基本的界面之一。一个用户登录界面一般包括背景图片、软件名称、中间框背景图、用户名和密码框、选择框、登录按钮等几个部分，如图 3-1 所示，部分物联网移动软件的用户登录界面还包括一个服务器 IP 地址框，方便用户填写物联网系统服务器 IP 地址。

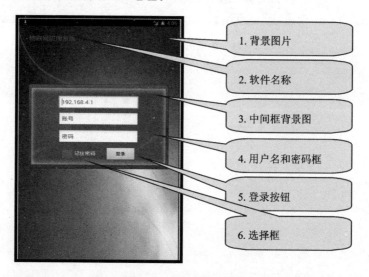

图 3-1 用户登录界面

用户登录界面布局结构如图 3-2 所示，采用线性布局来完成，共有四层进行嵌套。从图 3-1、图 3-2 的用户登录界面可以看出，其结构比较复杂，一是包括了多个控件；二是包括了复杂的布局。要实现用户登录界面的设计，首先得学习界面布局和常用控制的使用方法，因此，本章将重点介绍界面布局的设计方法。

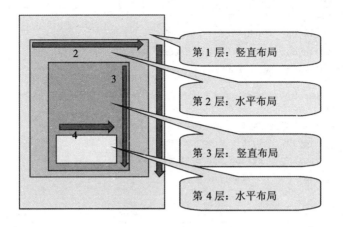

图 3-2 布局结构

3.2 安卓界面布局

3.2.1 界面布局文件

1. Android 界面布局文件

Android 提供了一种非常简单、方便的方法用于控制 UI 界面。该方法采用 XML 文件来进行界面布局,从而将布局界面的代码和逻辑控制的 Java 代码分离开来,使得程序的结构更加清晰明了。使用 XML 布局文件控制 UI 界面可以分为以下两个关键步骤:

(1) 在 Android 应用的 res/layout 目录下编写 XML 布局文件,可以是任何符合 Java 命名规则的文件名。创建后,R.java 会自动收录该布局资源。

(2) 在 Activity 中使用以下 Java 代码显示 XML 文件中布局的内容,setContentView(R.layout.main)。

这里"R.layout.main"之中的 main,对应 layout 目录下的 main.xml 布局文件。界面布局文件位于项目的\res\layout 子目录下。例如,包括一个文本显示和一个按钮的布局文件 main.xml 内容如下:

```
<? xml version = "1.0" encoding = "utf - 8"? >
<LinearLayout
xmlns:android = "http://schemas.android.com/apk/res/android"
    android:orientation = "vertical"
    android:layout_width = "fill_parent"
    android:layout_height = "fill_parent"
    >
```

```xml
<TextView
    android:layout_width = "fill_parent"
    android:layout_height = "wrap_content"
    android:text = "Hello World, hello"
/>
<Button
    android:layout_width = "fill_parent"
    android:layout_height = "wrap_content"
    android:text = "@string/str_button1"
    android:id = "@ + id/button1"
/>
</LinearLayout>
```

上述布局运行结果如图 3 - 3 所示。

图 3 - 3 布局运行效果

2. 界面布局类型

Android 的界面是由布局和组件协同完成,布局如建筑里的框架,而组件则相当于建筑里的砖瓦。组件按照布局的要求依次排列,就组成了用户所看见的界面。Android 具有以下常见布局:

- LinearLayout:线性布局;
- TableLayout:表格布局;
- RelativeLayout:相对布局;
- AbsoluteLayout:绝对布局;
- FrameLayout:帧布局。

3. 布局的属性

在安卓布局之中,位置相关的属性如图 3 - 4 所示。

布局常用属性:

(1) android:id 控件指定相应 ID。

(2) android:text 控件中显示文字。注意尽量使用 Strings.xml。

(3) android:gravity 控件中文字基本位置,如 center、left、right、center_horizontal 等。

(4) android:textsize 控件中字体大小,单位为 pt。

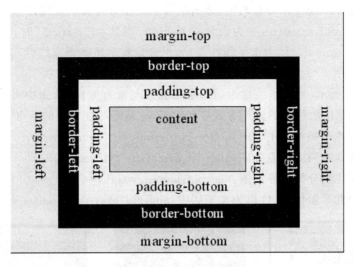

图3-4 位置相关的属性

(5) android:background 控件背景颜色。

(6) android:width 控件宽度。

(7) android:height 控件高度。

(8) android:padding 控件内边距,指控件当中内容到边界的距离。其中有 android:padding_left、android:padding_right 等。

(9) android:siglelise 如果设置为真,控件内容将在同一行显示。

(10) android:margin 外边距。

- android:layout_marginLeft 左外边距;
- android:layout_marginRight 右外边距;
- android:layout_marginTop 上外边距;
- android:layout_marginBottom 下外边距。

(11) android:orientation:线性布局的方式。

- vertical:垂直布局;
- horizontal:水平布局。

android:padding 和 android:layout_margin 的区别:padding 是内边距,是站在父类组件的角度描述问题,它规定里面的内容与这个父组件边界的距离;margin 则是外边距,是站在自己的角度描述问题,规定自己和其他(上下左右)的组件之间的距离。

3.2.2 线性布局

线性布局(LinearLayout)在单一方向上对齐所有的子视图——竖向或者横向,这依赖于怎么定义方向属性 orientation。所有子视图依次堆积,所以一个竖向列表每行只有一个子视图,不管它们有多宽,而一个横向列表将只有一行高(最高子视图的高度,加上填充)。

LinearLayout 线性布局有两种,分别是水平线性布局和垂直线性布局,LinearLayout 属性中 android:orientation 为设置线性布局,当其="vertical"时,为垂直线性布局,当其="horizontal"时,为水平线性布局,不管是水平还是垂直线性布局一行(列)只能放置一个控件。

如图 3-5 所示,使用了线性布局的水平方向与垂直方向,从图中可以清晰地看出来所有控件都是按照线性的排列方式显示出来,这就是线性布局的特点。

➤ 设置线性布局为水平方向:android:orientation="horizontal";
➤ 设置线性布局为垂直方向:android:orientation="vertical";
➤ 设置正比例分配控件范围:android:layout_weight="1";
➤ 设置控件显示位置,这里为水平居中:android:gravity="center_horizontal"。

图 3-5 线性布局

在下面的 xml 布局文件中使用了 LinearLayout 嵌套的方式配置了两个线性布局,一个水平显示,一个垂直显示。图 3-5 中间子图的程序代码如下:

```
<?xml version="1.0" encoding="utf-8"?>
<LinearLayout xmlns:android="http://schemas.android.com/apk/res/android"
    android:orientation="vertical"
    android:layout_width="fill_parent"
    android:layout_height="fill_parent"
    android:background="@drawable/login" >
    <LinearLayout
    android:gravity="center"
    android:layout_width="wrap_content"
    android:layout_height="wrap_content"
    android:layout_marginLeft="10.0dip"
```

```
        android:layout_marginRight = "10.0dip"
        android:orientation = "vertical"
        android:layout_marginTop = "110.0dip"
        android:background = "@drawable/bg3"    >
        <EditText android:layout_width = "fill_parent"
        android:layout_height = "wrap_content"
        android:id = "@ + id/edit1"
        android:hint = "请输入账号..."
        android:layout_marginTop = "30.0dip"
        android:layout_marginLeft = "65.0dip"
        android:layout_marginRight = "35.0dip" />
        <EditText android:layout_width = "fill_parent"
        android:layout_height = "wrap_content"
        android:id = "@ + id/edit2"
        android:hint = "输入密码..."
        android:password = "true"
        android:layout_marginTop = "20.0dip"
        android:layout_marginLeft = "65.0dip"
        android:layout_marginRight = "35.0dip" />
        <LinearLayout android:layout_width = "fill_parent"
        android:layout_height = "wrap_content"
        android:layout_weight = "1">
            <CheckBox android:layout_width = "wrap_content"
            android:layout_height = "wrap_content"
            android:text = "记住密码"
            android:id = "@ + id/checkbox1"
            android:layout_marginLeft = "10.0dip"
            android:layout_marginRight = "10.0dip" />
        <Button android:layout_width = "fill_parent"
            android:layout_height = "wrap_content"
            android:text = "登录"
            android:id = "@ + id/button1"
            android:layout_marginLeft = "10.0dip"
            android:layout_marginRight = "10.0dip" />
        </LinearLayout>
    </LinearLayout>
</LinearLayout>
```

(1) android:layout_weight 介绍

以水平的 LinearLayout 为例,它在分配子组件时使用的是 layout_width 和 layout_weight 参数的混合值。现在假设 LinearLayout 对应的 xml 布局如下:

```
<LinearLayout android:orientation = "horizontal"
    android:layout_width = "match_parent"
    android:layout_height = "wrap_content">
    <Button android:id = "@ + id/crime_date"
        android:layout_width = "wrap_content"
        android:layout_height = "wrap_content"
        android:layout_weight = "1"/>
    <CheckBox android:id = "@ + id/crime_solved"
        android:layout_width = "wrap_content"
        android:layout_height = "wrap_content"
        android:layout_weight = "1"/>
</LinearLayout>
```

根据上面的布局,LinearLayout 将如此安排布局:

第一步,LinearLayout 查看 layout_width 属性值(竖直则看 layout_height),由于上述的 Button 和 CheckBox 均是 wrap_content,因此,Button 和 Checkbox 获得的空间仅够绘制自身,LinearLayout 还会有一部分剩余空间。如果要占全部宽度 layout_width 设置为 fill_parent。

第二步,LinearLayout 按照 layout_weight 属性值进行剩余空间的分配,由于上述代码中 Button 和 Checkbox 的 weight 值一致,于是二者将平分额外的空间。

layout_weight 的值越大,组件获取剩余空间的比例越多。例如,假设 Button 的 layout_weight 为 2,Checkbox 的 layout_weight 的值仍为 1,那么 Button 将获得剩余空间的 2/3。

(2) **android:gravity 和 android:layout_gravity 介绍**

➢ android:gravity:设置的是控件自身上面的内容位置;

➢ android:layout_gravity:设置控件本身相对于父控件的显示位置。

android:gravity 指明了一个对象在其自己的边界内,以及在 X 轴和 Y 轴两个方向上,如何放置它的内容。它必须是表 3-1 所列的常量之一,或多个常量异或(|)在一起;与之对应的方法是:void setGravity(int gravity)。

表 3-1 gravity 常量

常量	值	描述
top	0x30	将对象放置在其容器的顶部(top),不改变其大小
bottom	0x50	将对象放置在其容器的底部(bottom),不改变其大小
left	0x03	将对象放置在其容器的左侧(left),不改变其大小
right	0x05	将对象放置在其容器的右侧(right),不改变其大小
center_vertical	0x10	将对象放置在其容器的垂直居中(vertical center),不改变其大小
fill_vertical	0x70	如果需要,增大对象的垂直大小;因此,它完全填满了其容器
center_horizontal	0x01	将对象放置在其容器的水平居中(horizontal center),不改变其大小

续表 3-1

常量	值	描述
fill_horizontal	0x07	如果需要,增大对象的水平大小;因此,它完全填满了其容器
center	0x11	将对象放置在其容器的垂直居中(vertical center)和水平居中(horizontal center),不改变其大小
fill	0x77	如果需要,增大对象的垂直大小和水平大小;因此,它完全填满了其容器
clip_vertical	0x80	附加选项,它被设置用于依据其容器的边界,裁剪子控件的顶部或/和底部的边缘; 裁剪区域将基于垂直对齐:靠顶部的,将裁剪底部边缘;靠底部的,将裁剪顶部边缘;两边都不靠的,同时裁剪顶部和底部的边缘
clip_horizontal	0x08	附加选项,它被设置用于依据其容器的边界,裁剪子控件的左侧或/和右侧的边缘; 裁剪区域基于水平对齐:靠左的裁剪右边缘;靠右的裁剪左边缘;左右都不靠的,同时裁剪左边缘和右边缘
start	0x00800003	将对象放置在其容器的开始处(beginning),不改变其大小
end	0x00800005	将对象放置在其容器的结束处(end),不改变其大小

3.2.3 相对布局

相对布局(RelativeLayout)是 android 布局中最为强大的,首先,它可以设置的属性是最多的,其次,它可以做的事情也是最多的。Android 手机屏幕的分辨率五花八门,所以为了考虑屏幕自适应的情况,在开发中建议大家都去使用相对布局,它的坐标取值范围都是相对的,使用它来做自适应屏幕是正确的,如图 3-6 所示。

图 3-6 相对布局

➢ 设置距父元素右对齐:android:layout_alignParentRight="true";
➢ 设置该控件在 id 为 re_edit_0 控件的下方:android:layout_below="@id/re_edit_0";

- 设置该控件在 id 为 img_0 控件的左边：android:layout_toLeftOf="@id/img_0";
- 设置当前控件与 id 为 name 控件的上方对齐：android:layout_alignTop="@id/name";
- 设置偏移的像素值：android:layout_marginRight="30dip"。

图 3-6 左边子图的程序代码如下：

```xml
<?xml version = "1.0" encoding = "utf-8"?>
<RelativeLayout xmlns:android = "http://schemas.android.com/apk/res/android"
    android:layout_width = "match_parent"
    android:layout_height = "match_parent" >
    <Button android:id = "@+id/cbut"
        android:layout_width = "wrap_content"
        android:layout_height = "wrap_content"
        android:layout_centerInParent = "true"
        android:text = "中间" />
    <Button android:id = "@+id/tbut"
        android:layout_width = "wrap_content"
        android:layout_height = "wrap_content"
        android:layout_centerInParent = "true"
        android:layout_above = "@id/cbut"
        android:text = "上面" />
    <Button android:id = "@+id/tlbut"
        android:layout_width = "wrap_content"
        android:layout_height = "wrap_content"
        android:layout_centerInParent = "true"
        android:layout_above = "@id/cbut"
        android:layout_toLeftOf = "@id/tbut"
        android:text = "左上" />
    <Button android:id = "@+id/trbut"
        android:layout_width = "wrap_content"
        android:layout_height = "wrap_content"
        android:layout_centerInParent = "true"
        android:layout_above = "@id/cbut"
        android:layout_toRightOf = "@id/tbut"
        android:text = "右上" />
    <Button android:id = "@+id/bbut"
        android:layout_width = "wrap_content"
        android:layout_height = "wrap_content"
        android:layout_centerInParent = "true"
        android:layout_below = "@id/cbut"
        android:text = "下面" />
    <Button android:id = "@+id/lbut"
```

```
        android:layout_width = "wrap_content"
        android:layout_height = "wrap_content"
        android:layout_centerInParent = "true"
        android:layout_toLeftOf = "@id/cbut"
        android:text = "左面" />
    <Button android:id = "@ + id/rbut"
        android:layout_width = "wrap_content"
        android:layout_height = "wrap_content"
        android:layout_centerInParent = "true"
        android:layout_toRightOf = "@id/cbut"
        android:text = "右面" />
    <Button android:id = "@ + id/blbut"
        android:layout_width = "wrap_content"
        android:layout_height = "wrap_content"
        android:layout_alignBottom = "@ + id/bbut"
        android:layout_alignLeft = "@ + id/lbut"
        android:text = "左下" />
    <Button android:id = "@ + id/brbut"
        android:layout_width = "wrap_content"
        android:layout_height = "wrap_content"
        android:layout_alignBottom = "@ + id/bbut"
        android:layout_alignLeft = "@ + id/rbut"
        android:text = "右下" />
</RelativeLayout>
```

3.2.4 帧布局

帧布局(FrameLayout)是在控件中绘制任何一个控件都可以被后绘制的控件覆盖,最后绘制的控件会盖住之前的控件,如图3-7所示,通常情况下带导航栏的布局一般都采用帧布局。帧布局可以说是五大布局中最为简单的一个布局,这个布局会默认把控件放在屏幕上的左上角的区域,后续添加的控件会覆盖前一个,如果控件的大小一样大的话,那么同一时刻就只能看到最上面的那个控件,如图3-7所示。

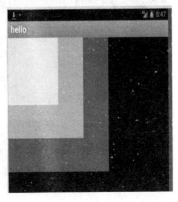

图3-7 帧布局

FrameLayout(帧布局)常用属性：
- android:foreground:设置该帧布局容器的前景图像；
- android:foregroundGravity:设置前景图像显示的位置。

图 3-7 左边子图的程序代码如下：

```
<?xml version = "1.0" encoding = "utf-8"?>
<FrameLayout xmlns:android = "http://schemas.android.com/apk/res/android"
    android:layout_width = "match_parent"
    android:layout_height = "match_parent" >

    <TextView
        android:layout_width = "200dp"
        android:layout_height = "200dp"
        android:background = "#FF0000" />
    <TextView
        android:layout_width = "150dp"
        android:layout_height = "150dp"
        android:background = "#00FFFF" />
    <TextView
        android:layout_width = "100dp"
        android:layout_height = "100dp"
        android:background = "#FFFF00" />
</FrameLayout>
```

3.2.5 绝对布局

绝对布局(AbsoluteLayout)可以设置任意控件在屏幕中 x、y 坐标点,和帧布局一样后绘制的控件会覆盖住之前绘制的控件,如图 3-8 所示,不建议大家使用绝对布局,因为 Android 的手机分辨率五花八门,所以使用绝对布局的话在其他分辨率的手机上就无法正常显示了。

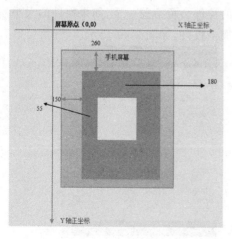
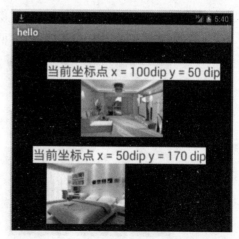

图 3-8 绝对布局

设置控件的显示坐标点的方法如下:
android:layout_x="50dip"
android:layout_y="30dip"
图3-8右边子图的用法程序代码如下:

```xml
<?xml version="1.0" encoding="utf-8"?>
<AbsoluteLayout xmlns:android="http://schemas.android.com/apk/res/android"
  android:layout_width="fill_parent"
  android:layout_height="fill_parent">
  <ImageView
    android:layout_width="wrap_content"
    android:layout_height="wrap_content"
    android:src="@drawable/alivingroom"
    android:layout_x="100dip"
    android:layout_y="50dip"    />
  <TextView
    android:layout_width="wrap_content"
    android:layout_height="wrap_content"
    android:text="当前坐标点 x = 100dip y = 50 dip"
    android:background="#FFFFFF"
    android:textColor="#FF0000"
    android:textSize="18dip"
    android:layout_x="50dip"
    android:layout_y="30dip"    />
  <ImageView
    android:layout_width="wrap_content"
    android:layout_height="wrap_content"
    android:src="@drawable/bedroom"
    android:layout_x="50dip"
    android:layout_y="170dip"    />
  <TextView
    android:layout_width="wrap_content"
    android:layout_height="wrap_content"
    android:text="当前坐标点 x = 50dip y = 170 dip"
    android:background="#FFFFFF"
    android:textColor="#FF0000"
    android:textSize="18dip"
    android:layout_x="30dip"
    android:layout_y="150dip"    />
</AbsoluteLayout>
```

3.2.6 表格布局

表格布局(TableLayout)可以设置 TableRow,可以设置表格中每一行显示的内容以及位置,可以设置显示的缩进、对齐的方式,如图 3-9 所示。

图 3-9 表格布局

图 3-9 的程序代码如下:

```
<?xml version = "1.0" encoding = "utf-8"?>
<TableLayout xmlns:android = "http://schemas.android.com/apk/res/android"
  android:layout_width = "fill_parent"
  android:layout_height = "fill_parent"
  android:stretchColumns = " * ">
  <EditText android:layout_width = "fill_parent"
    android:layout_height = "wrap_content"
    android:layout_marginRight = "30dip"
    android:layout_marginLeft = "30dip"
    android:layout_marginTop = "30dip"
    android:textSize = "20sp"
    android:textColor = "#F83333"
    android:text = "12345" />
  <TableRow android:layout_width = "fill_parent"
    android:layout_height = "fill_parent"
    android:padding = "10dip">
    <Button android:text = "1"/>
    <Button android:text = "2"/>
```

```
      <Button android:text = "3"/>
      <Button android:text = "4"/>
    </TableRow>
    <TableRow android:layout_width = "fill_parent"
      android:layout_height = "fill_parent"
      android:padding = "10dip">
      <Button android:text = "5" />
      <Button android:text = "6" />
      <Button android:text = "7" />
      <Button android:text = "8" />
    </TableRow>
    <TableRow android:layout_width = "fill_parent"
      android:layout_height = "fill_parent"
      android:padding = "10dip">
      <Button android:text = "9" />
      <Button android:text = "0" />
      <Button android:text = " + " />
      <Button android:text = " - " />
    </TableRow>
    <TableRow  android:layout_width = "fill_parent"
      android:layout_height = "fill_parent"
      android:padding = "10dip">
      <Button android:text = " * " />
      <Button android:text = "/" />
      <Button android:text = " # " />
      <Button android:text = "\@" />
    </TableRow > </TableLayout >
```

3.3　Android 常用控件

　　Android 常用控件包括 TextView、EditView、Button、Menu、RadioGroup、RadioButton、CheckBox、ProgressBar、ListView、TabWidget、SeekBar、ScrollView、GirdView、ImageSwitcher 等。

　　Android 控件常用属性如下：

　　（1）android:background 设置背景颜色/背景图片。可以通过以下两种方法设置背景为透明："@android:color/transparent"和"@null"。注意 TextView 默认是透明的，不用写此属性，但是 Buttom/ImageButton/ImageView 的透明效果就得写这个属性。

　　（2）android:clickable 是否响应单击事件。

　　（3）android:contentDescription 设置 View 的备注说明，作为一种辅助功能提供，

为一些没有文字描述的 View 提供说明,如 ImageButton。这里在界面上不会有效果,自己在程序中控制,可临时放一点字符串数据。

(4) android:drawingCacheQuality 设置绘图时半透明质量。有以下值可设置:auto(默认,由框架决定)/high(高质量,使用较高的颜色深度,消耗更多的内存)/low(低质量,使用较低的颜色深度,但是用更少的内存)。android:duplicateParentState 如果设置此属性,将直接从父容器中获取绘图状态(光标,按下等)。

(5) android:fadingEdge 设置拉滚动条时,边框渐变的方向。none(边框颜色不变),horizontal(水平方向颜色变淡),vertical(垂直方向颜色变淡)。

(6) android:fadingEdgeLength 设置边框渐变的长度。

(7) android:fitsSystemWindows 设置布局调整时是否考虑系统窗口(如状态栏)。

(8) android:focusable 设置是否获得焦点。若有 requestFocus() 被调用时,后者优先处理。注意在表单中想设置某一个(如 EditText)获取焦点,只设置这个还不行,需要将这个 EditText 前面的 focusable 都设置为 false 才行。在 Touch 模式下获取焦点需要设置 focusableInTouchMode 为 true。

(9) android:focusableInTouchMode 设置在 Touch 模式下 View 是否能取得焦点。

(10) android:hapticFeedbackEnabled 设置长按时是否接受其他触摸反馈事件。

(11) android:id 给当前 View 设置一个在当前 layout.xml 中的唯一编号,可以通过调用 View.findViewById() 或 Activity.findViewById() 根据这个编号查找到对应的 View。不同的 layout.xml 之间定义相同的 id 不会冲突。格式如"@+id/btnName"。

(12) android:isScrollContainer 设置当前 View 为滚动容器。

(13) android:keepScreenOnView 在可见的情况下是否保持唤醒状态。

(14) android:longClickable 设置是否响应长按事件。

(15) android:minHeight 设置视图最小高度。

(16) android:minWidth 设置视图最小宽度。

(17) android:nextFocusDown 设置下方指定视图获得下一个焦点。焦点移动是基于一个在给定方向查找最近邻居的算法。如果指定视图不存在,移动焦点时将报运行时错误。可以设置 imeOptions= actionDone,这样输入完即跳到下一个焦点。

(18) android:nextFocusLeft 设置左边指定视图获得下一个焦点。

(19) android:nextFocusRight 设置右边指定视图获得下一个焦点。

(20) android:nextFocusUp 设置上方指定视图获得下一个焦点。

(21) android:onClick 单击时从上下文中调用指定的方法。这里指定一个方法名称,一般在 Activity 定义符合如下参数和返回值的函数,并将方法名字符串指定为该值即可。

(22) android:padding 设置上下左右的边距,以像素为单位填充空白。android:paddingBottom 设置底部的边距,以像素为单位填充空白。android:paddingLeft 设置左边的边距,以像素为单位填充空白。android:paddingRight 设置右边的边距,以像素为单位填充空白。

(23) android:paddingTop 设置上方的边距,以像素为单位填充空白。android:saveEnabled 设置是否在窗口冻结时(如旋转屏幕)保存 View 的数据,默认为 true,但是前提是需要设置 id 才能自动保存。android:scrollX 以像素为单位设置水平方向滚动的偏移值,在 GridView 中可看的这个效果。

(24) android:scrollY 以像素为单位设置垂直方向滚动的偏移值 android:scrollbarAlwaysDrawHorizontalTrack 设置是否始终显示垂直滚动条。

(25) android:scrollbarAlwaysDrawVerticalTrack 设置是否始终显示垂直滚动条。

(26) android:scrollbarDefaultDelayBeforeFade 设置 N 毫秒后开始淡化,以毫秒为单位。

(27) android:scrollbarFadeDuration 设置滚动条淡出效果(从有到慢慢地变淡直至消失)时间,以毫秒为单位。

(28) android:scrollbarSize 设置滚动条的宽度。

(29) android:scrollbarStyle 设置滚动条的风格和位置。设置值:insideOverlay、insideInset、outsideOverlay、outsideInset。

(30) android:scrollbarThumbHorizontal 设置水平滚动条的 drawable(如颜色)。

(31) android:scrollbarThumbVertical 设置垂直滚动条的 drawable(如颜色)。

(32) android:scrollbarTrackHorizontal 设置水平滚动条背景(轨迹)的 drawable(如颜色)。

(33) android:scrollbarTrackVertical 设置垂直滚动条背景(轨迹)的 drawable。

(34) android:scrollbars 设置滚动条显示。none(隐藏),horizontal(水平),vertical(垂直)。

(35) android:soundEffectsEnabled 设置单击或触摸时是否有声音效果。

(36) android:tag 设置一个文本标签。可以通过 View.getTag()或 for with View.findViewWithTag()检索含有该标签字符串的 View。但一般最好通过 ID 来查询 View,因为它的速度更快,并且允许编译时类型检查。

(37) android:visibility 设置是否显示 View。设置值:visible(默认值,显示),invisible(不显示,但是仍然占用空间),gone(不显示,不占用空间)。

3.3.1 Button 控件

Android SDK 包含两个在布局中可以使用的简单按钮控件:Button(android.widget.Button)和 ImageButton(android.widget.ImageButton)。这两个控件的功能很相似。这两个控件不同之处是外观上,Button 控件有一个文本标签,而 ImageButton 使用一个可绘制的图像资源来代替。

典型 Button 控件标签写法:

```
<Button
    android:text = "button1"
    android:id = "@ + id/button1"
    android:layout_width = "wrap_content"
    android:layout_height = "wrap_content">
</Button>
```

1. 向布局添加 Button 控件

要添加一个 Button 控件到 main.xml 布局资源中,须要编辑布局文件。可使用 Eclipse 的布局资源设计器,或者直接编辑 XML。像按钮这样的控件也可以通过程序动态地创建并在运行时添加到屏幕上。对于简单的界面,可以不用布局文件,而是直接在类中创建合适的控件并将它添加到的活动界面之中。

要添加一个 Button 控件到布局资源文件,打开/res/layout/main.xml 布局文件,它是的 Android 项目的一部分。单击想要为其添加 Button 控件的 LinearLayout(或者父级布局控件,比如 RelativeLayout 或 FrameLayout)。在 Eclipse 中,可以单击 Outline 标签中的父级布局,然后使用绿色加号按钮添加一个新的控件。选择要添加的控件——在这个例子中是 Button 控件。

2. 实现 OnClickListener

实现单击事件处理,最常见的方法是使用 setOnClickListener()方法向的按钮控件注册一个新的 View.OnClickListener。

通常情况下通过 Activity 的 onCreate()方法来实现。使用 findViewById()方法找到控件,然后使用它的 setOnClickListener()方法来定义当它被单击时的行为——onClick()方法,例如:

```
Button myButton = (Button) findViewById(R.id.Button01);
    myButton.setOnClickListener(new View.OnClickListener() {
    public void onClick(View v) {
    Toast.makeText(BasicButtonActivity.this, "Button clicked!",
Toast.LENGTH_SHORT).show();
    }
});
```

3.3.2 CheckBox 控件

Android 提供了一个多选组件 CheckBox,实现多选操作。因为可以多选,所以它区别于 RadioButton,没有了组的概念,要监听用户操作时需要对每一个 CheckBox 进行监听。CheckBox 控件标签典型写法如下:

```
<CheckBox
    android:layout_width = "wrap_content"
    android:layout_height = "wrap_content"
    android:text = "checkbox1"
    android:id = "@ + id/checkbox1"/>
```

Checked 属性是 CheckBox 最重要的属性之一,改变它的方式有三种:
- XML 中申明;
- 代码动态改变;
- 用户触摸。

它的改变将会触发 OnCheckedChange 事件,可以使用 OnCheckedChange Listener 监听器来监听这个事件,例如:

```
//获取 CheckBox 实例
CheckBox cb = (CheckBox)this.findViewById(R.id.cb);
//绑定监听器
cb.setOnCheckedChangeListener(new OnCheckedChangeListener() {
    @Override
    public void onCheckedChanged(CompoundButton arg0, boolean arg1) {
        //TODO Auto - generated method stub
        Toast.makeText(MyActivity.this,
          arg1?"选中了":"取消了选中"  , Toast.LENGTH_LONG).show();
    }
});
```

3.3.3 EditText 控件

EditText 是一个非常重要的组件,可以说它是用户和 Android 应用进行数据传输的窗口,有了它就等于有了一扇和 Android 应用输入数据之门,通过它用户可以把数据传给 Android 应用。

EditText 是接受用户输入信息的最重要控件,可以利用 EditText.getText()获取它的文本。

典型控件标签写法如下:

```
<EditText
    android:id = "@ + id/edit_text"
    android:layout_width = "fill_parent"
    android:layout_height = "wrap_content"
    android:maxLength = "40"
    android:hint = "请输入用户名..."
    android:textColorHint = "#238745"/>
</LinearLayout>
```

Password 的实现方法如下：

```
<EditText
    android:id = "@ + id/edit_text"
    android:layout_width = "fill_parent"
    android:layout_height = "wrap_content"
    android:password = "true"/>
```

EditText 控件属性名称描述如下：

(1) android:gravity 设置文本位置，如设置成"center"，文本将居中显示。

(2) android:hintText 为空时显示的文字提示信息，可通过 textColorHint 设置提示信息的颜色。此属性在 EditView 中使用，但是这里也可以用。

(3) android:imeOptions 附加功能，设置右下角 IME 动作与编辑框相关的动作，如 actionDone 右下角将显示一个"完成"，而不设置默认是一个回车符号。这个在 EditView 中再详细说明，此处无用。

(4) android:imeActionId 设置 IME 动作 ID。

(5) android:imeActionLabel 设置 IME 动作标签。

(6) android:includeFontPadding 设置文本是否包含顶部和底部额外空白，默认为 true。

(7) android:inputMethod 为文本指定输入法，需要完全限定名(完整的包名)。例如：com. google. android. inputmethod. pinyin，但是这里报错找不到。

(8) android:inputType 设置文本的类型，用于帮助输入法显示合适的键盘类型。在 EditView 中再详细说明，这里无效果。

(9) android:linksClickable 设置链接是否单击连接，即使设置了 autoLink。

(10) android:marqueeRepeatLimit 在 ellipsize 指定 marquee 的情况下，设置重复滚动的次数，当设置为 marquee_forever 时表示无限次。

(11) android:ems 设置 TextView 的宽度为 N 个字符的宽度。

(12) android:maxEms 设置 TextView 的宽度最长为 N 个字符的宽度。与 ems 同时使用时覆盖 ems 选项。

(13) android:minEms 设置 TextView 的宽度最短为 N 个字符的宽度。与 ems 同时使用时覆盖 ems 选项。

(14) android:maxLength 限制显示的文本长度，超出部分不显示。

(15) android:lines 设置文本的行数，设置两行就显示两行，即使第二行没有数据。

(16) android:maxLines 设置文本的最大显示行数，与 width 或者 layout_width 结合使用，超出部分自动换行，超出行数将不显示。

(17) android:minLines 设置文本的最小行数，与 lines 类似。

(18) android:lineSpacingExtra 设置行间距。

(19) android:lineSpacingMultiplier 设置行间距的倍数，如"1.2"。

(20) android:numeric 如果被设置,该 TextView 有一个数字输入法。此处无用,设置后唯一效果是 TextView 有单击效果。

(21) android:password 以小点"."显示文本。

(22) android:phoneNumber 设置为电话号码的输入方式。

(23) android:privateImeOptions 设置输入法选项。

(24) android:scrollHorizontally 设置文本超出 TextView 的宽度的情况下,是否出现横拉条。

(25) android:selectAllOnFocus 如果文本是可选择的,让它获取焦点而不是将光标移动为文本的开始位置或者末尾位置。TextView 中设置后无效果。

(26) android:shadowColor 指定文本阴影的颜色,需要与 shadowRadius 一起使用。

(27) android:shadowDx 设置阴影横向坐标开始位置。

(28) android:shadowDy 设置阴影纵向坐标开始位置。

(29) android:shadowRadius 设置阴影的半径。设置为 0.1 就变成字体的颜色了,一般设置为 3.0 的效果比较好。

(30) android:singleLine 设置单行显示。如果和 layout_width 一起使用,当文本不能全部显示时,后面用""来表示。如 android:text="test_ singleLine " android:singleLine="true" android:layout_width="20dp"将只显示"t"。如果不设置 singleLine 或者设置为 false,文本将自动换行。

(31) android:text 设置显示文本。

(32) android:textSize 设置文字大小,推荐度量单位为"sp",如"15 sp"。

(33) android:textStyle 设置字形[bold(粗体) 0, italic(斜体) 1, bolditalic(又粗又斜) 2],可以设置一个或多个,用"|"隔开。

(34) android:typeface 设置文本字体,必须是以下常量值之一:normal 0, sans 1, serif 2, monospace(等宽字体) 3]。

(35) android:height 设置文本区域的高度,支持度量单位:px(像素)/dp/sp/in/mm(毫米)。

(36) android:maxHeight 设置文本区域的最大高度。

(37) android:minHeight 设置文本区域的最小高度。

(38) android:width 设置文本区域的宽度,支持度量单位:px(像素)/dp/sp/in/mm(毫米)。

(39) android:maxWidth 设置文本区域的最大宽度。

(40) android:minWidth 设置文本区域的最小宽度。

(41) android:textAppearance 设置文字外观。

使用 TextWatcher 类,这种方式是可以监听软键盘和硬键盘,只需要实现 onTextChanged 方法即可,另外,TextWatcher 还提供了 beforeTextChanged 和 afterTextChanged 方法,用于更加详细的输入监听处理,例如:

```
edittext.addTextChangedListener(new TextWatcher() {
    @Override
    public void onTextChanged(CharSequence s, int start, int before,
int count) {
        textview.setText(edittext.getText());
    }
    @Override
    public void beforeTextChanged(CharSequence s, int start,
int count, int after) {
    }
    @Override
    public void afterTextChanged(Editable s) {
    }
});
```

3.3.4 ImageButton 控件

ImageButton 显示一个可以被用户单击的图片按钮,默认情况下,ImageButton 看起来像一个普通的按钮,在不同状态(如按下)下改变背景颜色。按钮的图片可用通过 <ImageButton> XML 元素的 android:src 属性或 setImageResource(int)方法指定。ImageButton 使用方法与 Button 类似。

典型 ImageButton 控件标签写法如下:

```
<ImageButton
    android:id = "@ + id/ImageButton01"
    android:layout_width = "wrap_content"
    android:layout_height = "wrap_content"
    android:src = "@drawable/button1"
/>
```

Android 中 ImageButton 自定义按钮的按下效果的代码实现方法如下:

```
imageButton.setOnTouchListener(new OnTouchListener(){
    @Override
    public boolean onTouch(View v, MotionEvent event) {
    if(event.getAction() == MotionEvent.ACTION_DOWN){
        //更改为按下时的背景图片
        v.setBackgroundResource(R.drawable.pressed);
    }else if(event.getAction() == MotionEvent.ACTION_UP){
        //改为抬起时的图片
        v.setBackgroundResource(R.drawable.released);
    }
    return false;
    }
});
```

3.3.5　ImageView 控件

ImageView 控件显示任意图像,例如图标、图片等。ImageView 类可以加载各种来源的图片(如资源或图片库),需要计算图像的尺寸,以便它可以在其他布局中使用,并提供缩放、着色(渲染)等各种显示选项。

典型 ImageView 控件标签写法如下:

```
<ImageView android:paddingTop = "10px"
    android:background = "@android:color/white"
    android:scrollY = "10px"
    android:cropToPadding = "true"
    android:src = "@drawable/btn_mode_switch_bg"
    android:layout_width = "wrap_content"
    android:layout_height = "wrap_content">
</ImageView>
```

ImageView 常见 XML 属性如下:

(1) android:adjustViewBounds 是否保持宽高比。需要与 maxWidth、MaxHeight 一起使用,单独使用没有效果。

(2) android:cropToPadding 是否截取指定区域用空白代替。单独设置无效果,需要与 scrollY 一起使用。

(3) android:maxHeight 定义 View 的最大高度,需要与 AdjustViewBounds 一起使用,单独使用没有效果。如果想设置图片固定大小,又想保持图片宽高比,需要如下设置:

- 设置 AdjustViewBounds 为 true;
- 设置 maxWidth、MaxHeight;
- 设置 layout_width 和 layout_height 为 wrap_content。

(4) android:maxWidth 设置 View 的最大宽度;

(5) android:scaleType 设置图片的填充方式,参数:

- center:按图片原来的尺寸居中显示,当图片的长(宽)超过 View 的长(宽),则截取图片居中部分显示;
- centerCrop:按比例扩大图片的尺寸居中显示,使得图片长(宽)等于或大于 View 的长(宽);
- centerInside:将图片的内容完整居中显示,通过按比例缩小原来的尺寸使得图片长(宽)小于或等于 View 的长(宽);
- fitCenter:把图片按比例扩大/缩小到 View 的宽度,居中显示;
- fitEnd:把图片按比例扩大/缩小到 View 的宽度,显示在 View 的下半部分位置;
- fitStart:把图片按比例扩大/缩小到 View 的宽度,显示在 View 的上半部分位置；
- fitXY：把图片不按比例扩大/缩小到 View 的大小显示；
- matrix：用矩阵来绘制。

（6）android：src 设置 View 的图片或颜色。

（7）android：tint 将图片渲染成指定的颜色。

3.3.6 ListView 控件

ListView 是一个经常用到的控件，ListView 里面的每个子项 Item 可以是一个字符串，也可以是一个组合控件。典型 ListView 控件标签写法如下：

```
<ListView
    android:layout_width = "fill_parent"
    android:layout_height = "wrap_content"
    android:id = "@ + id/listview1"
> </ListView>
```

ListView 的实现：
- 准备 ListView 要显示的数据；
- 使用一维或多维动态数组保存数据；
- 构建适配器，适配器就是 Item 数组，动态数组有多少元素就生成多少个 Item；
- 把适配器添加到 ListView，并显示出来。

main.xml 文件之中使用 ListView 的代码如下：

```
<? xml version = "1.0" encoding = "utf - 8"? >
<LinearLayout
    android:id = "@ + id/LinearLayout01"
    android:layout_width = "fill_parent"
    android:layout_height = "fill_parent"
    xmlns:android = "http://schemas.android.com/apk/res/android">
    <ListView android:layout_width = "wrap_content"
        android:layout_height = "wrap_content"
        android:id = "@ + id/MyListView">
    </ListView>
</LinearLayout>
```

my_listitem.xml 用于设计 ListView 的 Item，my_listitem.xml 的代码如下：

```
<? xml version = "1.0" encoding = "utf - 8"? >
<LinearLayout
    android:layout_width = "fill_parent"
    xmlns:android = "http://schemas.android.com/apk/res/android"
    android:orientation = "vertical"
```

```xml
    android:layout_height = "wrap_content"
    android:id = "@ + id/MyListItem"
    android:paddingBottom = "3dip"
    android:paddingLeft = "10dip">
     <TextView
        android:layout_height = "wrap_content"
        android:layout_width = "fill_parent"
        android:id = "@ + id/ItemTitle"
        android:textSize = "30dip">
     </TextView>
     <TextView
        android:layout_height = "wrap_content"
        android:layout_width = "fill_parent"
        android:id = "@ + id/ItemText">
     </TextView>
</LinearLayout>
```

其中用到的一些属性：

- paddingBottom＝"3dip"，Layout 往底部留出 3 个像素的空白区域；
- paddingLeft＝"10dip"，Layout 往左边留出 10 个像素的空白区域；
- textSize＝"30dip"，TextView 的字体为 30 个像素。

在 Java 文件中为 ListView 动态添加选择项的代码如下：

```java
public void onCreate(Bundle savedInstanceState) {
    super.onCreate(savedInstanceState);
    setContentView(R.layout.main);
    //绑定 XML 中的 ListView, 作为 Item 的容器
    ListView list = (ListView) findViewById(R.id.MyListView);
    //生成动态数组，并且转载数据
    ArrayList <HashMap <String, String >>mylist
= new ArrayList <HashMap <String, String >>();
    for(int i = 0;i <30;i + + )
    {
    HashMap <String, String >map = new HashMap <String, String >();
    map.put("ItemTitle", "This is Title.....");
    map.put("ItemText", "This is text.....");
    mylist.add(map);
    }
    //生成适配器，数组 === 》ListItem
    SimpleAdapter mSchedule = new SimpleAdapter(this,mylist,//数据来源
    R.layout.my_listitem,//ListItem 的 XML 实现
    //动态数组与 ListItem 对应的子项
```

```
        new String[] {"ItemTitle","ItemText"},
        //ListItem 的 XML 文件里面的两个 TextView ID
        new int[] {R.id.ItemTitle,R.id.ItemText});
        //添加并且显示
        list.setAdapter(mSchedule);
}
```

3.3.7 ProgressBar 控件

当一个应用在后台执行时,前台界面就不会有什么信息,这时用户根本不知道程序是否在执行、执行进度如何、应用程序是否遇到错误终止等,这时需要使用进度条来提示用户后台程序执行的进度。

ProgressBar 用于某些操作的进度中的可视指示器,为用户呈现操作的进度,它还有一个次进度条,用来显示中间进度,如在流媒体播放的缓冲区的进度。一个进度条也可不确定其进度。在不确定模式下,进度条显示循环动画,这种模式常用于应用程序任务的长度是未知的情况。

Android 系统提供了两大类进度条样式,长形进度条(progressBarStyleHorizontal)和圆形进度条(progressBarStyleLarge)。

进度条用处很多,比如,应用程序装载资源和网络连接时,可以提示用户稍等,这一类进度条只是代表应用程序中某一部分的执行情况,而整个应用程序执行情况,则可以通过应用程序标题栏来显示一个进度条,这就需要先对窗口的显示风格进行设置"requestWindowFeature(Window.FEATURE_PROGRESS)"。

ProgressBarXML 常用属性:

- android:progressBarStyle:默认进度条样式
- android:progressBarStyleHorizontal:水平样式

ProgressBar 常用方法:

- getMax():返回这个进度条的范围的上限;
- getProgress():返回进度;
- getSecondaryProgress():返回次要进度;
- incrementProgressBy(int diff):指示增加的进度;
- isIndeterminate():指示进度条是否在不确定模式下;
- setIndeterminate(boolean indeterminate):设置不确定模式下;
- setVisibility(int v):设置该进度条是否可视。

ProgressBar 常用事件:

- onSizeChanged(int w, int h, int oldw, int oldh):当进度值改变时引发此事件。

典型 ProgressBar 控件标签写法如下:

```
<ProgressBar
    style = "? android:attr/progressBarStyleHorizontal"
    android:layout_width = "fill_parent"
    android:layout_height = "wrap_content"
    android:id = "@ + id/progress_bar"
    >
</ProgressBar>
```

3.3.8 RadioButton 控件

RadioButton 为单选按钮,如需实现多选功能,必须借助 RadioGroup 来实现。典型 RadioButton 控件标签写法如下:

```
<RadioButton
    android:text = "radio1"
    android:id = "@ + id/radio1"/>
```

RadioGroup 标签写法如下:

```
<RadioGroup
    android:layout_width = "fill_parent"
    android:layout_height = "wrap_content"
    android:orientation = "vertical"
    android:checkedButton = "@ + id/woman"
    android:id = "@ + id/sex">
    <RadioButton
    android:text = "man"
    android:id = "@ + id/man" />
    <RadioButton
    android:text = "woman"
    android:id = "@id/woman" />
</RadioGroup>
```

RadioGroup 控件事件代码如下:

```
RadioGroup rg = (RadioGroup)findViewById(R.id.radio_group_id);
rg.setOnCheckedChangeListener(new OnCheckedChangeListener() {
    @Override
    public void onCheckedChanged(RadioGroup group, int checkedId) {
        if(checkedId == R.id.boy_id){
            Toast toast = Toast.makeText(getApplicationContext(),
"选择了 男", Toast.LENGTH_LONG);
            toast.setGravity(Gravity.CENTER, 0, 0);
            toast.show();
```

```
}else if(checkedId == R.id.girl_id){
    Toast toast = Toast.makeText(getApplicationContext(),
"选择了 女", Toast.LENGTH_LONG);
    toast.setGravity(Gravity.CENTER, 0, 0);
    toast.show();
    }
    }
});
```

3.3.9 SeekBar 控件

SeekBar 可以通过滑块的位置来标识数值,允许用户拖动滑块来改变数值,因此,拖动条通常用于对系统的某种数值进行调节,比如调节音量等。

SeekBar 允许用户改变拖动条的滑块外观,改变滑块外观通常通过如下属性来指定:

android:thumb:指定一个 Drawable 对象,该对象将自定义滑块。

为了让程序能响应拖动条滑块位置的改变,程序可以考虑为它绑定一个 OnSeekBarChangeListener 监听器。

典型 SeekBar 控件标签写法如下:

```
<SeekBar
    android:layout_width = "fill_parent"
    android:layout_height = "wrap_content"
    android:id = "@ + id/seek_bar"
    />
```

SeekBar 控件事件代码如下:

```
public class Test_SeekBar extends Activity
implements SeekBar.OnSeekBarChangeListener{
    private SeekBar seekBar;
    private TextView textView1,textView2;
    @Override
    public void onCreate(Bundle savedInstanceState) {
        super.onCreate(savedInstanceState);
        setContentView(R.layout.main);
        seekBar = (SeekBar) this.findViewById(R.id.SeekBar01);
        textView1 = (TextView) this.findViewById(R.id.TextView1);
        textView2 = (TextView) this.findViewById(R.id.TextView2);
        seekBar.setOnSeekBarChangeListener(this);//添加事件监听
    }
    //拖动中
```

```
    @Override
    public void onProgressChanged(SeekBar seekBar, int progress,
boolean fromUser) {
        this.textView1.setText("当前值:" + progress);
    }
    //开始拖动
    @Override
    public void onStartTrackingTouch(SeekBar seekBar) {
        this.textView2.setText("拖动中...");
    }
    //结束拖动
    @Override
    public void onStopTrackingTouch(SeekBar seekBar) {
        this.textView2.setText("拖动完毕");
    }
}
```

3.3.10　Spinner 控件

Spinner 控件用来显示列表项,类似于一组单选框 RadioButton。Spinner 每次只能选择所有项其中之一。重要属性 android:prompt:当 Spinner 对话框关闭时显示该提示。

Spinner 控件常用方法:

- setPrompt(CharSequence prompt):设置当 Spinner 对话框关闭时显示的提示;
- performClick():如果它被定义就调用此视图的 OnClickListener;
- setOnItemClickListener(AdapterView.OnItemClickListener l):当项被单击时调用;
- onDetachedFromWindow():当 Spinner 脱离窗口时被调用。

典型 Spinner 控件标签写法如下:

```
<Spinner android:id = "@+id/spinner1"
    android:layout_width = "fill_parent"
    android:layout_height = "wrap_content"
    android:prompt = "title"
    android:entries = "@array/list"
/>
```

使用 ArrayAdapter 进行适配数据:

① 首先定义一个布局文件:

```
<span style="font-size:16px;">
<?xml version="1.0" encoding="utf-8"?>
<LinearLayout
    xmlns:android="http://schemas.android.com/apk/res/android"
    android:layout_width="fill_parent"
    android:layout_height="fill_parent"
    android:orientation="vertical" >

    <Spinner
      android:id="@+id/spinner1"
      android:layout_width="match_parent"
      android:layout_height="wrap_content"
      />
</LinearLayout></span>
```

上面的 Spinner 有两个属性：

- prompt 是初始的时候，Spinner 显示的数据，是一个引用类型；
- entries 是直接在 xml 布局文件中绑定数据源（可以不设置，即可以在 Activity 中动态绑定）。

② 建立数据源，使用数组，这些数据将会在 Spinner 下拉列表中进行显示：

```
<span style="font-size:16px;"><?xml version="1.0" encoding="utf-8"?>
    <resources>
        <string-array name="spinnername">
            <item>北京</item>
            <item>上海</item>
            <item>广州</item>
            <item>深圳</item>
        </string-array>
</resources></span>
```

③ 在 Activity 中加入如下的代码（使用了系统定义的下拉列表的布局文件，当然也可以自定义）：

```
// 初始化控件
mSpinner = (Spinner) findViewById(R.id.spinner1);
// 建立数据源
String[] mItems = getResources().getStringArray(R.array.spinnername);
// 建立 Adapter 并且绑定数据源
ArrayAdapter<String> _Adapter = new ArrayAdapter<String>(this,
    android.R.layout.simple_spinner_item, mItems);
//绑定 Adapter 到控件
mSpinner.setAdapter(_Adapter);
```

3.3.11 TabHost/TabWidget(切换卡)

1. TabHost/TabWidget 介绍

TabHost 是提供选项卡(Tab 页)的窗口视图容器。此对象包含两个子对象：一个是用户可以选择指定标签页的标签集合；另一个是用于显示标签页内容的 FrameLayout。选项卡中的个别元素一般通过其容器对象来控制，而不是直接设置子元素本身的值。

TabHost 是整个 Tab 的容器，包括两部分：TabWidget 和 FrameLayout。TabWidget 就是每个 Tab 标签，FrameLayout 则是 Tab 内容。Tab 标签页是界面设计时经常使用的界面控件，可以实现多个分页之间的快速切换，每个分页可以显示不同的内容。Tab 标签页的使用方法如下：

- 首先要设计所有分页的界面布局；
- 建立 Tab 标签页，给分页添加标识和标题，确定每个分页所显示的界面布局。

每个分页建立一个 XML 文件，用以编辑和保存分页的界面布局，使用的方法与设计普通用户界面没有什么区别。常用方法如下：

- addTab(TabHost.TabSpec tabSpec)：添加一项 Tab 页；
- clearAllTabs()：清除所有与之相关联的 Tab 页；
- getCurrentTab()：返回当前 Tab 页；
- getTabContentView()：返回包含内容的 FrameLayout；
- newTabSpec(String tag)：返回一个与之关联的新的 TabSpec。

2. TabHost/TabWidget 实例

建立一个"TabDemo"程序，包含三个 XML 文件，分别为 tab1.xml、tab2.xml 和 tab3.xml，这 3 个文件分别使用线性布局、相对布局和绝对布局示例中的 main.xml 的代码，并将布局的 ID 分别定义为 layout01、layout02 和 layout03。

(1) tab1.xml 文件代码如下：

```
<? xml version = "1.0" encoding = "utf-8"? >
<LinearLayout android:id = "@ + id/layout01"

</LinearLayout >
```

(2) tab2.xml 文件代码如下：

```
<? xml version = "1.0" encoding = "utf-8"? >
<AbsoluteLayout android:id = "@ + id/layout02"

</AbsoluteLayout >
```

（3）tab3.xml 文件代码如下：

```xml
<? xml version = "1.0" encoding = "utf-8"? >
<RelativeLayout android:id = "@ + id/layout03"

</RelativeLayout >
```

（4）TabDemo.java 文件：在 TabDemo.java 文件中键入下面的代码，创建 Tab 标签页，并建立子页与界面布局直接的关联关系，代码如下：

```java
public class TabDemo extends TabActivity {
    @Override
    public void onCreate(Bundle savedInstanceState) {
        super.onCreate(savedInstanceState);
        TabHost tabHost = getTabHost();
        LayoutInflater.from(this).inflate(R.layout.tab1,
        tabHost.getTabContentView(),true);
        LayoutInflater.from(this).inflate(R.layout.tab2,
        tabHost.getTabContentView(),true);
        LayoutInflater.from(this).inflate(R.layout.tab3,
            tabHost.getTabContentView(),true);
        tabHost.addTab(tabHost.newTabSpec("TAB1").
            setIndicator("线性布局").setContent(R.id.layout01));
        tabHost.addTab(tabHost.newTabSpec("TAB2").
            setIndicator("绝对布局").setContent(R.id.layout02));
        tabHost.addTab(tabHost.newTabSpec("TAB3").
            setIndicator("相对布局").setContent(R.id.layout03));
    }
}
```

第 1 行代码声明 TabDemo 类继承于 TabActivity，如同 ListActivity。

第 4 行代码通过 getTabHost()函数获得了 Tab 标签页的容器，用以承载可以单击的 Tab 标签和分页的界面布局。

第 5 行代码通过 LayoutInflater 将 tab1.xml 文件中的布局转换为 Tab 标签页可以使用的 View 对象。

第 7 行代码使用 addTab()函数添加了第 1 个分页，tabHost.newTabSpec（"TAB1"）表明在第 4 行代码中建立的 tabHost 上，添加一个标识为 TAB1 的 Tab 分页。

第 9 行代码使用 setIndicator()函数设定分页显示的标题，使用 setContent()函数设定分页所关联的界面布局。

TabHost/TabWidget 实例运行效果如图 3-10 所示。

图 3-10　运行效果

3.3.12　Gallery 与 ImageSwitcher

IPhone 曾经凭借这个效果吸引了无数的苹果粉丝，在 Android 平台上也可以实现这一效果。要实现这一效果，需要一个容器来存放 Gallery 显示的图片，这里使用一个继承自 BaseAdapter 类的派生类来装这些图片。

ImageSwitcher 类提供了图片切换功能，通过第三方的操作，设置当前 ImageSwitcher 显示的图片，同时设置图片变换的动画。

Gallery 与 ImageSwitcher 的结合使用，可以实现一个简单的浏览图片的功能。

Gallery 与 ImageSwitcher 实现图片浏览功能的运行效果如图 3-11 所示。

图 3-11　Gallery 与 ImageSwitcher 实现图片浏览功能运行效果

3.4　自定义按钮背景

3.4.1　Shape 介绍

在 Android 程序开发中，使用 Shape 可以很方便地设计出想要的背景，相对于 png

图片来说,使用 Shape 可以减小安装包的大小,而且能够更好地适配不同的手机。在 Android 程序开发中,经常会用到 Shape 定义各种各样的形状,首先了解一下 Shape 下面有哪些标签:

1. solid 填充

> android:color 指定填充的颜色。

2. gradient 渐变

> android:startColor 和 android:endColor 分别为起始和结束颜色。
> android:angle 是渐变角度,必须为 45 的整数倍。

另外,渐变默认的模式为 android:type="linear",即线性渐变,可以指定渐变为径向渐变 android:type="radial",径向渐变需要指定半径 android:gradientRadius="50"。angle 值对应的位置如图 3-12 所示。

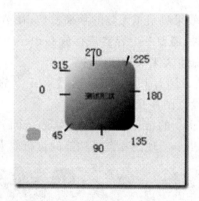

图 3-12　angle 值对应的位置

3. stroke 描边

> android:width="2dp" 描边的宽度,android:color 描边的颜色。

还可以把描边弄成虚线的形式,设置方式为:

> android:dashWidth="5dp";
> android:dashGap="3dp";

其中 android:dashWidth 表示"-"这样一个横线的宽度,android:dashGap 表示之间隔开的距离。

4. corners 圆角

> android:radius 为角的弧度,值越大角越圆。

还可以把四个角设定成不同的角度,同时设置五个属性,则 Radius 属性无效:

> android:Radius="20dp"　　　　　　　设置四个角的半径;
> android:topLeftRadius="20dp"　　　　设置左上角的半径;
> android:topRightRadius="20dp"　　　 设置右上角的半径;

> android:bottomLeftRadius="20dp"　　　设置右下角的半径；
> android:bottomRightRadius="20dp"　　设置左下角的半径。

3.4.2　Shape 使用步骤

1. 新建 Shape 文件

首先在 res/drawable 文件夹下，新建一个文件，命名为：shape_radius.xml，代码如下：

```
<?xml version="1.0" encoding="utf-8"?>
<shape xmlns:android="http://schemas.android.com/apk/res/android">
    <corners android:radius="20dip"/>
    <solid android:color="#ff00ff"/>
</shape>
```

2. 添加到控件中

在定义好 Shape 文件后，下一步就是将其添加到控件中，一般是使用设置 background 属性，将其设置为控件背景，下面将其添加到 MainActivity 对应的布局中（activity_main.xml），将其设为 TextView 的背景，代码如下：

```
<RelativeLayout xmlns:android="http://schemas.android.com/apk/res/android"
    xmlns:tools="http://schemas.android.com/tools"
    android:layout_width="match_parent"
    android:layout_height="match_parent"
    tools:context="com.harvic.tryshape.MainActivity">
    <TextView
        android:layout_width="wrap_content"
        android:layout_height="wrap_content"
        android:layout_margin="50dip"
        android:text="@string/hello_world"
        android:background="@drawable/shape_radius"/>
</RelativeLayout>
```

显示出来的效果如图 3-13 所示。

图 3-13　简单的 Shape 圆角效果图

3.4.3 Shape 常用属性

Shape 用于设定形状，可以在 selector、layout 等里面使用，有 6 个子标签，默认为矩形，可以设置为矩形（rectangle）、椭圆形（oval）、线性形状（line）、环形（ring）。各属性如下：

android:shape=["rectangle" | "oval" | "line" | "ring"]

1. 圆 角

- android:radius 整型，半径；
- android:topLeftRadius 整型，左上角半径；
- android:topRightRadius 整型，右上角半径；
- android:bottomLeftRadius 整型，左下角半径；
- android:bottomRightRadius 整型，右下角半径。

2. 渐变色

- android:startColor，颜色值，起始颜色；
- android:endColor 颜色值，结束颜色；
- android:centerColor 整型，渐变中间颜色，即开始颜色与结束颜色之间的颜色；
- android:angle 整型，渐变角度（PS：当 angle＝0 时，渐变色是从左向右。然后逆时针方向转，当 angle＝90 时为从下往上。angle 必须为 45 的整数倍）；
- android:type["linear" | "radial" | "sweep"]，渐变类型（取值：linear、radial、sweep）linear 线性渐变，这是默认设置；radial 放射性渐变，以开始色为中心；sweep 扫描线式的渐变；
- android:useLevel["true" | "false"]，如果要使用 LevelListDrawable 对象，就要设置为 true。设置为 true 无渐变。false 有渐变色；
- android:gradientRadius，整型，渐变色半径。当 android:type＝"radial" 时才使用。单独使用 android:type＝"radial"会报错；
- android:centerX 整型，渐变中心 x 点坐标的相对位置；
- android:center 整型，渐变中心 y 点坐标的相对位置。

3. 内边距，即内容与边的距离

- android:left 整型，左内边距；
- android:top 整型，上内边距；
- android:right 整型，右内边距；
- android:bottom 整型，下内边距。

4. size 大小

- android:width 整型，宽度；
- android:height 整型，高度；

5. 内部填充

- android:color 颜色值,填充颜色。

6. 描边

- android:width 整型,描边的宽度;
- android:color 颜色值,描边的颜色;
- android:dashWidth 整型,表示描边的样式是虚线的宽度,值为 0 时,表示为实线。值大于 0 则为虚线;
- android:dashGap 整型,表示描边为虚线时,虚线之间的间隔。

7. 其他属性

下面的属性只有在 android:shape="ring"时可用:

- android:innerRadius 尺寸,内环的半径;
- android:innerRadiusRatio 浮点型,以环的宽度比率来表示内环的半径,例如,如果 android:innerRadiusRatio"=5",表示内环半径等于环的宽度除以 5,这个值是可以被覆盖的,默认为 9;
- android:thickness 尺寸,环的厚度;
- android:thicknessRatio 浮点型,以环的宽度比率来表示环的厚度,例如,如果 android:thicknessRatio=2",那么环的厚度就等于环的宽度除以 2。这个值是可以被 android:thickness 覆盖的,默认值是 3;
- android:useLevel boolean 值,如果当做是 LevelListDrawable 使用时值为 true,否则为 false。

3.4.4 常见 Shape 标签的种类

Shape 标签能定义多种类型的 Drawable 图片。Shape 可以定义四种类型的几何图形,由 android:shape 属性指定:

- line——线;
- rectangle——矩形(圆角矩形);
- oval——椭圆,圆;
- ring——圆环。

Shape 可以定义边框属性:有边框、无边框、虚线边框、实线边框。

Shape 可以实现矩形圆角效果,可以指定其中一个角或者多个角设置圆角效果,指定圆角半径设置圆角的大小。

Shape 可以实现三种渐变,由子标签 gradient 实现:

- linear——线性渐变(水平,垂直,对角线三个渐变);
- sweep——扫描渐变(只支持顺时针方向,其实颜色反过来就跟逆时针一样);
- radial——径向渐变(由指定的中心点开始向外渐变,指定半径);

xml 实现只支持三个颜色,startColor,CenterColor,endColor。由上面的组合可以定义很多 Drawable,下面依次进行介绍:

1. 线(实线＋虚线)

线(实线＋虚线)的效果如图 3-14 所示。

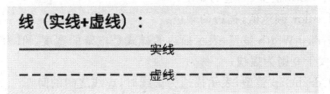

图 3-14 线(实线＋虚线)的效果图

实线:line_solid.xml

```
<?xml version="1.0" encoding="utf-8"?>
<!-- 实线 -->
<shape xmlns:android="http://schemas.android.com/apk/res/android"
    android:shape="line"
    android:useLevel="true">
<stroke
    android:width="2dp"
    android:color="#ffff0000"/>
</shape>
```

虚线:line_dashed.xml

```
<?xml version="1.0" encoding="utf-8"?>
<!-- 虚线 设置类型会 line
    需要关闭硬件加速虚线才能绘制出来,布局文件中使用的时候需要设置
android:layerType="software"
    android:width 线宽,布局文件中的 View 的高度需要比这个值大才可以绘制出来
    android:dashWidth 每段破折线的长度
    android:dashGap="5dp"每段破折线之间的间隔 -->
<shape xmlns:android="http://schemas.android.com/apk/res/android"
    android:shape="line"
    android:useLevel="true">
    <stroke
        android:width="2dp"
        android:dashGap="5dp"
        android:dashWidth="10dp"
        android:color="#ffff0000"/>
</shape>
```

2. 矩形(边框+填充)

矩形(边框+填充)的效果如图 3-15 所示。

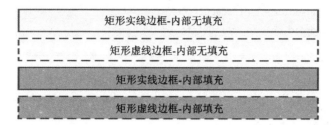

图 3-15 矩形(边框+填充)效果图

矩形实线边框-内部无填充:rect_solid_border.xml

```
<?xml version="1.0" encoding="utf-8"?>
<!-- 实线边框 -->
<shape xmlns:android="http://schemas.android.com/apk/res/android"
    android:shape="rectangle"
    android:useLevel="true">
    <stroke
        android:width="2dp"
        android:color="#ffff0000" />
</shape>
```

矩形虚线边框-内部无填充:rect_dashed_border.xml

```
<?xml version="1.0" encoding="utf-8"?>
<!-- 虚线边框 -->
<shape xmlns:android="http://schemas.android.com/apk/res/android"
    android:shape="rectangle"
    android:useLevel="true">
    <stroke
        android:width="2dp"
        android:color="#ffff0000"
        android:dashGap="5dp"
        android:dashWidth="10dp" />
</shape>
```

矩形实线边框-内部填充:rect_solid_border_and_fill.xml

```
<?xml version="1.0" encoding="utf-8"?>
<!-- 实线边框+内部填充 -->
<shape xmlns:android="http://schemas.android.com/apk/res/android"
    android:shape="rectangle"
```

```
        android:useLevel = "true">
    <stroke
        android:width = "2dp"
        android:color = "#ffff0000" />
    <solid android:color = "#ff00ffff" />
</shape>
```

矩形虚线边框-内部填充:rect_dashed_border_and_fill.xml

```
<?xml version = "1.0" encoding = "utf-8"?>
<!--虚线边框+内部填充 -->
<shape xmlns:android = "http://schemas.android.com/apk/res/android"
    android:shape = "rectangle"
    android:useLevel = "true">
    <stroke
        android:width = "2dp"
        android:color = "#ffff0000"
        android:dashGap = "5dp"
        android:dashWidth = "10dp" />
    <solid android:color = "#ff00ffff" />
</shape>
```

3. 圆角矩形

圆角矩形的效果如图3-16所示。

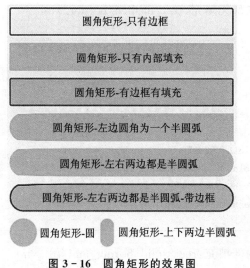

图3-16 圆角矩形的效果图

圆角矩形-只有边框:rect_rounded_border.xml

```
<?xml version = "1.0" encoding = "utf-8"?>
<!--矩形边框圆角 -->
```

```xml
<shape xmlns:android = "http://schemas.android.com/apk/res/android"
    android:shape = "rectangle"
    android:useLevel = "true">
    <size android:height = "100dp"
        android:width = "100dp"/>
    <stroke
        android:width = "2dp"
        android:color = "#ffff0000" />
    <corners android:bottomLeftRadius = "2dp"
        android:bottomRightRadius = "2dp"
        android:topLeftRadius = "2dp"
        android:topRightRadius = "2dp" />
</shape>
```

圆角矩形-只有内部填充：rect_rounded_fill.xml

```xml
<?xml version = "1.0" encoding = "utf-8"?>
<!-- 圆角矩形 -->
<shape xmlns:android = "http://schemas.android.com/apk/res/android"
    android:shape = "rectangle"
    android:useLevel = "true">
    <size android:height = "100dp"
        android:width = "100dp"/>
    <solid android:color = "#8000ff00" />
    <corners android:bottomLeftRadius = "2dp"
        android:bottomRightRadius = "2dp"
        android:topLeftRadius = "2dp"
        android:topRightRadius = "2dp" />
</shape>
```

圆角矩形-有边框有填充：rect_rounded_border_and_fill.xml

```xml
<?xml version = "1.0" encoding = "utf-8"?>
<!-- 矩形边框 + 填充 圆角 -->
<shape xmlns:android = "http://schemas.android.com/apk/res/android"
    android:shape = "rectangle"
    android:useLevel = "true">
    <size android:height = "100dp"
        android:width = "100dp"/>
    <stroke
        android:width = "2dp"
        android:color = "#ffff0000" />
    <solid android:color = "#8000ff00" />
```

```xml
    <corners android:bottomLeftRadius = "2dp"
        android:bottomRightRadius = "2dp"
        android:topLeftRadius = "2dp"
        android:topRightRadius = "2dp" />
</shape>
```

圆角矩形-左边圆角为一个半圆弧:rect_rounded_left_arc.xml

```xml
<?xml version = "1.0" encoding = "utf-8"?>
<!-- 矩形圆角+左边为一个半圆弧 -->
<shape xmlns:android = "http://schemas.android.com/apk/res/android"
    android:shape = "rectangle"
    android:useLevel = "true">
    <size android:width = "50dp"
        android:height = "10dp" />
    <solid android:color = "#8000ff00" />
    <!-- 圆角半径是高度的一半就是一个圆弧了 -->
    <corners
        android:bottomLeftRadius = "20dp"
        android:topLeftRadius = "20dp" />
</shape>
```

圆角矩形-左右两边都是半圆弧:rect_rounded_left_right_arc.xml

```xml
<?xml version = "1.0" encoding = "utf-8"?>
<!-- 矩形圆角+左右两边为一个半圆弧 -->
<shape xmlns:android = "http://schemas.android.com/apk/res/android"
    android:shape = "rectangle"
    android:useLevel = "true">
    <size android:width = "50dp"
        android:height = "10dp" />
    <solid android:color = "#8000ff00" />
    <!-- 圆角半径是高度的一半就是一个圆弧了 -->
    <corners android:radius = "20dp" />
</shape>
```

圆角矩形-左右两边都是半圆弧-带边框:rect_rounded_left_right_arc_border.xml

```xml
<?xml version = "1.0" encoding = "utf-8"?>
<!-- 矩形圆角+左右两边为一个半圆弧 -->
<shape xmlns:android = "http://schemas.android.com/apk/res/android"
    android:shape = "rectangle"
    android:useLevel = "true">
```

```
<size   android:width = "50dp"
     android:height = "10dp" />
<stroke android:color = "#ffff0000"
     android:width = "2dp"/>
<solid android:color = "#8000ff00" />
<!-- 圆角半径是高度的一半就是一个圆弧了 -->
<corners android:radius = "20dp" />
</shape>
```

圆角矩形-圆:rect_rounded_arc.xml

```
<? xml version = "1.0" encoding = "utf - 8"? >
<!-- 矩形圆角 + 圆出一个圆弧 -->
<shape xmlns:android = "http://schemas.android.com/apk/res/android"
    android:shape = "rectangle"
    android:useLevel = "true">
    <size android:height = "10dp"
        android:width = "10dp"/>
    <solid android:color = "#8000ff00" />
    <corners android:radius = "20dp" />
</shape>
```

圆角矩形-上下两边半圆弧:rect_rounded_top_bottom_arc.xml

```
<? xml version = "1.0" encoding = "utf - 8"? >
<!-- 矩形圆角 + 上下两边为一个圆弧 -->
<shape xmlns:android = "http://schemas.android.com/apk/res/android"
    android:shape = "rectangle"
    android:useLevel = "true">
    <size   android:width = "10dp"
        android:height = "60dp" />
    <solid android:color = "#8000ff00" />
    <!-- 圆角半径是宽度的一半就是一个圆弧了 -->
    <corners android:radius = "10dp" />
</shape>
```

4. 渐变效果(以矩形为例)

渐变效果(以矩形为例)如图 3-17 所示。

垂直线性渐变:rect_gradient_linear_vertical.xml

```
<? xml version = "1.0" encoding = "utf - 8"? >
<!-- 矩形内部填充-垂直线性渐变 -->
<shape xmlns:android = "http://schemas.android.com/apk/res/android"
```

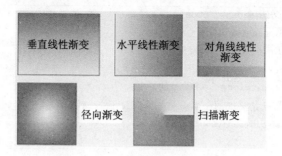

图 3-17 渐变效果(以矩形为例)图

```
    android:shape = "rectangle"
    android:useLevel = "true">
    <size android:width = "@dimen/shape_size"
        android:height = "@dimen/shape_size" />
    <stroke
        android:width = "1px"
        android:color = "#ffff00ff" />
    <!-- 调整 angle 实现水平渐变,垂直渐变或者对角渐变 -->
    <gradient
        android:angle = " - 45"
        android:centerX = "0.5"
        android:centerY = "0.4"
        android:centerColor = "#8000ff00"
        android:endColor = "#1000ff00"
        android:startColor = "#ff00ff00"
        android:type = "linear" />
</shape>
```

水平线性渐变:rect_gradient_linear_horizon.xml

```
<? xml version = "1.0" encoding = "utf - 8"? >
<!-- 矩形内部填充 - 水平线性渐变 -->
<shape xmlns:android = "http://schemas.android.com/apk/res/android"
    android:shape = "rectangle"
    android:useLevel = "true">
    <size android:width = "@dimen/shape_size"
        android:height = "@dimen/shape_size" />
    <stroke
        android:width = "1px"
        android:color = "#ffff00ff" />
    <!-- 调整 angle 实现水平渐变,垂直渐变或者对角渐变 -->
    <gradient
```

```
        android:angle = "0"
        android:centerX = "0.5"
        android:centerY = "0.5"
        android:centerColor = "#8000ff00"
        android:endColor = "#ff00ff00"
        android:startColor = "#1000ff00"
        android:type = "linear" />
</shape>
```

对角线线性渐变:rect_gradient_linear_diagonal.xml

```
<?xml version = "1.0" encoding = "utf-8"?>
<!-- 矩形内部填充 - 对角线线性渐变 -->
<shape xmlns:android = "http://schemas.android.com/apk/res/android"
    android:shape = "rectangle"
    android:useLevel = "true">
    <size
        android:width = "@dimen/shape_size"
        android:height = "@dimen/shape_size" />
    <stroke
        android:width = "1px"
        android:color = "#ffff00ff" />
    <!-- 调整 angle 实现水平渐变,垂直渐变或者对角渐变 -->
    <gradient
        android:angle = "45"
        android:centerX = "0.5"
        android:centerY = "0.5"
        android:centerColor = "#8000ff00"
        android:endColor = "#1000ff00"
        android:startColor = "#ff00ff00"
        android:type = "linear" />
</shape>
```

径向渐变:rect_gradient_radial.xml

```
<?xml version = "1.0" encoding = "utf-8"?>
<!-- 矩形内部填充 - 径向渐变,一般不用在 rect 上,用到圆或者椭圆上 -->
<shape xmlns:android = "http://schemas.android.com/apk/res/android"
    android:shape = "rectangle"
    android:useLevel = "true">
    <size android:width = "@dimen/shape_size"
        android:height = "@dimen/shape_size" />
    <stroke
```

```xml
        android:width = "1px"
        android:color = "#ffff00ff" />
    <!-- 径向渐变 angle 无效 -->
    <gradient
        android:angle = "0"
        android:centerX = "0.5"
        android:centerY = "0.5"
        android:startColor = "#0000ff00"
        android:endColor = "#ff00ff00"
        android:gradientRadius = "40dp"
        android:type = "radial" />
</shape>
```

扫描渐变:rect_gradient_sweep.xml

```xml
<?xml version = "1.0" encoding = "utf-8"?>
<!-- 矩形内部填充-扫描渐变 -->
<shape xmlns:android = "http://schemas.android.com/apk/res/android"
    android:shape = "rectangle"
    android:useLevel = "true">
    <!-- 如果布局中没有设置 View 的大小,会以 size 设置的大小为默认值 -->
    <size android:width = "20dp"
        android:height = "20dp" />
    <stroke
        android:width = "1px"
        android:color = "#ffff00ff" />
    <!-- 调整 angle 不能实现角度变化
        centerX,centerY 是中心点的位置,这里用的是百分比值(0-1)
        在 rect 中 gradientRadius 无效 -->
    <gradient
        android:angle = "0"
        android:centerX = "0.5"
        android:centerY = "0.5"
        android:startColor = "#ff00ff00"
        android:gradientRadius = "20dp"
        android:type = "sweep" />
</shape>
```

5. 圆(边框+填充+渐变)

圆(边框+填充+渐变)效果如图 3-18 所示。

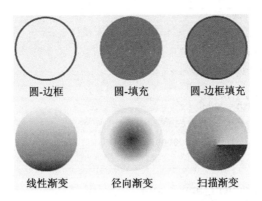

图 3-18 圆(边框＋填充＋渐变)效果图

圆-边框：circle_border.xml

```
<?xml version="1.0" encoding="utf-8"?>
<!-- 圆形边框 -->
<shape xmlns:android="http://schemas.android.com/apk/res/android"
    android:shape="oval"
    android:useLevel="true">
    <size android:width="80dp"
        android:height="80dp" />
    <stroke
        android:width="2dp"
        android:color="#ffff0000" />
</shape>
```

圆-填充：circle_fill.xml

```
<?xml version="1.0" encoding="utf-8"?>
<!-- 圆形+填充 -->
<shape xmlns:android="http://schemas.android.com/apk/res/android"
    android:shape="oval"
    android:useLevel="true">
    <size android:width="80dp"
        android:height="80dp" />
    <solid android:color="#800000ff" />
</shape>
```

圆-边框填充：circle_border_and_fill.xml

```
<?xml version="1.0" encoding="utf-8"?>
<!-- 圆形边框 + 填充 -->
<shape xmlns:android="http://schemas.android.com/apk/res/android"
```

```
    android:shape = "oval"
    android:useLevel = "true">
     <size android:width = "80dp"
        android:height = "80dp" />
     <stroke
        android:width = "2dp"
        android:color = "#ffff0000" />
     <solid android:color = "#800000ff" />
</shape>
```

线性渐变:circle_gradient_linear.xml

```
<?xml version = "1.0" encoding = "utf-8"?>
<!-- 圆形内部填充-线性渐变 -->
<shape xmlns:android = "http://schemas.android.com/apk/res/android"
    android:shape = "oval"
    android:useLevel = "true">
     <size android:width = "@dimen/shape_size"
        android:height = "@dimen/shape_size" />
<!-- angle 调整渐变角度,只能是 45 的倍数,centerX,centerY 是百分百(0-1) -->
    <gradient
        android:angle = "-90"
        android:centerX = "0.5"
        android:centerY = "0.8"
        android:centerColor = "#80ff0000"
        android:endColor = "#ffff0000"
        android:startColor = "#00ff0000"
        android:type = "linear" />
</shape>
```

径向渐变:circle_gradient_radial.xml

```
<?xml version = "1.0" encoding = "utf-8"?>
<!-- 圆形内部填充-径向渐变 -->
<shape xmlns:android = "http://schemas.android.com/apk/res/android"
    android:shape = "oval"
    android:useLevel = "true">
     <size
        android:width = "40dp"
        android:height = "40dp" />
    <!-- centerX,centerY 是百分百(0-1) -->
    <gradient
        android:centerX = "0.5"
        android:centerY = "0.5"
```

```
        android:startColor = "#ffff0000"
        android:centerColor = "#80ff0000"
        android:endColor = "#10ff0000"
        android:gradientRadius = "30dp"
        android:type = "radial" />
</shape>
```

扫描渐变:circle_gradient_sweep.xml

```
<?xml version = "1.0" encoding = "utf-8"?>
<!-- 圆形内部填充-扫描渐变 -->
<shape xmlns:android = "http://schemas.android.com/apk/res/android"
    android:shape = "oval"
    android:useLevel = "true">
    <size android:width = "@dimen/shape_size"
        android:height = "@dimen/shape_size" />
<!-- sweep 类型 angle,gradientRadius 无效,centerX, centerY 是百分百 0-1 -->
    <gradient
        android:centerX = "0.5"
        android:centerY = "0.6"
        android:startColor = "#ffff0000"
        android:centerColor = "#80ff0000"
        android:endColor = "#20ff0000"
        android:gradientRadius = "20dp"
        android:type = "sweep" />
</shape>
```

6. 椭圆(边框+填充+渐变)

椭圆(边框+填充+渐变)效果如图 3-19 所示。

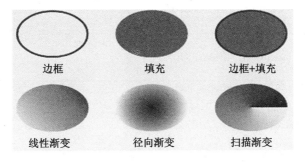

图 3-19　椭圆(边框+填充+渐变)效果图

边框:oval_border.xml

```xml
<?xml version="1.0" encoding="utf-8"?>
<!-- 椭圆边框 -->
<shape xmlns:android="http://schemas.android.com/apk/res/android"
    android:shape="oval"
    android:useLevel="true">
    <stroke
        android:width="2dp"
        android:color="#ffff0000" />
</shape>
```

填充:oval_fill.xml

```xml
<?xml version="1.0" encoding="utf-8"?>
<!-- 椭圆填充 -->
<shape xmlns:android="http://schemas.android.com/apk/res/android"
    android:shape="oval"
    android:useLevel="true">
    <solid android:color="#800000ff" />
</shape>
```

边框+填充:oval_border_and_fill.xml

```xml
<?xml version="1.0" encoding="utf-8"?>
<!-- 椭圆边框 + 填充 -->
<shape xmlns:android="http://schemas.android.com/apk/res/android"
    android:shape="oval"
    android:useLevel="true">
    <stroke
        android:width="2dp"
        android:color="#ffff0000" />
    <solid android:color="#800000ff" />
</shape>
```

线性渐变:oval_gradient_linear.xml

```xml
<?xml version="1.0" encoding="utf-8"?>
<!-- 椭圆内部填充-线性渐变 -->
<shape xmlns:android="http://schemas.android.com/apk/res/android"
    android:shape="oval"
    android:useLevel="true">
    <size android:width="80dp"
        android:height="60dp" />
    <gradient
        android:angle="45"
```

```
        android:centerX = "0.5"
        android:centerY = "0.7"
        android:centerColor = "#80ff0000"
        android:endColor = "#ffff0000"
        android:startColor = "#00ff0000"
        android:type = "linear" />
</shape>
```

径向渐变:oval_gradient_radial.xml

```
<?xml version = "1.0" encoding = "utf-8"?>
<!-- 椭圆内部填充-径向渐变 -->
<shape xmlns:android = "http://schemas.android.com/apk/res/android"
    android:shape = "oval"
    android:useLevel = "true" >
    <size android:width = "80dp"
        android:height = "60dp" />
    <gradient
        android:centerX = "0.5"
        android:centerY = "0.5"
        android:centerColor = "#80ff0000"
        android:endColor = "#00ff0000"
        android:startColor = "#ffff0000"
        android:gradientRadius = "40dp"
        android:type = "radial" />
</shape>
```

扫描渐变:oval_gradient_sweep.xml

```
<?xml version = "1.0" encoding = "utf-8"?>
<!-- 椭圆内部填充-扫描渐变 -->
<shape xmlns:android = "http://schemas.android.com/apk/res/android"
    android:shape = "oval"
    android:useLevel = "true" >
    <size android:width = "80dp"
        android:height = "60dp" />
    <gradient
        android:centerX = "0.5"
        android:centerY = "0.5"
        android:centerColor = "#80ff0000"
        android:endColor = "#ffff0000"
        android:startColor = "#00ff0000"
        android:type = "sweep" />
</shape>
```

7. 圆环(边框＋填充＋渐变)

圆环(边框＋填充＋渐变)效果如图 3-20 所示。

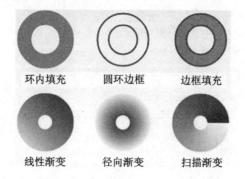

图 3-20 圆环(边框＋填充＋渐变)效果图

环内填充：ring_fill.xml

```
<? xml version = "1.0" encoding = "utf-8"? > <!-- 圆环 -->
<shape xmlns:android = "http://schemas.android.com/apk/res/android"
    android:innerRadiusRatio = "4"
    android:shape = "ring"
    android:thicknessRatio = "4"
    android:useLevel = "false">
    <!-- android:useLevel = "false"必须是 false -->
    <size android:width = "80dp"
        android:height = "80dp" />
    <solid android:color = "#80ff0000" />
</shape>
```

圆环边框：ring_border.xml

```
<? xml version = "1.0" encoding = "utf-8"? >
<!-- 圆环-仅有边框 -->
<shape xmlns:android = "http://schemas.android.com/apk/res/android"
    android:innerRadius = "20dp"
    android:shape = "ring"
    android:thickness = "16dp"
    android:useLevel = "false">
    <!-- android:useLevel = "false"必须是 false -->
    <size
        android:width = "80dp"
        android:height = "80dp" />
    <stroke
```

```
        android:width = "2dp"
        android:color = "#ffff00ff" />
</shape>
```

边框填充：ring_border_and_fill.xml

```
<?xml version = "1.0" encoding = "utf-8"?>
<!-- 圆环 -->
<shape xmlns:android = "http://schemas.android.com/apk/res/android"
    android:innerRadius = "20dp"
    android:shape = "ring"
    android:thickness = "16dp"
    android:useLevel = "false">
    <!-- android:useLevel = "false"必须是 false -->
    <size android:width = "80dp"
        android:height = "80dp" />
    <solid android:color = "#80ff0000" />
    <stroke
        android:width = "2dp"
        android:color = "#ffff00ff" />
</shape>
```

线性渐变：ring_gradient_linear.xml

```
<?xml version = "1.0" encoding = "utf-8"?>
<!-- 圆环-线性渐变 -->
<shape xmlns:android = "http://schemas.android.com/apk/res/android"
    android:shape = "ring"
    android:innerRadius = "10dp"
    android:thickness = "30dp"
    android:useLevel = "false">
    <!-- android:useLevel = "false"必须是 false -->
    <size android:width = "80dp"
        android:height = "80dp" />
    <gradient
        android:angle = "45"
        android:centerColor = "#80ff0000"
        android:endColor = "#ffff0000"
        android:startColor = "#00ff0000"
        android:type = "linear" />
</shape>
```

径向渐变：ring_gradient_radial.xml

```xml
<?xml version="1.0" encoding="utf-8"?>
<!--圆环-径向渐变-->
<shape xmlns:android="http://schemas.android.com/apk/res/android"
    android:shape="ring"
    android:innerRadius="10dp"
    android:thickness="30dp"
    android:useLevel="false">
    <!--android:useLevel="false"必须是false-->
    <size
        android:width="80dp"
        android:height="80dp" />
    <!--设置径向渐变半径,渐变从圆心开始-->
    <gradient
        android:centerX="0.5"
        android:centerY="0.5"
        android:centerColor="#80ff0000"
        android:endColor="#00ff0000"
        android:startColor="#ffff0000"
        android:gradientRadius="40dp"
        android:type="radial" />
</shape>
```

扫描渐变:ring_gradient_sweep.xml

```xml
<?xml version="1.0" encoding="utf-8"?>
<!--圆环-扫描渐变-->
<shape xmlns:android="http://schemas.android.com/apk/res/android"
    android:shape="ring"
    android:innerRadius="10dp"
    android:thickness="30dp"
    android:useLevel="false">
    <!--android:useLevel="false"必须是false-->
    <size android:width="80dp"
        android:height="80dp" />
    <!--扫描渐变shape不能设置角度-->
    <gradient
        android:centerColor="#80ff0000"
        android:endColor="#ffff0000"
        android:startColor="#00ff0000"
        android:type="sweep" />
</shape>
```

3.4.5 自定义背景的按钮

自定义背景的按钮目前有两种方式实现:矢量和位图。

1. 矢量图形绘制的方式

矢量图形绘制的方式实现简单,适合对按钮形状和图案要求不高的场合,下面介绍其设计步骤。

(1) 使用 xml 定义一个圆角矩形,外围轮廓线实线、内填充渐变色,xml 代码如下:

```xml
//bg_alibuybutton_default.xml
<?xml version = "1.0" encoding = "utf-8"?>
<layer-list
        xmlns:android = "http://schemas.android.com/apk/res/android">
  <item>
    <shape android:shape = "rectangle">
     <solid android:color = "#FFEC7600" />
     <corners
      android:topLeftRadius = "5dip"
      android:topRightRadius = "5dip"
      android:bottomLeftRadius = "5dip"
      android:bottomRightRadius = "5dip" />
    </shape>
  </item>
  <item android:top = "1px" android:bottom = "1px"
android:left = "1px" android:right = "1px">
    <shape>
     <gradient
      android:startColor = "#FFEC7600" android:endColor = "#FFFED69E"
      android:type = "linear" android:angle = "90"
      android:centerX = "0.5" android:centerY = "0.5" />
     <corners
      android:topLeftRadius = "5dip"
      android:topRightRadius = "5dip"
      android:bottomLeftRadius = "5dip"
      android:bottomRightRadius = "5dip" />
    </shape>
  </item>
</layer-list>
```

定义 bg_alibuybutton_pressed.xml 和 bg_alibuybutton_selected.xml,内容相同,就是渐变颜色不同,用于按钮按下后的背景变化效果。

（2）定义按钮按下后的效果变化描述文件 drawable/bg_alibuy button.xml，代码如下：

```xml
<?xml version = "1.0" encoding = "UTF-8"?>
<selector
    xmlns:android = "http://schemas.android.com/apk/res/android">
  <item android:state_pressed = "true"
    android:drawable = "@drawable/bg_alibuybutton_pressed" />
  <item android:state_focused = "true"
    android:drawable = "@drawable/bg_alibuybutton_selected" />
  <item android:drawable = "@drawable/bg_alibuybutton_default" />
</selector>
```

（3）在界面定义文件中，如 layout/main.xml 中定义一个 Button 控件，代码如下：

```xml
<Button
    android:layout_width = "120dip"
    android:layout_height = "40dip"
    android:text = "矢量背景按钮"
    android:background = "@drawable/bg_alibuybutton" />
```

这样，自定义背景的按钮就可以使用了，在实现 onClick 方法后就可以响应操作。

2. 9-patch 图片背景方式

9-patch 图片背景方式相对复杂，但可以制作出更多、更复杂的按钮样式。9-patch 格式是 Android 中特有的一种 PNG 图片格式，以"***.9.png"结尾。该格式的图片定义了可以伸缩拉伸的区域和文字显示区域，如图 3-21 所示，这样就可以在 Android 开发中对非矢量图进行拉伸而仍然保持美观。如果使用位图而没有经过 9-patch 处理的话，图片就会被整体拉伸，影响效果。Android 中大量使用了该技术，默认的按钮背景就是用了类似的方法实现。

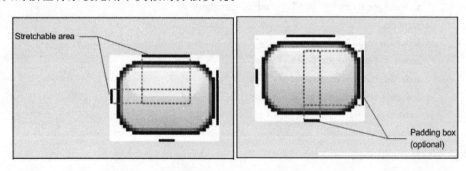

图 3-21 9-patch 图

该格式相对于一般 PNG 图片来说，上下左右各多了一条 1px 的黑线。左、上黑线隔开了 9 个格子，当中一个格子（Strechable area 区域）声明为可以进行拉伸。右、下两条黑线所定义的 Padding box 区域,是在该图片当作背景时,能够在图片上填写文字的区域。每条黑线都可以是不连续的,这样就可以定义出很多自动拉伸的规格。Android sdk 中提供了设置工具,启动命令位于:$ANDROID_SDK/ tools/draw9patch.bat,使用它对于原始 PNG 进行设置 9-patch 格式,非常方便,如图 3-22 所示。

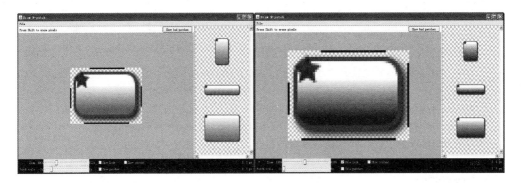

图 3-22 9-patch 图制作界面

draw9patch 工具的右侧能够看到各方向拉伸后的效果图,所要做的就是在图上最外侧一圈 1px 宽的像素上涂黑线。

注意,在 draw9patch.bat 第一次运行时,sdk2.2 版本上会报错:java.lang.NoClassDefFoundError:org/jdesktop/swingworker/SwingWorker。需要下载 swing-worker-1.1.jar,放到$android_sdk/tools/lib 路径下,才能成功运行。下面介绍其实现步骤。

(1) 使用 draw9patch.bat 画完图片后,得到两张按钮背景,分别是正常和按下状态下的图片,命名为 bg_btn.9.png 和 bg_btn_2.9.png。

(2) 编写图片使用描述文件 bg_9patchbutton.xml,代码如下:

```xml
// in bg_9patchbutton.xml
<?xml version = "1.0" encoding = "UTF-8" ?>
<selector  xmlns:android =
        "http://schemas.android.com/apk/res/android" >
  <item  android:state_pressed = "true"
    android:drawable = "@drawable/bg_btn_2"  />
  <item  android:state_focused = "true"
    android:drawable = "@drawable/bg_btn_2"  />
  <item  android:drawable = "@drawable/bg_btn"  />
</selector>
```

(3) 在界面定义文件 layout/main.xml 中添加 Button、ImageButton 按钮控件的定义。Button、ImageButton 都可以使用背景属性,代码如下:

```
<Button
  android:layout_width = "120dip"
  android:layout_height = "40dip"
  android:text = "9-patch 图片背景按钮"
  android:background = "@drawable/bg_9patchbutton"    />
<Button
  android:layout_width = "200dip"
  android:layout_height = "40dip"
  android:text = "9-patch 图片背景按钮"
  android:background = "@drawable/bg_9patchbutton"    />
<Button
  android:layout_width = "120dip"
  android:layout_height = "80dip"
  android:text = "9-patch 图片背景按钮"
  android:background = "@drawable/bg_9patchbutton"    />
<ImageButton
  android:layout_width = "120dip"
  android:layout_height = "40dip"
  android:src = "@drawable/bg_9patchbutton"
  android:scaleType = "fitXY"
  android:background = "@android:color/transparent"    />
```

3. 自定义形状、颜色、图样按钮的实现

下面介绍自定义形状、颜色、图样按钮的实现步骤。

（1）设计一张自定义形状风格的背景图片。

（2）未单击和按下后的状态各做一张，形成一套图片（两个图片文件）。

　　forward.png　　forward2.png

（3）创建和编写不同状态的按钮图片，描述文件 drawable/ib_forward.xml 如下：

```
// ib_forward.xml
<? xml   version = "1.0"   encoding = "UTF-8" ? >
<selector   xmlns:android
    = "http://schemas.android.com/apk/res/android" >
 <item   android:state_pressed = "true"
    android:drawable = "@drawable/forward2"   />
 <item   android:state_focused = "true"
    android:drawable = "@drawable/forward2"   />
 <item   android:drawable = "@drawable/forward"   />
</selector>
```

（4）在界面定义文件 layout/main.xml 中添加 ImageButton 按钮控件的定义：

```
// in layout/main.xml
  <ImageButton
    android:layout_width = "80dip"
    android:layout_height = "40dip"
    android:src = "@drawable/ib_forword"
    android:scaleType = "fitXY"
    android:background = "@android:color/transparent"   />
```

3.5 Selector 的使用

Android 中的 Selector 主要是用来改变 ListView 和 Button 控件的默认背景。Selector 中文的意思选择器，在 Android 中常用来作组件的背景，这样做的好处是省去了用代码控制实现组件在不同状态下不同的背景颜色或图片的变换，使用十分方便。

前面介绍的 Shape 主要用于自定义矩形、圆形、线形和环形，以及有哪些需要注意的地方。不过，Shape 只能定义单一的形状，而实际应用中，很多地方比如按钮、Tab、ListItem 等都是不同状态有不同的展示形状。举个例子，一个按钮的背景，默认时是一个形状，按下时是一个形状，不可操作时又是另一个形状。有时候，不同状态下改变的不只是背景、图片等，文字颜色也会相应改变。而要处理这些不同状态下展示什么的问题，就要用 Selector 来实现了。

Selector 标签，可以添加一个或多个 item 子标签，而相应的状态是在 item 标签中定义的。定义的 xml 文件可以作为两种资源使用：drawable 和 color。作为 drawable 资源使用时，一般和 Shape 一样放于 drawable 目录下，item 必须指定 android:drawable 属性；作为 color 资源使用时，则放于 color 目录下，item 必须指定 android:color 属性。

Selector 常用属性：
- android:state_enabled：设置触摸或点击事件是否可用状态，一般只在 false 时设置该属性，表示不可用状态；
- android:state_pressed：设置是否按压状态，一般在 true 时设置该属性，表示已按压状态，默认为 false；
- android:state_selected：设置是否选中状态，true 表示已选中，false 表示未选中；
- android:state_checked：设置是否勾选状态，主要用于 CheckBox 和 RadioButton，true 表示已被勾选，false 表示未被勾选；
- android:state_checkable：设置勾选是否可用状态，类似 state_enabled，只是 state_enabled 会影响触摸或点击事件，而 state_checkable 影响勾选事件；
- android:state_focused：设置是否获得焦点状态，true 表示获得焦点，默认为 false，表示未获得焦点；

- android:state_window_focused：设置当前窗口是否获得焦点状态，true 表示获得焦点，false 表示未获得焦点，例如，拉下通知栏或弹出对话框时，当前界面就会失去焦点；另外，ListView 的 ListItem 获得焦点时也会触发 true 状态，可以理解为当前窗口就是 ListItem 本身；
- android:state_activated：设置是否被激活状态，true 表示被激活，false 表示未激活，API Level 11 及以上才支持，可通过代码调用控件的 setActivated(boolean)方法设置是否激活该控件；
- android:state_hovered：设置是否鼠标在上面滑动的状态，true 表示鼠标在上面滑动，默认为 false，API Level 14 及以上才支持。

Selector 是在 drawable/xxx.xml 中配置，相关图片放在同目录下。先看一下 listview 中的状态，把下面的 XML 文件保存成你自己命名的.xml 文件（比如 list_item_bg.xml），在系统使用时根据 ListView 中的列表项的状态来使用相应的背景图片。

```xml
<?xml version="1.0" encoding="utf-8"?>
<selector xmlns:android="http://schemas.android.com/apk/res/android">
    <!-- 默认时的背景图片 -->
    <item android:drawable="@drawable/pic1" />

    <!-- 没有焦点时的背景图片 -->
    <item android:state_window_focused="false"
        android:drawable="@drawable/pic1" />

    <!-- 非触摸模式下获得焦点并单击时的背景图片 -->
    <item android:state_focused="true"
        android:state_pressed="true"
        android:drawable="@drawable/pic2" />

    <!-- 触摸模式下单击时的背景图片 -->
    <item android:state_focused="false"
        android:state_pressed="true"
        android:drawable="@drawable/pic3" />

    <!-- 选中时的图片背景 -->
    <item android:state_selected="true"
        android:drawable="@drawable/pic4" />

    <!-- 获得焦点时的图片背景 -->
    <item android:state_focused="true"
        android:drawable="@drawable/pic5" />
</selector>
```

- android:state_selected 是选中；
- android:state_focused 是获得焦点；
- android:state_pressed 是点击；
- android:state_enabled 是设置是否响应事件，指所有事件。

根据这些状态同样可以设置 button 的 selector 效果。也可以设置 selector 改变 button 中的文字状态。使用 xml 文件：

（1）方法一：

在 listview 中配置 android:listSelector="@drawable/xxx"

或者在 listview 的 item 中添加属性 android:background="@drawable/xxx"

（2）方法二：

Drawable drawable = getResources().getDrawable(R.drawable.xxx);

ListView.setSelector(drawable);

但是这样会出现列表有时候为黑的情况，需要加上：android:cacheColorHint="@android:color/transparent"使其透明。以下是配置 Button 中的文字效果，drawable/button_font.xml：

```xml
<?xml version="1.0" encoding="utf-8"?>
<selector xmlns:android="http://schemas.android.com/apk/res/android">
    <item android:state_selected="true" android:color="#FFF" />
    <item android:state_focused="true" android:color="#FFF" />
    <item android:state_pressed="true" android:color="#FFF" />
    <item android:color="#000" />
</selector>
```

Button 还可以实现更复杂的效果，例如渐变，drawable/button_color.xml：

```xml
<?xml version="1.0" encoding="utf-8"?>
<selector xmlns:android="http://schemas.android.com/apk/res/android">   /
  <item android:state_pressed="true">//定义当 Button 处于 pressed 状态时的形态。
  <shape>
    <gradient android:startColor="#8600ff" />
    <stroke  android:width="2dp" android:color="#000000" />
    <corners android:radius="5dp" />
    <padding android:left="10dp" android:top="10dp"
        android:bottom="10dp" android:right="10dp"/>
  </shape>
  </item>
  <item android:state_focused="true">//定义当 Button 获得 focus 时的形态
  <shape>
    <gradient android:startColor="#eac100"/>
    <stroke android:width="2dp" android:color="#333333"
```

```
        color = "#ffffff"/>
    <corners android:radius = "8dp" />
    <padding android:left = "10dp" android:top = "10dp"
      android:bottom = "10dp" android:right = "10dp"/>
  </shape>
  </item>
</selector>
```

最后,需要在包含 Button 的 XML 文件里添加两项。假如是 main.xml 文件,需要在 <Button /> 里加两项。

android:focusable="true"

android:background="@drawable/button_color"

这样,使用 Button 的时候就可以甩掉系统自带的黄颜色的背景了,实现个性化的背景,配合应用的整体布局非常有用。

3.6　Android 沉浸式状态栏及悬浮效果

现在大多数的电商 App 的详情页长得几乎都差不多,当滑动时,会有 Tab 悬浮在上面,这样做用户体验确实不错,如果 Tab 滑上去,用户可能还需要滑下来,再来点击 Tab,这样确实很麻烦。沉浸式状态栏对比如图 3-23 所示。

图 3-23　沉浸式状态栏对比图

实现步骤如下:

(1) 下载 SystemBarTint 开源库,拷贝或引入其中的 SystemBarTintManager.java 至本项目中,下载地址:https://github.com/jgilfelt/SystemBarTint。

(2) 在项目 res 目录下新建 values－v19 文件夹,新建 style.xml 文件,内容为:

```xml
<resources xmlns:android = "http://schemas.android.com/apk/res/android">
  <style name = "AppBaseTheme" parent = "android:Theme.Holo.Light.NoActionBar">
    <item name = "android:windowContentOverlay">@null</item>
  </style>
</resources>
```

此处,确认 manifest 文件中 application 的 android:theme="AppBaseTheme"或继承自"AppBaseTheme",以保证在 Android 4.4 以上该样式生效。

(3) 在需要体现沉浸式效果的页面 Activity 中添加以下代码:

```java
protected void onCreate(Bundle savedInstanceState) {
    super.onCreate(savedInstanceState);
    if (Build.VERSION.SDK_INT >= Build.VERSION_CODES.KITKAT) {
        Window window = getWindow();
        //半透明状态栏
        window.setFlags(WindowManager.LayoutParams.FLAG_TRANSLUCENT_STATUS,
            WindowManager.LayoutParams.FLAG_TRANSLUCENT_STATUS);
        SystemBarTintManager tintManager = new SystemBarTintManager(this);
        tintManager.setStatusBarTintEnabled(true);
        //通知栏所需颜色
        tintManager.setStatusBarTintResource(R.color.main_bg_blue);
    }
}
```

此时,会发现 App 的整个界面充满了屏幕,UI 顶部的部分原色被通知栏遮盖了一点,透明或半透明,效果如图 3-24 所示。

图 3-24 沉浸式状态栏效果图

（4）实现 UI 顶部的自适应，从通知栏下方开始显示。在界面根布局文件中添加 android:fitsSystemWindows="true" 属性，让界面布局适应系统屏幕空间，给状态栏让出位子。例如：

```
<LinearLayout xmlns:android = "http://schemas.android.com/apk/res/android"
    android:layout_width = "match_parent"
    android:layout_height = "match_parent"
    android:fitsSystemWindows = "true"
    android:background = "@color/white"
    android:orientation = "vertical" >
    ...
</LinearLayout>
```

沉浸式状态栏最终效果如图 3-25 所示。

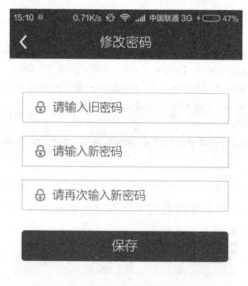

图 3-25　沉浸式状态栏最终效果图

其实，不用 SystemBarTintManager 库也可以实现上述效果，不过需要对界面布局的背景色做特定设置（界面布局需要与状态栏背景色一致），这样工作量有点大，而 SystemBarTint Manager 类可以灵活地设置背景色，其变更不会影响界面其他元素的背景色。Android 沉浸式状态栏及悬浮效果实例下载地址如下：

https://github.com/xiaoyuanandroid/ProductPage。

3.7　登录界面布局的实现

本章的登录界面布局的实现，在第 2 章欢迎界面设计项目"IoT_IS_App_001"的基础上完成。在 Obtain_Studio 中打开 IoT_IS_App_001 项目。在与 welcome.xml 文件

第3章 登录界面布局设计

相同的目录\Obtain_Studio\WorkDir\IoT_IS_App_001\app\src\main\res\layout 下新建一个 login.xml 文件，代码内容如下：

```xml
<? xml version = "1.0" encoding = "utf - 8"? >
<LinearLayout xmlns:android = "http://schemas.android.com/apk/res/android"
    android:orientation = "vertical"
    android:layout_width = "fill_parent"
    android:layout_height = "fill_parent"
    android:background = "@drawable/login" >
    <TextView
    android:layout_width = "fill_parent"
    android:layout_height = "wrap_content"
    android:layout_marginRight = "30dip"
    android:layout_marginLeft = "30dip"
    android:layout_marginTop = "30dip"
    android:textSize = "20sp"
    android:textColor = "#F83333"
    android:text = "物联网应用系统"  />
    <LinearLayout
    android:orientation = "vertical"
    android:layout_width = "fill_parent"
    android:layout_height = "300dip"
    android:background = "@drawable/bg3"
    android:layout_marginRight = "30dip"
    android:layout_marginLeft = "30dip"
    android:layout_marginTop = "150dip" >
        <EditText
            android:id = "@ + id/ip"
            android:layout_marginTop = "40dip"
            android:layout_marginRight = "90dip"
            android:layout_marginLeft = "90dip"
            android:layout_width = "fill_parent"
            android:layout_height = "50dip"
            android:maxLength = "40"
            android:hint = "192.168.4.1"
            android:textColorHint = "#238745"
            android:background = "#eeeeee"/>
    <EditText
        android:layout_width = "fill_parent"
        android:layout_height = "wrap_content"
        android:id = "@ + id/edit1"
        android:layout_marginTop = "10dip"
```

```xml
        android:layout_marginRight = "90dip"
        android:layout_marginLeft = "90dip"
        android:layout_marginBottom = "5dip"
        android:textColorHint = "#238745"
        android:background = "#eeeeee"
        android:hint = "请输入用户名..." />
    <EditText
        android:layout_width = "fill_parent"
        android:layout_height = "wrap_content"
        android:id = "@+id/edit2"
        android:layout_marginTop = "10dip"
        android:layout_marginRight = "90dip"
        android:layout_marginLeft = "90dip"
        android:textColorHint = "#238745"
        android:background = "#eeeeee"
        android:hint = "请输入密码..." />
    <LinearLayout
        android:layout_width = "fill_parent"
        android:layout_height = "wrap_content"
        android:layout_marginTop = "10dip"
        android:layout_marginRight = "20dip"
        android:layout_marginLeft = "90dip"  >
        <CheckBox
            android:layout_width = "wrap_content"
            android:layout_height = "wrap_content"
            android:layout_marginLeft = "10dip"
            android:text = "记住密码"
            android:id = "@+id/checkbox1"/>
        <Button
            android:layout_width = "90dip"
            android:layout_height = "wrap_content"
            android:text = "登录"
            android:id = "@+id/button1"
            android:layout_marginRight = "30dip"
            android:layout_marginLeft = "30dip"/>
    </LinearLayout>
  </LinearLayout>
</LinearLayout>
```

在与 WelcomeActivity.java 文件相同的目录\Obtain_Studio\WorkDir\IoT_IS_App_001\app\src\main\java\com\example\administrator\myapplication 下创建新文件 login.java，代码内容如下：

```java
package com.example.administrator.myapplication;
import android.app.Activity;
import android.os.Bundle;
import android.view.View;
import android.widget.Button;
import android.widget.EditText;
import android.widget.Toast;
import android.view.Gravity;
import android.content.Intent;
public class login extends Activity
{
    @Override
    public void onCreate(Bundle savedInstanceState)
    {
        super.onCreate(savedInstanceState);
        setContentView(R.layout.login);
        Button button = (Button)findViewById(R.id.button1);
        button.setOnClickListener(new View.OnClickListener(){
            // @Override
            public void onClick(View v)
            {
                EditText ipadders = (EditText) findViewById(R.id.ip);
                EditText username = (EditText) findViewById(R.id.edit1);
                EditText password = (EditText) findViewById(R.id.edit2);
                String ipaddersStr = ipadders.getText().toString();
                String usernameStr = username.getText().toString();
                String passwordStr = password.getText().toString();
                if("lbl".equals(usernameStr) && "133".equals(passwordStr) )
                {
                    Intent intent = new Intent();
                    intent.putExtra("IP",ipaddersStr);
                    intent.setClass(login.this, MainFragmentPagerActivity.class);
                    startActivity(intent);
                }
                //else
                {
                    Toast toast = Toast.makeText(getApplicationContext()," error",Toast.LENGTH_LONG);
                    toast.setGravity(Gravity.CENTER, 0, 0);
                    toast.show();
                } }
        });
    }
}
```

在 Drawable 目录下保存上述两个布局文件之中使用到的图片文件 login.jpg、

bg3.png。

在 AndroidManifest.xml 文件中,找到 <activity android:name=".WelcomeActivity">…. </activity> 一项的后面添加 login 一项的 Activity 注册,代码如下:

```
<activity android:name=".WelcomeActivity">
    <intent-filter>
        <action android:name="android.intent.action.MAIN" />
        <category android:name="android.intent.category.LAUNCHER" />
    </intent-filter>
</activity>
<activity android:name=".login">
</activity>
```

修改 IoT_IS_App_001.prj 文件中的运行 Activity 类为 login,代码如下:

```
<run>
cd app\build\outputs\apk\debug
adb -s emulator-5554 uninstall com.example.administrator.myapplication
adb -s emulator-5554 install app-debug.apk
adb -s emulator-5554 shell am start -n com.example.administrator.myapplication/.login
</run>
```

启动模拟器,最后编译和运行该项目,可以显示上述登录界面,在模拟器上的运行效果如图 3-26 所示。

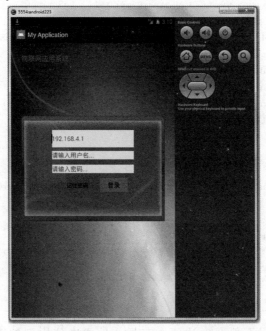

图 3-26 登录界面运行效果图

第 4 章 界面切换设计

4.1 界面切换设计目标

第2章和第3章分别介绍了欢迎界面和登录界面设计方法,特别是详细介绍了界面的布局与控件的应用,但还没有介绍如何实现欢迎界面和登录界面之间的切换。登录界面和主界面之间的切换过程与效果如图4-1所示,包括从欢迎界面自动切换到登录界面,再从登录界面通过输入用户名和密码再切换到主界面。

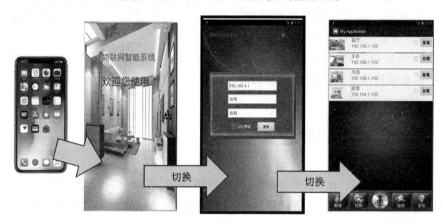

图 4-1 登录界面与主界面之间的切换

1. 主界面中 4 项地点列表

主界面中4项地点列表如主界面布局图4-2所示。全部采用线性布局,其中,总体结构采用垂直布局,每一项列表采用水平布局,两列文字采用垂直布局,右边的选择框和查看按钮采用水平布局。

2. 从欢迎界面自动切换到登录界面

从欢迎界面自动切换到登录界面的几个关键点:
(1) 设置隐藏状态栏。
(2) 设置隐藏标题栏。
(3) 设置定时器,到时间自动切换到登录界面。

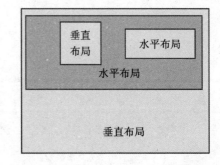

图 4-2 主界面布局

(4) 两个 Activity 之间的切换。两个 Activity 之间的切换需要使用到 Intents 类来实现,因此,Intents 是本章的重点内容之一。

3. 从登录界面通过输入用户名和密码再切换到主界面

从登录界面通过输入用户名和密码再切换到主界面的几个关键点:

(1) 读取用户输入的用户名和密码。

(2) 判断用户输入的用户名和密码是否正确。需要进行 Activity 消息的处理,因此,Activity 消息的处理也是本章的重点内容之一。

(3) 如果密码不对,要弹出一个对话框,提示"输入的用户名或者密码不正确!",因此,本章还要重点讲解对话框的实现方法。

(4) 两个 Activity 之间的切换。

4.2 安卓应用程序组件

Android 应用程序由一些松散联系的组件构成,遵守一个应用程序清单(manifest) 的规则,这个清单描述了每个组件以及它们如何交互,还包含了应用程序的硬件和平台需求的元数据(metadata)。

Android 应用程序组件包括 Activites、Services、Content Providers、Intents、Broadcast Receivers、Widgets、Notifications 等几个部分。

1. Activites

应用程序的每个界面都将是 Activity 类的扩展。Acitvities 用视图(View)构成 GUI 来显示信息、响应用户操作。就桌面开发而言,一个活动(Activity)相当于一个窗体(Form)。

2. Services

Service 组件在后台运行,后台更新数据源,可见 Activities 以及触发通知(Notification)。在应用程序的 Activities 不激活或不可见时,Service 组件用于执行依然需要继续的长期处理。

3. Content Providers

Content Providers 用于管理和共享应用程序数据库。是跨应用程序边界数据共享的优先方式。这表示可以配置自己的 Content Providers 以允许其他应用程序的访问，用他人提供的 Content Providers 来访问他人存储的数据。Android 设备包括几个本地 Content Providers，提供了像媒体库和联系人明细这样有用的数据库。

4. Intents

一个应用程序间（inter－application）的消息传递框架。使用 Intents 可以在系统范围内广播消息或者对一个目标 Activity 或 Service 发送消息，来表示要执行一个动作。系统将辨别出相应要执行活动的目标（target）。

5. Broadcast Receivers

Intent 广播的消费者。如果创建并注册了一个 Broadcase Receiver，应用程序就可以监听匹配了特定过滤标准的广播 Intent。Broadcase Receiver 会自动开启应用程序以响应一个收到的 Intent，使得可以用它们完美地创建事件驱动的应用程序。

6. Widgets

可以添加到主屏幕界面（home screen）的可视应用程序组件。作为 Broadcase Receiver 的特殊变种，Widgets 可以为用户创建可嵌入到主屏幕界面的动态的、交互的应用程序组件。

7. Notifications

一个用户通知框架。Notification 让不必窃取焦点或中断当前 Activities 就能通知用户。这是在 Service 和 Broadcast Receiver 中获取用户注意的推荐技术。例如，当设备收到一条短消息或一个电话，它会通过闪光灯、发出声音、显示图标或显示消息来提醒。可以在应用程序中使用 Notifications 触发相同的事件。

4.3 Activity

4.3.1 Activity 类

回顾第 2 章介绍的 Android 项目，那个项目中包含了一个 Java 文件，实现了一个 Activity 类的派生类。

1. Activity 类

HelloWorld 类继承自 Activity 类且重写了 onCreate 方法。Activity 是用户唯一可以看得到的东西。几乎所有的 Activity 都与用户进行交互，所以，Activity 主要负责的就是创建显示窗口，可以在这些窗口里使用 setContentView(View) 来显示自己的 UI。Activity 展现在用户面前的经常是全屏窗口，也可以将 Activity 作为浮动窗口来使用(使用设置了 windowIsFloating 的主题),或者嵌入到其他的 Activity(使用 ActivityGroup)中。onCreate 和 onPause 是几乎所有 Activity 子类都实现的两个方法。

onCreate(Bundle) 这个方法是初始化 Activity 的地方。最重要的是,经常需要在这里使用 setContentView(int) 来设置 UI 布局所使用的 layout 资源,当需要使用程序控制 UI 中的组件时,可以使用 findViewById(int) 来获得对应的视图。

当用户离开 Activity 时,可以在 onPause() 进行相应的操作。更重要的是,用户做的任何改变都应该在该点上提交(经常提交到 ContentProvider,这里保存数据)。

如果要使用 Context. startActivity() 来启动 Activity,Activity 必须在启动者应用包的 AndroidManifest. xml 文件中有对应的 <activity> 定义。

2. onCreate 函数

初始化活动(Activity),比如完成一些图形的绘制。最重要的是,在这个方法里通常将用布局资源(layout resource)调用 setContentView(int) 方法定义 UI,和用 findViewById(int) 在 UI 中检索需要编程交互的小部件(widgets)。setContentView 指定由哪个文件指定布局(main. xml),可以将这个界面显示出来,然后进行相关操作,操作会被包装成为一个意图,然后这个意图对应有相关的 Activity 进行处理。

(1) 重写方法

在重写父类的 onCreate 方法时,在方法前面加上 @Override 系可以检查方法的正确性。例如:

public void onCreate(Bundle savedInstanceState){…….}

这种写法是正确的,如果写成:

public void oncreate(Bundle savedInstanceState){…….}

这样编译器会报如下错误——The method oncreate(Bundle) of type HelloWorld must override or implement a supertype method,以确保正确重写 onCreate 方法。因为"oncreate"应该为"onCreate"。

而如果不加 @Override,则编译器将不会检测出错误,而是会认为新定义了一个方法 oncreate。

(2) android. os. Bundle 类

Bundle 类用作携带数据,它类似于 Map,用于存放 key - value 形式的值。相对于 Map,它提供了各种常用类型的 putXxx()/getXxx() 方法,如: putString()/getString() 和 putInt()/getInt(),putXxx() 用于往 Bundle 对象放入数据,getXxx() 方法用于从 Bundle 对象里获取数据。Bundle 的内部实际上是使用了 HashMap 类型的变量来存放 putXxx() 方法放入的值。

onCreate 方法的完整定义如下:

```
public void onCreate(Bundle saveInsanceState){
super.onCreate(saveInsanceState);
```

从上面的代码可以看出,onCreate 方法的参数是一个 Bundle 类型的参数。Bundle 类型的数据与 Map 类型的数据相似,都是以 key‐value 的形式存储数据的。

saveInsanceState 保存实例状态。实际上,saveInsanceState 也就是保存 Activity 的状态。

onsaveInsanceState 方法是用来保存 Activity 的状态。当一个 Activity 在生命周期结束前,会调用该方法保存状态。这个方法有一个参数名称与 onCreate 方法参数名称相同,如下所示:

```
public void onSaveInsanceState(Bundle saveInsanceState){
super.onSaveInsanceState(saveInsanceState);
}
```

在实际应用中,当一个 Activity 结束前,如果需要保存状态,就在 onsaveInsanceState 中将状态数据以 key‐value 的形式放入到 saveInsanceState 中。这样,当一个 Activity 被创建时,就能从 onCreate 的参数 saveInsanceState 中获得状态数据。

状态数据参数在实际应用中有很大作用,比如,一个游戏在退出前,保存一下当前游戏运行的状态,当下次开启时能接着上次的继续玩下去。

3. setOnClickListener 函数

View.OnClickListener()是 View 定义的点击事件的监听器接口,并在接口中仅定义了 onClick()函数。当 Button 从 Android 界面框架中接收到事件后,首先检查这个事件是否是点击事件,如果是点击事件,同时 Button 又注册了监听器,则会调用该监听器中的 onClick()函数。每个 View 仅可以注册一个点击事件的监听器,如果使用 setOnClickListener()函数注册第二个点击事件的监听器,之前注册的监听器将被自动注销。Android 事件侦听器更加详细的内容见本章"Android 事件侦听器"小节的介绍。

4. setContentView 函数

setContentView 方法实例化了一个 layout,当然也可以用 Layout Inflater.inflate()来实现相同的功能。可以使用 findViewById 方法来找到其中的某项 View。

R.layout.main 是个布局文件,描述控件都是如何摆放、如何显示,setContentView 就是设置一个 Activity 的布局文件,这样 Activity 就可以依照该布局文件绘制界面了。

使用 setContentView 可以在 Activity 中动态切换显示的 View。这样,不需要多个 Activity 就可以显示不同的界面,因此,不再需要在 Activity 间传送数据,变量可以直接引用。

但是,在 Android SDK 给我们建的默认 Hello World 程序中,调用的是 setContentView(int layoutResID)方法,如果使用该方法切换 View,在切换后再切换回,无法显示切换前修改后的样子,也就是说,相当于重新显示一个 View,并非是把原来的 View 隐藏后再显示。

其实 setContentView 是个多态方法,可以先用 LayoutInflater 把布局 XML 文件引入成 View 对象,再通过 setContentView(View view) 方法来切换视图。因为所有对 View 的修改都保存在 View 对象里,所以当切换回原来的 View 时,就可以直接显示原来修改后的样子。

5. findViewById 函数

findViewById 函数是用来查找当前 Activity 对应 layout 中的某项 View,使用了 R. Id. XXX 作为参数来索引。也就是说,要先 inflate 了一个 layout,再用 findViewById 来找 layout 实例中的某项 View。

4.3.2 Android 事件侦听器

1. Android 事件侦听器介绍

Android 事件侦听器是视图 View 类的接口,包含一个单独的回调方法。这些方法将在视图中注册的侦听器被用户界面操作触发时由 Android 框架调用。回调方法被包含在 Android 事件侦听器接口中。

例如,Android 的 View 对象都含有一个命名为 OnClickListener 接口成员变量,用户的点击操作都会交给 OnClickListener 的 OnClick() 方法进行处理。

开发者如果需要对点击事件作处理,可以定义一个 OnClickListener 接口对象,赋给需要被点击的 View 接口成员变量 OnClickListener,一般是用 View 的 setOnClickListener() 函数来完成这一操作。当有用户点击事件时,系统就会回调被点击 View 的 OnClickListener 接口成员的 OnClick() 方法。

2. Android 事件监听器接口中常用的回调方法

下面是一些常用的 Android 事件监听器接口中的回调方法。

(1) onClick()

包含于 View.OnClickListener。当用户触摸这个 item(在触摸模式下),或者通过浏览键或跟踪球聚焦在这个 item 上,然后按下"确认"键或者按下跟踪球时被调用。

(2) onLongClick()

包含于 View.OnLongClickListener。当用户触摸并控制住这个 item(在触摸模式下),或者通过浏览键或跟踪球聚焦在这个 item 上,然后保持按下"确认"键或者按下跟踪球 1 s 时被调用。

(3) onFocusChange()

包含于 View.OnFocusChangeListener。当用户使用浏览键或跟踪球浏览进入或离开这个 item 时被调用。

(4) onKey()

包含于 View.OnKeyListener。当用户聚焦在这个 item 上并按下或释放设备上的一个按键时被调用。

(5) onTouch( )

包含于 View.OnTouchListener。当用户执行的动作被当作一个触摸事件时被调用,包括按下、释放,或者屏幕上任何的移动手势(在这个 item 的边界内)。

(6) onCreateContextMenu( )

包含于 View.OnCreateContextMenuListener。当创建一个上下文菜单时被调用。

3. Listener 使用方法

Listener 在使用上有多种写法,例如:

方法 1:使用用户类对象,这样要求使用继承 OnClickListener 接口。调用方法如下:m_button1.setOnClickListener(this);

其中的 this 相当于 new OnClickListener()对象,即用户类中的一个对象。如果要用这种方式,public void onClick 方法必须写在该用户类中,并且在开头使用 implements OnClickListener。然后 this 对象可以直接调用该方法。

方法 2:使用对象 clickListener:

    m_button2.setOnClickListener(clickListener);

方法 3:使用匿名对象创建监听,同方法 2,可以看作另一种写法:

```
m_button3.setOnClickListener(new Button.OnClickListener() {
    @Override
    public void onClick(View v) {
        String strTmp = "点击 Button03";
        ET.setText(strTmp);
    }
});
```

方法 4:使用 XML 文件创建时绑定方法 Btn4OnClick。例如:

```
<Button
    android:id = "@ + id/button4"
    android:layout_width = "wrap_content"
    android:layout_height = "wrap_content"
    android:text = "Button4"
    android:onClick = "Btn4OnClick" />
```

方法 5:自己设计个监听类,监听的方法引用 OnClickListener 接口中的方法,创建的是匿名对象:

    m_button5.setOnClickListener(new clickListener2());

方法 6:外部类实现事件监听器接口,很少用,方法如下:

m_button6.setOnClickListener(new callOut(this));

4.4 Intent

4.4.1 Intent 简介

Android 中提供了 Intent 机制来协助应用间的交互与通信，Intent 负责对应用中一次操作的动作、动作涉及的数据、附加的数据进行描述，Android 则根据此 Intent 的描述，负责找到对应的组件，将 Intent 传递给调用的组件，并完成组件的调用。Intent 不仅可用于应用程序之间，也可用于应用程序内部的 Activity/Service 之间的交互。

可以在 Intent 中指定程序要执行的动作（比如：view，edit，dial），以及程序执行到该动作时所需要的资料。都指定好后，只要调用 startActivity()，Android 系统会自动寻找最符合指定要求的应用程序，并执行该程序。

1. 显示网页

打开浏览器，并且显示网页内容，代码如下：

```
Uri uri = Uri.parse("http://google.com");
Intent it = new Intent(Intent.ACTION_VIEW, uri);
startActivity(it);
```

2. 显示地图

通过 geo-uri 方式调用外部程序，可以启动 google 地图、百度地图等，代码如下：

```
Uri uri = Uri.parse("geo:34.899533,-77.036476");
Intent it = new Intent(Intent.ACTION_VIEW, uri);
startActivity(it);
//其他 geo URI 例子
//geo:latitude,longitude
//geo:latitude,longitude? z = zoom
//geo:0,0? q = my + street + address
//geo:0,0? q = business + near + city
//google.streetview:cbll = lat,lng&cbp = 1,yaw,,pitch,zoom&mz = mapZoom
```

3. 路径规划

在浏览器中使用 google 地图实现路径规划，代码如下：

```
Uri uri = Uri.parse("http://maps.google.com/maps?
        f = d&saddr = startLat % 20startLng&daddr = endLat % 20endLng&hl = en");
Intent it = new Intent(Intent.ACTION_VIEW, uri);
startActivity(it);
//where startLat, startLng, endLat, endLng are a long with 6 decimals like: 50.123456
```

4. 打电话

启动打电话功能,包括显示打电话窗口和直接打电话两种方法,代码如下:

```
//叫出拨号程序
Uri uri = Uri.parse("tel:0800000123");
Intent it = new Intent(Intent.ACTION_DIAL, uri);
startActivity(it);

//直接打电话出去
Uri uri = Uri.parse("tel:0800000123");
Intent it = new Intent(Intent.ACTION_CALL, uri);
startActivity(it);
//用这个,要在 AndroidManifest.xml 中,加上
// <uses-permission id = "android.permission.CALL_PHONE" />
```

5. 传送 SMS/MMS

实现传送短信、传送 MMS 的功能,代码如下:

```
//调用短信程序
Intent it = new Intent(Intent.ACTION_VIEW, uri);
it.putExtra("sms_body", "The SMS text");
it.setType("vnd.android-dir/mms-sms");
startActivity(it);

//传送消息
Uri uri = Uri.parse("smsto://0800000123");
Intent it = new Intent(Intent.ACTION_SENDTO, uri);
it.putExtra("sms_body", "The SMS text");
startActivity(it);

//传送 MMS
Uri uri = Uri.parse("content://media/external/images/media/23");
Intent it = new Intent(Intent.ACTION_SEND);
it.putExtra("sms_body", "some text");
it.putExtra(Intent.EXTRA_STREAM, uri);
it.setType("image/png");
startActivity(it);
```

6. 传送 Email

发电子邮件功能,包括传送电子邮件附件功能,代码如下:

```
Uri uri = Uri.parse("mailto:xxx@abc.com");
Intent it = new Intent(Intent.ACTION_SENDTO, uri);
startActivity(it);

Intent it = new Intent(Intent.ACTION_SEND);
it.putExtra(Intent.EXTRA_EMAIL, "me@abc.com");
it.putExtra(Intent.EXTRA_TEXT, "The email body text");
it.setType("text/plain");
startActivity(Intent.createChooser(it, "Choose Email Client"));

Intent it = new Intent(Intent.ACTION_SEND);
String[] tos = {"me@abc.com"};
String[] ccs = {"you@abc.com"};
it.putExtra(Intent.EXTRA_EMAIL, tos);
it.putExtra(Intent.EXTRA_CC, ccs);
it.putExtra(Intent.EXTRA_TEXT, "The email body text");
it.putExtra(Intent.EXTRA_SUBJECT, "The email subject text");
it.setType("message/rfc822");
startActivity(Intent.createChooser(it, "Choose Email Client"));

//传送附件
Intent it = new Intent(Intent.ACTION_SEND);
it.putExtra(Intent.EXTRA_SUBJECT, "The email subject text");
it.putExtra(Intent.EXTRA_STREAM, "file:///sdcard/mysong.mp3");
sendIntent.setType("audio/mp3");
startActivity(Intent.createChooser(it, "Choose Email Client"));
```

7. 播放多媒体

打开媒体播放器播放多媒体文件的功能,代码如下:

```
Uri uri = Uri.parse("file:///sdcard/song.mp3");
Intent it = new Intent(Intent.ACTION_VIEW, uri);
it.setType("audio/mp3");
startActivity(it);
Uri uri = Uri.withAppendedPath(MediaStore.Audio.Media.INTERNAL_CONTENT_URI, "1");
Intent it = new Intent(Intent.ACTION_VIEW, uri);
startActivity(it);
```

4.4.2 Intent 实现两个 Activity 之间切换

Android 中每个 Activity 通常描述了一个屏幕上的所有画面(窗口级别的 Activity

第4章 界面切换设计

除外),因此,通常手机屏幕两个界面(准确地说是整个屏幕)之间的切换就涉及到了Activity的切换。

假定有两个Activity,分别是Activity01和Activity02,现在Activity01页面中有一个按钮,点下之后会切换到Activity02。并且在Activity切换时,Activity01给Activity02传递了一个参数。intent对象可以在切换Activity时使用,且能传递数据。方法如下:

(1) 在Activity01中设置一个可触发的空间,并添加一个触发器;

(2) 在Activity01的触发器添加listener;

(3) 在listener的接口实现中,设置一个Intent,让这个Intent能够将Activity01和Activity02绑定起来,并且通过putExtra将要传输的值放到Intent对象中存储;

(4) listener接口实现结尾,通过Activity01启动调用这个Intent对象,通过调用来切换到Activity02;

(5) 在Activity02中,使用getIntent来获取上下文切换中自己启动了的那个Intent对象实例;

(6) 通过获取到Intent对象实例,使用getStringExtra来获取先前putExtra的值。

下面是一个Android的窗口切换程序设计的小实验,实现两个窗口,单击第一个窗口的其中一个按钮,然后使用Intent,切换到第二个窗口。要求两个窗口的界面不同。实验步骤如下:

(1) 创建项目,项目名称为姓名拼音开头字母__学号最后三位__03。例如"lyk_110_03"。

(2) 在视图最上方面显示"实验三",按钮上显示姓名。

(3) 重新建立一个Java文件,扩展名是".java"。可以从hello.java拷贝。

(4) 重新建立一个界面布局文件,扩展名是".xml"。可以从main.xml拷贝。窗口中显示班级、姓名和学号。

(5) 在hello.java中的按钮单击事件监听函数中,创建Intent对象,启动新的界面。

1) import 数据包

import android.content.Intent;

2) Intent 实例及启动新的界面

Intent intent=newIntent(hello.this, kui.class);

startActivity(intent);

(6) 在AndroidManifest.xml文件中添加新的Active。

<activity android:name="kui" android:label="@string/app_name">

</activity>

(7) 判断密码方法,代码如下:

```
EditText txtName = (EditText)findViewById(R.id.用户名文本框ID);
String name = txtName.getText().toString();
if( "110".equals(name)){
    //空
```

```
}else{
    //非空
}
```

4.4.3 Intent 实现两个 Activity 之间传递数据

1. 在 Intent 中夹带数据传给新 Activity

可以使用 Intent 的 putExtra(数据名称,数据)方法将数据附加到 Intent 中,其中的参数说明如下:

- 第一个参数为要传入字符串类型的数据名称(可以称之为键值),以便后面以此名称来读出数据。
- 第二个参数为要实际附加的数据,其类型可以是任何常用类型,或者是常用类型的数组。

实例如下:

```
String[] favor = {"骨头","肉","狗粮"};
Intent it = new Intent(this,second.class);
it.putExtra("编号",1);
it.putExtra("说明",狗狗);
it.putExtra("爱吃",favor);
startActivity(it);
```

2. Intent( )与 getXxxExtra( ):从 Intent 中读取数据

在新的 Activity 中,则可以使用 getIntent()来获取传入的 Intent 对象,然后再用 Intent 的 getXxxExtra(数据名称,默认值)方法来读取数据,其中 Xxx 为数据的类型名称。

第二个参数则是默认值,当 Intent 中找不到指定名称的数据时就会将这个默认值返回。如果存入数组则要用 getXxxArrayExtra(数据名称)来读取。实例如下:

```
Intent it = getIntent();
int num = it.getIntExtra("编号",0);           //当获取数据失败时返回 0
String da = it.getStringExtra("说明");         //当获取数据失败时返回 null
String[] a = it.getStringArrayExtra("爱吃");   //当获取数据失败时返回 null
```

3. 要求新的 Activity 返回数据

如果想让新启动的 Activity 返回数据,步骤如下:

(1) 在主 Activity 中改用 startActivityForResult()来启动新 Activity:
startActivityForResult(Intent it,int 标识符);

标识符作为一个自定义的值,当新 Activity 返回数据时,也会一并返回此标识符以辨别。

(2) 新 Activity 在结束前使用 setResult() 返回执行的结果与数据：
　　setResult(int 结果码, Intent, it);

结果码可以设置为 Activity 类中定义的 RESULT_OK 或是 RESULT_CANCELED 常数, it 为 Intent 对象用来夹带数据可以为 null。例如：

```
@Override
public void onBackPressed() {
    Intent intent = new Intent();
    intent.putExtra("img_file","update");
    intent.putExtra("id", m_hotel_id.getText().toString());
    setResult(2, intent);
    super.onBackPressed();
}
```

在往 Activity 中加入 onActivityResult() 方法接收返回的数据：
　　protect void onActivityRestlt(int 标识符, int 结果码, Intent it)

在这个方法中应检查标识符是否与步骤 1 的相符, 然后按结果码而进行不同的处理, 并由 it 中读取返回的数据。例如：

```
@Override
protected void onActivityResult(int requestCode, int resultCode,
    Intent data) {
        Bundle bundle = data.getExtras();
        String str = (String) bundle.get("img_file");
        String id = (String) bundle.get("id");
}
```

4.5　Service

4.5.1　Service 介绍

　　Android 支持服务的概念。服务是在后台运行组件, 没有用户界面。可以将这些组件想象为 Windows 服务或 UNIX 服务。与这些服务类型类似, Android 服务始终可用, 但无须主动执行某些操作。

　　Android 支持两种服务类型的服务：本地服务和远程服务。本地服务无法供在设备上运行其他应用程序访问。一般而言, 这些服务类型仅支持承载该服务的应用程序。而对于远程服务, 除了可以承载服务的应用程序访问, 还可以从其他应用程序访问。远程服务使用 AIDL(Android Interface Definition Language, Android 接口定义语言)向客户端定义其自身。

Service 是 Android 系统中的四大组件(Activity、Service、Broadcast Receiver、ContentProvider)之一,它跟 Activity 的级别差不多,但不能自己运行,只能后台运行,可以和其他组件进行交互。

Service 可以在很多应用场合中使用,比如播放多媒体的时候用户启动了其他 Activity,这时程序要在后台继续播放;例如检测 SD 卡上文件的变化;例如在后台记录你地理信息位置的改变,等等。

Service 的启动有两种方式:
- context.startService();
- context.bindService()。

4.5.2 Service 启动流程

context.startService()启动流程如下:
  context.startService()—>onCreate()—>onStart()—>Servicerunning
    —>context.stopService()—>onDestroy()—>Servicestop

如果 Service 还没有运行,则 Android 先调用 onCreate(),然后调用 onStart()。

如果 Service 已经运行,则只调用 onStart(),所以一个 Service 的 onStart 方法可能会重复多次调用。

如果是调用者直接退出而没有调用 stopService 的话,Service 会一直在后台运行,该 Service 的调用者再启动起来后可以通过 stopService 关闭 Service。

所以,调用 startService 的生命周期为:
  onCreate——onStart(可多次调用)——onDestroy

context.bindService()启动流程如下:
  context.bindService()—>onCreate()—>onBind()—>Servicerunning
    —>onUnbind()—>onDestroy()—>Servicestop

4.6 消息提示框和对话框

4.6.1 Toast 消息提示框

Android 中的 Toast 是一种简易的消息提示框。和 Dialog 不一样的是,Toast 是没有焦点的,Toast 提示框不能被用户点击,而且 Toast 显示的时间有限,Toast 会在用户设置的显示时间后自动消失。

1. 默认的显示方式

默认的显示方式 Java 代码如下:

第4章 界面切换设计

```
//第一个参数:当前的上下文环境。可用 getApplicationContext()或 this
//第二个参数:要显示的字符串。也可是 R.string 中字符串 ID
//第三个参数:显示的时间长短。Toast 默认的有两个 LENGTH_LONG(长)和 LENGTH_SHORT(短),
也可以使用毫秒,如 2 000 ms
Toasttoast = Toast.makeText(getApplicationContext(),
                "默认的 Toast",Toast.LENGTH_SHORT);
//显示 Toast 信息
toast.show();
```

2. 自定义显示位置

自定义显示位置的 Java 代码如下:

```
Toast toast = Toast.makeText(getApplicationContext(), "自定义显示位置的 Toast", Toast.
LENGTH_SHORT);
//第一个参数:设置 Toast 在屏幕中显示的位置。现在的设置是居中靠顶
//第二个参数:相对于第一个参数设置 Toast 位置的横向 X 轴的偏移量,正数向右偏移,负数向
左偏移
//第三个参数:同第二个参数道理一样
//如果设置的偏移量超过了屏幕的范围,Toast 将在屏幕内靠近超出的那个边界显示
toast.setGravity(Gravity.TOP|Gravity.CENTER, -50, 100);
//屏幕居中显示,X 轴和 Y 轴偏移量都是 0
//toast.setGravity(Gravity.CENTER, 0, 0);
toast.show();
```

3. 带图片的 Toast

带图片 Toast 的 Java 代码如下:

```
Toast toast = Toast.makeText(getApplicationContext(),
"显示带图片的 toast", 3000);
toast.setGravity(Gravity.CENTER, 0, 0);
//创建图片视图对象
ImageView imageView = new ImageView(getApplicationContext());
//设置图片
imageView.setImageResource(R.drawable.ic_launcher);
//获得 Toast 的布局
LinearLayout toastView = (LinearLayout) toast.getView();
//设置此布局为横向的
toastView.setOrientation(LinearLayout.HORIZONTAL);
//将 ImageView 在加入到此布局中的第一个位置
toastView.addView(imageView, 0);
toast.show();
```

4. 完全自定义显示方式

完全自定义显示方式的 Java 代码如下：

```
//Inflater 意思是填充
//LayoutInflater 这个类用来实例化 XML 文件到其相应的视图对象的布局
LayoutInflater inflater = getLayoutInflater();
//通过制定 XML 文件及布局 ID 来填充一个视图对象
View layout = inflater.inflate(R.layout.custom2,
(ViewGroup) findViewById(R.id.llToast));
ImageView image = (ImageView) layout.findViewById(R.id.tvImageToast);
//设置布局中图片视图中图片
image.setImageResource(R.drawable.ic_launcher);
TextView title = (TextView) layout.findViewById(R.id.tvTitleToast);
//设置标题
title.setText("标题栏");
TextView text = (TextView) layout.findViewById(R.id.tvTextToast);
//设置内容
text.setText("完全自定义 Toast");
Toast toast = new Toast(getApplicationContext());
toast.setGravity(Gravity.CENTER, 0, 0);
toast.setDuration(Toast.LENGTH_LONG);
toast.setView(layout);
toast.show();
```

5. 其他线程通过 Handler 的调用

其他线程通过 Handler 的调用的 Java 代码如下：

```
//调用方法 1
//Thread th = new Thread(this);
//th.start();
//调用方法 2
handler.post(new Runnable() {
    @Override
    public void run() {
        showToast();
    }
});
```

showToast 函数代码如下：

```
public void showToast(){
    Toast toast = Toast.makeText(getApplicationContext(),
"Toast 在其他线程中调用显示", Toast.LENGTH_SHORT);
    toast.show();
}
```

Handler 线程代码如下:

```
Handler handler = new Handler(){
    @Override
    public void handleMessage(Message msg) {
        int what = msg.what;
        switch (what) {
        case 1:
            showToast();
            break;
        default:
            break;
        }
        super.handleMessage(msg);
    }
};
```

run 函数代码如下:

```
@Override
public void run() {
    handler.sendEmptyMessage(1);
}
```

4.6.2 对话框

1. 分 类

Android 对话框有如下几种类型:

(1) AlertDialog:它能够管理 0~3 个按钮和一个包含 Radio 或者 Checkbox 的可选项列表。

(2) ProgressDialog:一个用于显示进度圈或者进度条的 Dialog,继承自 AlertDialog,所以它也支持按钮。

(3) DatePickerDialog:用于让用户选择日期的 Dialog。

(4) TimePickerDialog:用于让用户选择时间的 Dialog。

如果需要定义自己的 Dialog,只需要继承 Dialog 或者上面提到四个组件之一,并且为新的 Dialog 定义自己的布局就可以了。

2. AlertDialog. Builder

Android 中 AlertDialog 的创建一般是通过其内嵌类 AlertDialog.Builder 来实现。Builder 所提供的方法如下:

- setTitle():给对话框设置 Title;

- setIcon()：给对话框设置图标；
- setMessage()：设置对话框的提示信息；
- setItems()：设置对话框要显示的一个List，一般用于要显示几个命令时；
- setSingleChoiceItems()：设置对话框显示一个单选的List；
- setMultiChoiceItems()：用来设置对话框显示一系列的复选框；
- setPositiveButton()：给对话框添加"Yes"按钮；
- setNegativeButton()：给对话框添加"No"按钮。

3. 创建 Dialog 的方法

Dialog 的创建方式有两种：

（1）直接新建一个 Dialog 对象，然后调用 Dialog 对象的 Show 和 Dismiss 方法来控制对话框的显示和隐藏。

（2）在 Activity 的 onCreateDialog(int id)方法中创建 Dialog 对象并返回，然后调用 Activty 的 showDialog(int id)和 dismissDialog(int id)来显示和隐藏对话框。

区别在于通过第二种方式创建的对话框会继承 Activity 的属性，比如获得 Activity 的 menu 事件等。

使用 AlertDialog 可以创建普通对话框、带列表的对话框，以及带单选按钮和多选按钮的对话框。

4. 最简单的 AlertDialog 对话框

（1）导入的包：

```
import android.app.AlertDialog;
```

（2）在需要显示对话框的地方加入以下代码：

```
new AlertDialog.Builder(this).setTitle("标题")
    .setMessage("消息内容")
    .setPositiveButton("确定", null)
    .show();
```

最简单对话框运行效果如图 4-3 所示。

5. AlertDialog 对话框实例

AlertDialog 中的对话框形式有四种，分别是一般对话框形式、列表对话框形式、单选按钮对话框、多选按钮对话框。一般对话框形式之中，可以定义多个按钮，用于实现不同的功能选择，例如，下面的例子，实现一个确认是否退出的对话框功能。代码如下：

图 4-3 最简单对话框运行效果图

```
protected void dialog() {
    AlertDialog.Builder builder = new Builder(Main.this);
    builder.setMessage("确认退出吗?");
    builder.setTitle("提示");
    builder.setPositiveButton("确认", new OnClickListener() {
      @Override
      public void onClick(DialogInterface dialog, int which) {
        dialog.dismiss();
        Main.this.finish();
      }
    });
    builder.setNegativeButton("取消", new OnClickListener() {
      @Override
      public void onClick(DialogInterface dialog, int which) {
        dialog.dismiss();
      }
    });
    builder.create().show();
}
```

上述确认对话框代码实现的效果如图 4-4 所示。

图 4-4　确认对话框效果图

4.7　Android 程序生命周期

1. Android 程序生命周期

与大多数传统的操作系统环境不同,Android 应用程序并不能控制自己的生命周期。所以,应用的各个组件(Activity,Service……)就得时刻小心地监听应用的状态变化对它们的影响,防止在不适当的时机被终止掉。

在 Android 中,每个应用都具有独立的进程运行在独立的 Dalvik(Android 特有的虚拟机)。各个应用在运行时的进程管理和内存管理都是相对独立的。使用一种"侵占性"的方式管理系统资源,这意味着为了释放资源给高优先级的程序(通常情况下是正在与用户进行直接交互的程序),某些进程及其宿主程序将会在没有任何提示警告的情况下被结束。

2. 理解应用的优先级和进程状态

结束一些进程是为了释放回收资源,那么哪些进程会被结束?这决定于宿主应用程序的优先级了。一个应用的优先级等同于具有最高优先级的组件的优先级。如果此刻两个程序具有相同的优先级,那么曾经哪个进程处于低优先级的时间较长,就会被结束掉。

进程的优先级受到进程间附属关系的影响,比如 A 应用依赖的 Service 或者 Content Provider 是由 B 应用提供的,那么 B 将会具有更高的优先级。所有的 Android 应用都会遗留在内存中运行,直到系统需要释放回收资源才会被结束掉。

3. Activity 状态

Activity 有三种基本状态:

(1) Active:处于屏幕前景(当前 Task 的栈顶 Activity 处于活动 Active 状态),同一时刻只能有一个 Activity 处于活动 Active 状态。

(2) Paused 状态:处于背景画面状态,失去了焦点,但依然是活动状态。

(3) stopped:不可见,但依然保持所有的状态和内存信息。

(4) 可以调用 finish()结束处理 Paused 或者 Stopped 状态的 Activity。

各种状态之间通过下列函数调用转换:

- void onCreate(Bundle savedInstanceState);
- void onStart();
- void onRestart();
- void onResume();
- void onPause();
- void onStop();
- void onDestroy()。

4. Android 程序命周期的过程

Android 程序命周期的过程如图 4-5 所示。图中各函数的说明如下:

(1) onCreate():在窗体创建的时候执行,只执行一次。

(2) onStart():在窗体启动的时候执行。

(3) onResume():在窗体的所有内容都被加载完成之后执行。例如,加载一个网页,文字的加载优先级要高于图片,当图片加载完成之后就是执行 onResume()方法的时机。

(4) onPause():当用户对窗体有数据交互时,并且窗体程序会对该数据有处理方式的时候。就是说,比方说有个按钮,程序为这个按钮设置了功能。用户点击了它,就会执行 onPause()方法。若是按钮没有设置功能的时候,是不会执行 onPause()方法的。

(5) onStop()：当整个页面窗体被其他的窗体完全覆盖的时候，就会执行 onStop() 的方法。来停止当前程序的后台数据交互，整个页面会被保存到系统的堆栈里。当用户在该页面的下一个页面中点击回退键的时候，就会在堆栈里提取出该页面（堆栈空间没有被清空的时候）。

(6) onRestart()：回退键会调用堆栈中即将被显示的页面的 onRestart() 方法。或者当你执行程序时，点击了 HOME 键，再次运行该程序（回到该程序的运行页面）时，也会执行 onRestart() 方法。

(7) onDestory()：这个方法和 onCreate() 一样，只执行一次。在退出程序的时候执行。会清空未以文件形式保存的所有缓存记录。

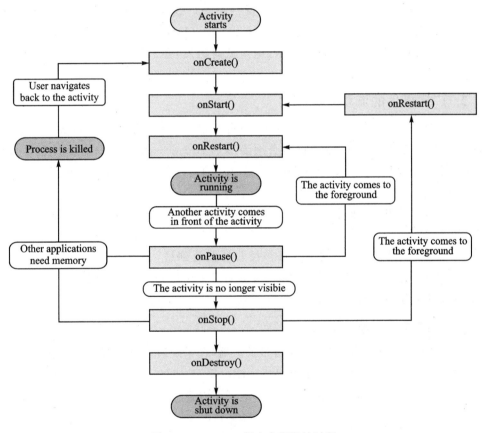

图 4-5 Android 程序命周期的过程

4.8 广播接收器

1. 广播接收器介绍

广播接收器 BroadcastReceiver 是 Android 中的四大组件之一。广播接收器是一个专注于接收广播通知信息,并做出对应处理的组件。很多广播是源自于系统代码,比如通知时区改变、电池电量低、拍摄了一张照片或者用户改变了语言选项。应用程序也可以进行广播,比如通知其他应用程序一些数据下载完成并处于可用状态。

应用程序可以拥有任意数量的广播接收器以对所有它感兴趣的通知信息予以响应。所有的接收器均继承自 BroadcastReceiver 基类。

广播接收器没有用户界面。然而,它们可以启动一个 Activity 来响应它们收到的信息,或者用 NotificationManager 来通知用户。通知可以用很多种方式来吸引用户的注意力,比如闪动背灯、震动、播放声音等。一般来说是在状态栏上放一个持久的图标,用户可以打开它并获取消息。

Android 中的广播事件有两种,一种就是系统广播事件,比如 ACTION_BOOT_COMPLETED(系统启动完成后触发)、ACTION_TIME_CHANGED(系统时间改变时触发)、ACTION_BATTERY_LOW(电量低时触发)等。另一种是我们自定义的广播事件。

2. 广播事件的流程

① 注册广播事件:注册方式有两种,一种是静态注册,就是在 AndroidManifest.xml 文件中定义,注册的广播接收器必须要继承 BroadcastReceiver;另一种是动态注册,是在程序中使用 Context. registerReceiver 注册,注册的广播接收器相当于一个匿名类。两种方式都需要 IntentFIlter。

② 发送广播事件:通过 Context. sendBroadcast 来发送,由 Intent 来传递注册时用到的 Action。

③ 接收广播事件:当发送的广播被接收器监听到后,会调用它的 onReceive()方法,并将包含消息的 Intent 对象传给它。onReceive 中代码的执行时间不要超过 5 s,否则,Android 会弹出超时 Dialog。

3. 广播接收器应用实例

广播接收器应用实例:接收系统自带的广播,在系统启动时播放一首音乐。

(1) 建立一个项目 Lesson21_BroadcastReceiver,拷贝一首音乐进 res/raw 目录;

(2) 建立 HelloBroadcastReceiver.java,程序内容如下:

```
Codepackage android.basic.lesson21;
import android.content.BroadcastReceiver;
import android.content.Context;
import android.content.Intent;
import android.media.MediaPlayer;
import android.util.Log;
public class HelloBroadReciever extends BroadcastReceiver {
    //如果接收的事件发生
    @Override
```

第4章 界面切换设计

```
        public void onReceive(Context context, Intent intent) {
            //则输出日志
            Log.e("HelloBroadReciever", "BOOT_COMPLETED!!!!! ");
            Log.e("HelloBroadReciever", "" + intent.getAction());
            //则播放一首音乐
            MediaPlayer.create(context, R.raw.babayetu).start();
        }
    }
```

（3）在 AndroidManifest.xml 中注册此 Receiver，程序内容如下：

```
Code <? xml version = "1.0" encoding = "utf - 8"? >
<manifest
xmlns:android = "http://schemas.android.com/apk/res/android"
android:versionname = "1.0" android:versioncode = "1"
 package = "android.basic.lesson21">
 <application android:icon = "@drawable/icon" android:label = "@string/app_name">
    <activity android:label = "@string/app_name"
android:name = ".MainBroadcastReceiver">
    < intent - filter = "">
        <action android:name = "android.intent.action.MAIN">
        <category android:name = "android.intent.category.LAUNCHER">
    </category> </action> </intent>
   </activity>
<!-- 定义 Broadcast Receiver 指定监听的 Action -->
<receiver android:name = "HelloBroadReciever">
   < intent - filter = "">
      <action android:name = "android.intent.action.BOOT_COMPLETED">
   </action> </intent>
</receiver>
</application> </manifest>
```

（4）运行应用程序，在系统启动时能听到音乐播放的声音，说明程序接收到了系统启动的广播事件，并做出了响应。

4.9 界面切换的实现

1. 添加主界面

本章的界面切换的实现，在第3章登录界面布局项目"IoT_IS_App_001"的基础上完成。在 Obtain_Studio 中打开 IoT_IS_App_001 项目。在与 welcome.xml 文件相同的目录\Obtain_Studio\WorkDir\IoT_IS_App_001\app\src\main\res\layout 中新建

一个 main.xml 文件。做一个简单的物联网地点列表布局,程序内容如下:

```xml
<?xml version="1.0" encoding="utf-8"?>
<LinearLayout xmlns:android="http://schemas.android.com/apk/res/android"
    android:orientation="vertical"
    android:layout_width="fill_parent"
    android:layout_height="fill_parent"
    android:background="@drawable/main" >
  <LinearLayout  android:layout_width="fill_parent"
      android:layout_height="70px"
      android:background="#eeeeee"
      android:layout_marginTop="2px" >
    <ImageView android:paddingTop="10px"
        android:src="@drawable/alivingroom"
        android:layout_width="70px"
        android:layout_height="60px"/>
  <LinearLayout  android:orientation="vertical"
      android:layout_width="wrap_content"
      android:layout_height="wrap_content"
      android:layout_marginLeft="30dip" >
  <TextView android:layout_width="wrap_content"
    android:layout_height="wrap_content"
    android:textSize="20sp"
    android:textColor="#F83333"
    android:text="客厅" />
    <TextView android:layout_width="wrap_content"
    android:layout_height="wrap_content"
    android:textSize="20sp"
    android:textColor="#F83333"
    android:text="192.168.1.100" />
  </LinearLayout>
    <LinearLayout
        android:layout_width="fill_parent"
        android:layout_height="fill_parent"
        android:gravity="right" >
    <CheckBox android:layout_width="wrap_content"
    android:layout_height="wrap_content"
    android:text=""
    android:id="@+id/checkbox1"/>
      <Button android:text="查看"
        android:id="@+id/button1"
        android:layout_width="wrap_content"
        android:layout_height="wrap_content"/>
    </LinearLayout>
  </LinearLayout>
  (下面还有三个跟上面结构完全相同的布局,复制上面代码三份,然后修改一下图片资料和地点名称即可)
  </LinearLayout>
```

第 4 章 界面切换设计

在与 WelcomeActivity.java 文件相同的目录 \Obtain_Studio\WorkDir\IoT_IS_App_001\app\src\main\java\com\example\administrator\myapplication 中创建新文件 main.java,程序内容如下:

```java
package com.example.administrator.myapplication;
import android.app.Activity;
import android.os.Bundle;
import android.view.View;
import android.widget.Button;
import android.widget.EditText;
import android.widget.Toast;
import android.view.Gravity;
import android.content.Intent;
public class main extends Activity
{
    @Override
    public void onCreate(Bundle savedInstanceState)
    {
        super.onCreate(savedInstanceState);
        setContentView(R.layout.main);
    }
}
```

2. 从欢迎界面自动切换到登录界面

从欢迎界面自动切换到登录界面的实现方案:

(1) 设置隐藏状态栏,设置隐藏标题栏,采用 Window、WindowManager 类实现,需要导入这两个包:

import android.view.Window;

import android.view.WindowManager;

(2) 设置定时器,到时间自动切换到登录界面,需要使用 Handler、Message 类创建线程,需要导入这两个包:

import android.os.Handler;

import android.os.Message;

(3) 两个 Activity 之间的切换,采用 Intent 类,需要导入包:

import android.content.Intent;

完整的程序内容如下:

```java
import android.os.Bundle;
import android.os.Handler;
import android.os.Message;
```

```java
import android.app.Activity;
import android.content.Intent;
import android.view.Window;
import android.view.WindowManager;
public class WelcomeActivity extends Activity {
    @Override
    protected void onCreate(Bundle savedInstanceState)
    {
        super.onCreate(savedInstanceState);
        final Window window = getWindow();// 获取当前的窗体对象
        window.setFlags(WindowManager.LayoutParams.FLAG_FULLSCREEN, WindowManager.LayoutParams.FLAG_FULLSCREEN);// 隐藏了状态栏
        requestWindowFeature(Window.FEATURE_NO_TITLE);// 隐藏了标题栏
        setContentView(R.layout.welcome);
        welcomeUI();
    }
    private void welcomeUI()
    {
        new Thread(new Runnable()
        {
            @Override
            public void run()
            {
                try
                {
                    Thread.sleep(2000);
                    Message message = new Message();
                    welHandler.sendMessage(message);//接收的函数不需要改参数
                } catch (InterruptedException e)
                {
                    e.printStackTrace();
                }
            }
        }).start();
    }
    Handler welHandler = new Handler()
    {
        @Override
        public void handleMessage(Message msg)
        {
            welcomeFunction();
        }
```

第4章 界面切换设计

```
    };
    public void welcomeFunction()
    {
        Intent intent = new Intent();
        intent.setClass(WelcomeActivity.this,login.class);
        startActivity(intent);
        WelcomeActivity.this.finish();
    }
}
```

3. 从登录界面通过输入用户名和密码再切换到主界面

从登录界面通过输入用户名和密码再切换到主界面完整的程序内容如下：

```
import android.app.Activity;
import android.os.Bundle;
import android.view.View;
import android.widget.Button;
import android.widget.EditText;
import android.widget.Toast;
import android.view.Gravity;
import android.content.Intent;
import android.app.AlertDialog;
import android.content.DialogInterface;
public class login extends Activity
{
    @Override
    public void onCreate(Bundle savedInstanceState)
    {
        super.onCreate(savedInstanceState);
        setContentView(R.layout.login);
        Button button = (Button)findViewById(R.id.button1);
        button.setOnClickListener(new View.OnClickListener(){
          // @Override
            public void onClick(View v)
            {
                EditText ipadders = (EditText) findViewById(R.id.ip);
                EditText username = (EditText) findViewById(R.id.edit1);
                EditText password = (EditText) findViewById(R.id.edit2);
                String ipaddersStr = ipadders.getText().toString();
                String usernameStr = username.getText().toString();
                String passwordStr = password.getText().toString();
                if("123".equals(usernameStr) && "123".equals(passwordStr))
```

```
                    {
                        Intent intent = new Intent();
                        intent.putExtra("IP",ipaddersStr);
                        intent.setClass(login.this, MainFragmentPagerActivity.class);
                        startActivity(intent);
                    }
                    else
                    {
                        new AlertDialog.Builder(login.this).setTitle("标题")
                            .setMessage("用户名或者密码不对!")
                            .setPositiveButton("确定", null)
                            .show();
                    }
                }
            });
        }
    }
```

4. 配置与运行

在 Drawable 目录中保存上述布局文件之中使用到的图片文件 main.jpg。在 AndroidManifest.xml 文件中,找到 <activity android:name=".WelcomeActivity">…. </activity> 一项的后面添加 main 一项的 Activity 注册,程序内容如下:

```
<activity android:name=".WelcomeActivity">
    <intent-filter>
        <action android:name="android.intent.action.MAIN" />
        <category android:name="android.intent.category.LAUNCHER" />
    </intent-filter>
</activity>
<activity android:name=".login"></activity>
<activity android:name=".main"></activity>
```

修改 IoT_IS_App_001.prj 文件中的运行 Activity 类为 WelcomeActivity。程序内容如下:

```
<run>
cd app\build\outputs\apk\debug
adb -s emulator-5554 uninstall com.example.administrator.myapplication
adb -s emulator-5554 install app-debug.apk
adb -s emulator-5554 shell am start -n com.example.administrator.myapplication/.WelcomeActivity
</run>
```

启动模拟器,最后编译和运行该项目,在模拟器上可以演示上述从欢迎界面自动切换到登录界面以及从登录界面通过输入用户名和密码再切换到主界面的切换过程。

第 5 章 列表视图界面设计

5.1 列表视图界面设计目标

在移动软件 App 主界面里,一般都会以列表的形式显示各种信息,例如电子商务 App 显示产品、社交软件显示联系人或者消息等。对于物联网智能系统,则需要在主界面上以列表的形式显示地点(位置)、设备等。

对于一个物联网智能系统 App,除了登录界面、主界面之外,其核心主要包括设备界面、控制界面、设置界面等界面与功能,如图 5-1 所示。

图 5-1 物联网智能系统 App 核心界面与功能

本章的设计目标是实现一个类似于图 5-1 左边所示的地点列表视图界面的简化版本,不带导航栏,只在主界面里显示一个物联网智能系统的地点列表,如图 5-2 所示。

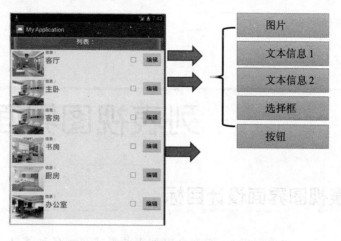

图 5-2 物联网智能系统地点列表界面

主要包括：
(1) 在列表上面包括一个标题和背景图片；
(2) 每一项列表包括一个图片、两个文本、一个选择框和一个按钮。

这些列表信息，在安卓里以列表视图 ListView 的形式实现，因此，列表视图是物联网智能系统主界面的核心内容之一。

5.2 ListView 应用

5.2.1 ListView 列表视图的工作原理

ListView 列表视图的工作原理如图 5-3 所示，主要包括 Adapter 适配器和 ListView 视图对象两个主要部分。Adapter 适配器是沟通数据与视图的桥梁，用于对要进行显示的数据进行处理，并通过与 ListView 视图对象的绑定将数据显示到视图对象中。

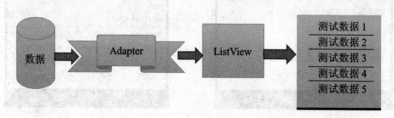

图 5-3 ListView 列表视图的工作原理

下面是最简单的 ListView 程序，采用最简单的 ArrayAdapter 适配器，代码如下：

```
public class hello extends Activity
{
    private ListView listView;
```

```
    @Override
    public void onCreate(Bundle savedInstanceState)
    {
        super.onCreate(savedInstanceState);
        listView = new ListView(this);
        listView.setAdapter(new ArrayAdapter<String>(this,
          android.R.layout.simple_expandable_list_item_1,getData()));
        setContentView(listView);
    }
    private List<String>getData(){
            List<String>data = new ArrayList<String>();
            data.add("测试数据 1");
            data.add("测试数据 2");
            data.add("测试数据 3");
            data.add("测试数据 4");
            return data;
        }
}
```

在上述程序中,用到了安卓系统自带的 List View 布局 simple_list_item_1。自带布局还包括 simple_list_item_1、simple_list_item_2、two_line_list_item 等。

- simple_list_item_1,此布局显示最为简单,其中只有一个 TextView,id 为 android.R.id.text1。
- simple_list_item_1 和 two_line_list_item 都有两个 TextView:android.R.id.text1 和 android.R.id.text2,不同之处在于,前者两行字是不一样大小的,而后者中两行字体一样大小。
- simple_list_item_single_choice、simple_list_item_multiple_choice、simple_list_item_checked(不同的呈现方式),这三种布局增加了选项,有单选和多选模式。常用方法为 setChoiceMode( ),getCheckedItemPositions( ),getCheckedItemIds( )。

上述 ArrayAdapter 适配器程序运行效果如图 5-4 所示。

图 5-4　ArrayAdapter 适配器程序运行效果图

5.2.2　SimpleCursorAdapter

列表的显示需要三个元素:
(1) ListVeiw,用来展示列表的 View。
(2) 适配器,用来把数据映射到 ListView 上的中介。

(3) 数据,具体的将被映射的字符串、图片,或者基本组件。

根据列表的适配器类型,列表分为三种,ArrayAdapter,SimpleAdapter 和 SimpleCursorAdapter。

其中以 ArrayAdapter 最为简单,只能展示一行字。SimpleAdapter 有最好的扩充性,可以自定义出各种效果。上面代码使用了 ArrayAdapter(Context context,int textViewResourceId,List < T > objects)来装配数据,要装配这些数据就需要一个连接 ListView 视图对象和数组数据的适配器来完成两者的适配工作,ArrayAdapter 的构造需要三个参数,依次为 this,布局文件(注意这里的布局文件描述的是列表的每一行的布局),android.R.layout.simple_list_item_1 是系统定义好的布局文件只显示一行文字),数据源(一个 List 集合)。同时用 setAdapter()完成适配的最后工作。

SimpleCursorAdapter 可以认为是 SimpleAdapter 对数据库的简单结合,可以方便地把数据库的内容以列表的形式展示出来。SimpleCursorAdapter 简单地说就是方便地把从游标得到的数据进行列表显示,并可以把指定的列映射到对应的 TextView 中。

下面的程序是从电话簿中把联系人显示到列表中。先在通讯录中添加一个联系人作为数据库的数据,然后获得一个指向数据库的 Cursor 并且定义一个布局文件(当然也可以使用系统自带的)。SimpleCursorAdapter 的实例程序如下:

```java
public class MyListView2 extends Activity {
    private ListView listView;
    @Override
    public void onCreate(Bundle savedInstanceState){
        super.onCreate(savedInstanceState);
        listView = new ListView(this);
        Cursor cursor = getContentResolver().query(People.CONTENT_URI, null, null, null, null);
        startManagingCursor(cursor);
        ListAdapter listAdapter = new SimpleCursorAdapter(this,
            android.R.layout.simple_expandable_list_item_1,
                cursor,
                new String[]{People.NAME},
                new int[]{android.R.id.text1});
        listView.setAdapter(listAdapter);
        setContentView(listView);
    }
}
```

Cursor cursor = getContentResolver().query(People.CONTENT_URI, null, null, null, null);先获得一个指向系统通讯录数据库的 Cursor 对象获得数据来源。

startManagingCursor(cursor);将获得的 Cursor 对象交由 Activity 管理,这样 Cursor 的生命周期和 Activity 便能够自动同步,省去手动管理 Cursor。

SimpleCursorAdapter 构造函数前面 3 个参数和 ArrayAdapter 是一样的,最后两

个参数:一个包含数据库的列的 String 型数组,一个包含布局文件中对应组件 id 的 int 型数组。其作用是自动地将 String 型数组所表示的每一列数据映射到布局文件对应 id 的组件上。上面的代码,将 NAME 列的数据一次映射到布局文件的 id 为 text1 的组件上。

注意:需要在 AndroidManifest.xml 中加权限:< uses – permission android:name = "android.permission.READ_CONTACTS"> </uses – permission >

运行后效果如图 5 – 5 所示。

图 5 – 5 SimpleCursorAdapter 实例效果图

5.2.3 SimpleAdapter

SimpleAdapter 的扩展性最好,可以定义各种各样的布局,可以放上 ImageView (图片),还可以放上 Button(按钮)、CheckBox(复选框)等。下面的程序是实现一个带有图片的列表,首先,需要定义好一个用来显示每一个列内容,vlist.xml:

```xml
<? xml version = "1.0" encoding = "utf - 8"? >
<LinearLayout xmlns:android = "http://schemas.android.com/apk/res/android"
        android:layout_width = "fill_parent"
        android:layout_height = "fill_parent" >
    <ImageView   android:paddingTop = "10px"
            android:layout_width = "80px"
            android:layout_height = "70px"
            android:id = "@ + id/img"/>
    <TextView android:layout_width = "wrap_content"
            android:layout_height = "wrap_content"
            android:id = "@ + id/title"      />
    <TextView android:layout_width = "wrap_content"
            android:layout_height = "wrap_content"
            android:id = "@ + id/info"      />
</LinearLayout>   </LinearLayout>
```

下面是实现代码:

```java
public class   main extends Activity
{
  private ListView listView;
    @Override
    public void onCreate(Bundle savedInstanceState)
    {
        super.onCreate(savedInstanceState);
        listView = new ListView(this);
```

```java
        SimpleAdapter adapter = new SimpleAdapter(
                this,getData(),R.layout.list_item,
                new String[]{"title","info","img"},
                new int[]{R.id.title,R.id.info,R.id.img});
        listView.setAdapter(adapter);
        setContentView(listView);
    }
    private List<Map<String,Object>>getData() {
    List<Map<String,Object>>list = new ArrayList<Map<String,Object>>();
        Map<String,Object>map = new HashMap<String,Object>();//1
        map.put("title", "客厅");
        map.put("info", "192.168.1.100");
        map.put("img", R.drawable.alivingroom);
        list.add(map);
        map = new HashMap<String,Object>();/////////////////////2
        map.put("title", "卧室");
        map.put("info", "192.168.1.100");
        map.put("img", R.drawable.bedroom);
        list.add(map);
        map = new HashMap<String,Object>();/////////////////////3
        map.put("title", "厨房");
        map.put("info", "192.168.1.100");
        map.put("img", R.drawable.kitchen);
        list.add(map);
        return list;
    }
}
```

使用 SimpleAdapter 一般都是 HashMap 构成 List 数据，List 每一节对应 ListView 每一行。HashMap 每个键值数据映射到布局文件中对应 id 组件上。因为系统没有对应的布局文件可用，因此，上述代码定义一个布局 list_item.xml，运行效果如图 5-6 所示。

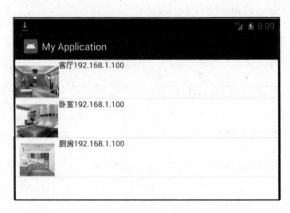

图 5-6　SimpleAdapter 实例运行效果图

5.2.4　有按钮的 ListView

有时候列表不仅用来做显示用,同样可以在上面添加按钮。添加按钮首先要写一个有按钮的 XML 文件,然后,自然会想到用上面的方法定义一个适配器,将数据映射到布局文件上。但是事实并非这样,因为按钮是无法映射的,即使成功地用布局文件显示出了按钮也无法添加按钮的响应,这时就要研究一下 ListView 是如何现实的了,而且必须要重写一个类继承 BaseAdapter。下面的示例将显示一个按钮和一个图片,两行字,如果单击按钮将删除此按钮的所在行,布局 vlist2.xml：

```xml
<?xml version="1.0" encoding="utf-8"?>
<LinearLayout xmlns:android="http://schemas.android.com/apk/res/android"
    android:orientation="horizontal"
    android:layout_width="fill_parent"
    android:layout_height="fill_parent"
    android:background="#eeeeee"
    >
    <ImageView android:id="@+id/img"
        android:layout_width="80px"
        android:layout_height="70px"
        android:layout_margin="5px"/>
    <LinearLayout android:orientation="vertical"
        android:layout_width="wrap_content"
        android:layout_height="wrap_content"
        android:layout_margin="15px">
        <TextView android:id="@+id/title"
            android:layout_width="wrap_content"
            android:layout_height="wrap_content"
            android:textColor="#FF0000"
            android:textSize="22px" />
        <TextView android:id="@+id/info"
            android:layout_width="wrap_content"
            android:layout_height="wrap_content"
            android:textColor="#FF0000"
            android:textSize="13px" />
    </LinearLayout>
    <Button android:id="@+id/view_btn"
        android:layout_width="wrap_content"
        android:layout_height="wrap_content"
        android:text="查看"
        android:layout_gravity="bottom|right" />
</LinearLayout>
```

程序代码如下:

```java
public class main extends Activity
{
   private ListView listView;
     @Override
     public void onCreate(Bundle savedInstanceState)
     {
          super.onCreate(savedInstanceState);
          listView = new ListView(this);
          MyAdapter adapter = new MyAdapter(this,getData());
        listView.setAdapter(adapter);
          setContentView(listView);
     }
     private List <Map <String, Object>>getData() {
     List <Map <String, Object>>list = new ArrayList <Map <String, Object>>();
     Map <String, Object>map = new HashMap <String, Object>();//1
     map.put("title", "客厅");
     map.put("info", "192.168.1.100");
     map.put("img", R.drawable.alivingroom);
     list.add(map);
     map = new HashMap <String, Object>();/////////////////////2
     map.put("title", "卧室");
     map.put("info", "192.168.1.100");
     map.put("img", R.drawable.bedroom);
     list.add(map);
     map = new HashMap <String, Object>();/////////////////////3
     map.put("title", "厨房");
     map.put("info", "192.168.1.100");
     map.put("img", R.drawable.kitchen);
     list.add(map);
     return list;
}

public final class ViewHolder{
     public ImageView img;
     public TextView title;
     public TextView info;
     public Button viewBtn;
}
public class MyAdapter extends BaseAdapter{
     private List <Map <String, Object>>mData;
```

```java
        private LayoutInflater mInflater;
        private Context context;
        public MyAdapter(Context m_context,List <Map <String, Object >>data){
            context = m_context;
            mData = data;
            this.mInflater = LayoutInflater.from(m_context);
        }
        @Override
        public int getCount() {return mData.size();}
        @Override
        public Object getItem(int arg0) {return null; }
        @Override
        public long getItemId(int arg0) {return 0;}
        @Override
        public View getView(int position, View convertView,ViewGroup parent){
          ViewHolder holder = null;
          if (convertView == null) {
             holder = new ViewHolder();
             convertView = mInflater.inflate(R.layout.list_item, null);
             holder.img = (ImageView)convertView.findViewById(R.id.img);
             holder.title = (TextView)convertView.findViewById(R.id.title);
             holder.info = (TextView)convertView.findViewById(R.id.info);
             holder.viewBtn = (Button)convertView.findViewById(R.id.view_btn);
             convertView.setTag(holder);
        }else {

              holder = (ViewHolder)convertView.getTag();
            }
holder.img.setBackgroundResource((Integer)mData.get(position).get("img"));
            holder.title.setText((String)mData.get(position).get("title"));
            holder.info.setText((String)mData.get(position).get("info"));
            final int index = position;
            holder.viewBtn.setOnClickListener(new View.OnClickListener()
            {
                @Override
                public void onClick(View v) {
                    showInfo(index);
                }
            });
            return convertView;
        }
```

```
    public void showInfo(int index){
    new AlertDialog.Builder(context)
    .setTitle((String)mData.get(index).get("title"))
    .setMessage("介绍..." + (String)mData.get(index).get("info"))
    .setPositiveButton("确定", new DialogInterface.OnClickListener() {
        @Override
        public void onClick(DialogInterface dialog, int which) {
        }
    }).show();
    }
    }
}
```

有按钮的 ListView 实例运行效果如图 5-7 所示。

图 5-7 有按钮的 ListView 实例运行效果图

ListView 在开始绘制的时候,系统首先调用 getCount() 函数,根据函数的返回值得到 ListView 的长度(这也是为什么在开始的第一张图特别的标出列表长度),然后根据这个长度,调用 getView() 逐一绘制每一行。如果 getCount() 返回值是 0,列表将不显示。如果返回值是 1,就只显示一行。

系统显示列表时,首先实例化一个适配器(这里将实例化自定义的适配器)。手动完成适配时,必须手动映射数据,这需要重写 getView() 方法。系统在绘制列表的每一行的时候将调用此方法。getView() 有三个参数,Position 表示将显示的是第几行,CovertView 是从布局文件通过 Inflate 获取的布局。用 LayoutInflater 的方法将定义好的 vlist2.xml 文件提取成 View 实例用来显示,然后将 XML 文件中的各个组件实例化(简单的 findViewById() 方法),这样便可以将数据对应到各个组件上。为了响应按钮点击事件,需要为它添加点击监听器,这样就能捕获点击事件。

至此一个自定义的 ListView 就完成了,现在从新审视这个过程。系统要绘制 ListView,首先获得要绘制的这个列表的长度,然后开始绘制第一行,怎么绘制呢?调用 getView() 函数。在这个函数里面首先获得一个 View(实际上是一个 ViewGroup),然后再实例化并设置各个组件,显示出来。

在实际的运行过程中会发现 ListView 的每一行没有焦点了,这是因为 Button 抢夺了 ListView 的焦点,在布局文件中将 Button 设置为没有焦点,运行效果如图 5-8 所示。

图 5-8 simpleAdapter 综合实例效果图

5.2.5 getView 应用

getView 工作原理如下:

(1) ListView 针对 List 中每个 Item,要求 Adapter "给我一个视图"(getView)。

(2) 一个新的视图被返回并显示。

如果有上亿个项目要显示怎么办?为每个项目创建一个新视图?这不可能。实际上 Android 缓存了视图。Android 中有个叫做 Recycler 的构件,工作原理如图 5-9 所示。

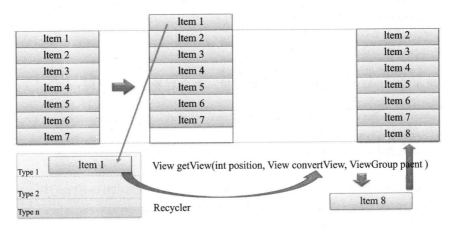

图 5-9 Recycler 构件工作原理图

① 如果有10亿个项目(item)，其中只有可见的项目存在内存中，其他的在Recycler中。

② ListView先请求一个type1视图(getView)，然后请求其他可见的项目。convertView在getView中是空(null)的。

③ 当item1滚出屏幕，并且一个新的项目从屏幕低端上来时，ListView再请求一个type1视图。convertView此时不是空值了，它的值是item1。只需设定新的数据，然后返回convertView，不必重新创建一个视图。

ListView的Adapter的作用如下图5-10所示。

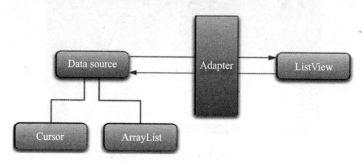

图5-10　ListView的Adapter的作用

Adapter的作用就是ListView界面与数据之间的桥梁，当列表里的每一项显示到页面时，都会调用Adapter的getView方法返回一个View。如果列表有1 000 000项时会是什么样的？是不是会占用极大的系统资源？先看看下面的代码：

```
public View getView(int position, View convertView, ViewGroup parent) {
    View item = mInflater.inflate(R.layout.list_item_icon_text, null);
    ((TextView)item.findViewById(R.id.text)).setText(DATA[position]);
    ((ImageView)item.findViewById(R.id.icon)).setImageBitmap(
        (position & 1) == 1 ? mIcon1 : mIcon2);
    return item;
}
```

如果超过1 000 000项时，后果不堪设想。再来看看下面的代码：

```
public View getView(int position, View convertView, ViewGroup parent) {
    if (convertView == null) {
        convertView = mInflater.inflate(R.layout.item, null);
    }
    ((TextView)convertView.findViewById(R.id.text)).setText(DATA[position]);
    ((ImageView)convertView.findViewById(R.id.icon)).setImageBitmap(
        (position & 1) == 1 ? mIcon1 : mIcon2);
    return convertView;
}
```

上面的代码,系统将会减少创建很多 View,性能得到了很大的提升。还有没有优化的方法呢? 答案是肯定的代码如下:

```java
public View getView(int position, View convertView, ViewGroup parent) {
    ViewHolder holder;
    if (convertView == null) {
    convertView = mInflater.inflate(R.layout.list_item_icon_text, null);
    holder = new ViewHolder();
    holder.text = (TextView)convertView.findViewById(R.id.text);
    holder.icon = (ImageView)convertView.findViewById(R.id.icon);
    convertView.setTag(holder);
    } else {
    holder = (ViewHolder) convertView.getTag();
    }
    holder.text.setText(DATA[position]);
    holder.icon.setImageBitmap((position & 1) == 1 ? mIcon1 : mIcon2);
    return convertView;
    }
    static class ViewHolder {
    TextView text;
    ImageView icon;
}
```

会不会又给系统带来很大的提升呢? 三种方式的性能对比如图 5-11 所示。

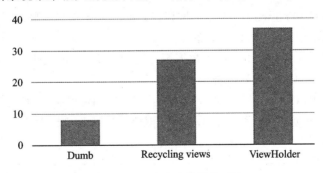

图 5-11 三种方式的性能对比

5.3 GridView 应用

1. GridView 简介

GridView 是 ViewGroup 子类,主要用于把内容显示在一个二维可滚动的网格,比如九宫格。使用 Gridview 和 Listview 类似,都是使用 ListAdapter 来填充数据加载布局。

2. XML Attributes

(1) android:columnWidth 主要用来指定每一列的宽度,使用该属性时应该注意和后面的拉伸模式相关联。如果拉伸模式为 spacingWidth,则该属性必须指定;如果拉伸模式为 columnWidth,指定该属性无效。

(2) android:gravity 用于指定每个 Ltem 的对其方式起始位置,当有多个时可以使用'|'隔开。

(3) android:horizontalSpacing 用于指定每一列之间的间距。注意:如果拉伸模式为 spacingWidth,则该值设定无效。

(4) android:numColumns 每一列的 Ltem 数目。

(5) android:verticalSpacing 用于指定每一行之间的间距。

(6) android:stretchMode(***)如果以列间距拉伸,则不需要指定列之间的间距;如果以列的等宽度拉伸,则不需要指定列的宽度。

> none,值等于 0,拉伸被禁用,不可以被拉伸;
> spacingWidth,值等于 1,每一列之间的间距会被拉伸,因此,使用该拉伸模式时,必须指定 columnWidth,而指定 horizontalSpacing 就会无效,即不需要先考虑 horizontalSpacing;
> columnWidth,值等于 2,每一列是等宽度,只需要指定 numColumns 和 horizontalSpacing,即先指定 columnWidth 无效;
> spacingWidthUniform,值等于 3,每一列的间距均匀拉伸,拉伸被禁用,不可以被拉伸。

Android 的 GridView 以二维滚动网格(行和列)显示项目,网格项目不一定是预定的,但它们会自动使用 ListAdapter 布局插入。

一个适配器实际上是 UI 组件和数据源之间的桥梁,填充数据到 UI 组件。适配器可以用来提供数据,如微调、列表视图、网格视图等。

ListView 和 GridView 是 AdapterView 的子类,它们可以绑定填充到一个适配器,从外部源检索数据,并创建一个视图表示每个数据项。GridView 属性如表 5-1 所列。

表 5-1 GridView 的属性表

属性	描述
android:id	这是唯一标识的布局的 ID
android:columnWidth	这指定了固定的宽度为每列。这可能是 px、dp、sp、in 或 mm
android:gravity	指定每个单元内的重力。可能的值是 top、bottom、left、right、center、center_vertical、center_horizontal 等
android:horizontalSpacing	定义列之间的默认水平间距。可能形式为 px、dp、sp、in 或 mm
android:numColumns	定义了要显示多少列。可以是一个整数值,例如"100"或 auto_fit,这意味着显示尽可能多的列可能填补可用空间

第5章 列表视图界面设计

续表 5-1

属 性	描 述
android:stretchMode	定义列应如何拉伸以填充可用的空白,如果有的话,值必须是: • none:延长被禁止。 • spacingWidth:每一列之间的间距被拉伸。 • columnWidth:每列被均等地拉伸。 • spacingWidthUniform:每一列之间的间距被均匀拉伸
android:verticalSpacing	定义行之间的缺省垂直间距。这可能是 px、dp、sp、in 或 mm

下面的例子将通过简单的步骤显示如何使用 GridView 创建自己的 Android 应用程序。按照下面的步骤来 创建 Android 应用程序 GridView:

Step 1 使用 Eclipse IDE 创建 Android 应用程序,并将其命名为 GridView,存在包 com.yiibai.gridview 下

Step 2 修改 res/layout/activity_main.xml 文件的默认内容,包括 GridView 的内容以及它的属性

Step 3 在 res/values/strings.xml 文件中定义所需的常量

Step 4 把几张照片放在 res/drawable-hdpi 文件夹。这几张照片如下:sample0.jpg,sample1.jpg,sample2.jpg,sample3.jpg,sample4.jpg,sample5.jpg,sample6.jpg 和 sample7.jpg

Step 5 在包 com.yiibai.helloworld 下创建一个新类 ImageAdapter 扩展 BaseAdapter。将用于填充视图,这个类将实现一个适配器的功能

Step 6 运行该应用程序,启动 Android 模拟器,并验证应用程序所运行的结果

以下是内容是主活动文件 src/com.yiibai.gridview/MainActivity.java。这个文件可以包括每个的基本生命周期方法。

```
package com.yiibai.gridview;
import android.os.Bundle;
import android.app.Activity;
import android.view.Menu;
import android.widget.GridView;
public class MainActivity extends Activity {
    @Override
    protected void onCreate(Bundle savedInstanceState) {
        super.onCreate(savedInstanceState);
        setContentView(R.layout.activity_main);
        GridView gridview = (GridView) findViewById(R.id.gridview);
        gridview.setAdapter(new ImageAdapter(this));
    }
    @Override
    public boolean onCreateOptionsMenu(Menu menu) {
```

```
        getMenuInflater().inflate(R.menu.main, menu);
        return true;
    }
}
```

以下是文件 res/layout/activity_main.xml 的内容:

```xml
<?xml version = "1.0" encoding = "utf-8"?>
<GridView xmlns:android = "http://schemas.android.com/apk/res/android"
    android:id = "@+id/gridview"
    android:layout_width = "fill_parent"
    android:layout_height = "fill_parent"
    android:columnWidth = "90dp"
    android:numColumns = "auto_fit"
    android:verticalSpacing = "10dp"
    android:horizontalSpacing = "10dp"
    android:stretchMode = "columnWidth"
    android:gravity = "center"
/>
```

以下是文件 res/values/strings.xml 的内容,定义了两个常量:

```xml
<?xml version = "1.0" encoding = "utf-8"?>
<resources>
    <string name = "app_name">HelloWorld</string>
    <string name = "action_settings">Settings</string>
</resources>
```

以下是文件 src/com.yiibai.gridview/ImageAdapter.java 的内容:

```java
package com.yiibai.gridview;
import android.content.Context;
import android.view.View;
import android.view.ViewGroup;
import android.widget.BaseAdapter;
import android.widget.GridView;
import android.widget.ImageView;
public class ImageAdapter extends BaseAdapter {
    private Context mContext;
    // Constructor
    public ImageAdapter(Context c) {
        mContext = c;
    }
    public int getCount() {
```

```
        return mThumbIds.length;
    }
    public Object getItem(int position) {
        return null;
    }
    public long getItemId(int position) {
        return 0;
    }
    // create a new ImageView for each item referenced by the Adapter
    public View getView(int position, View convertView, ViewGroup parent) {
        ImageView imageView;
        if (convertView == null) {
        imageView = new ImageView(mContext);
        imageView.setLayoutParams(new GridView.LayoutParams(85, 85));
        imageView.setScaleType(ImageView.ScaleType.CENTER_CROP);
        imageView.setPadding(8, 8, 8, 8);
        } else {
        imageView = (ImageView) convertView;
        }

        imageView.setImageResource(mThumbIds[position]);
        return imageView;
    }

    // Keep all Images in array
    public Integer[] mThumbIds = {
        R.drawable.sample_2, R.drawable.sample_3,
        R.drawable.sample_4, R.drawable.sample_5,
        R.drawable.sample_6, R.drawable.sample_7,
        R.drawable.sample_0, R.drawable.sample_1,
        R.drawable.sample_2, R.drawable.sample_3,
        R.drawable.sample_4, R.drawable.sample_5,
        R.drawable.sample_6, R.drawable.sample_7,
        R.drawable.sample_0, R.drawable.sample_1,
        R.drawable.sample_2, R.drawable.sample_3,
        R.drawable.sample_4, R.drawable.sample_5,
        R.drawable.sample_6, R.drawable.sample_7
    };
}
```

GridView 实例运行效果如图 5-12 所示。

图 5-12　GridView 实例运行效果

5.4　RecyclerView 应用

自 Android 5.0 之后,谷歌公司推出了 RecylerView 控件,据官方的介绍,该控件用于在有限的窗口中展示大量数据集,其实这样功能的控件并不陌生,例如 ListView、GridView。

那么有了 ListView、GridView 为什么还需要 RecyclerView 这样的控件呢?整体上看 RecyclerView 架构,提供了一种插拔式的体验,高度的解耦,异常的灵活,通过设置它提供的不同 LayoutManager、ItemDecoration、ItemAnimator 实现绚丽的效果。

RecyclerView 是 Android 5.0 版本中新添加的一个用来取代 ListView 的 SDK,它的灵活性与可替代性比 ListView 更好。RecyclerView 与 ListView 原理是类似的,都是仅仅维护少量的 View 并且可以展示大量的数据集。在 RecyclerView 标准化了 ViewHolder 类似于 ListView 中 convertView 用来做视图缓存。

1. RecyclerView 是什么?

RecylerView 是 support-v7 包中的新组件,是一个强大的滑动组件,与经典的 ListView 相比,同样拥有 Ltem 回收复用的功能,这一点从它的名字 RecyclerView 即回收 View 也可以看出。看到这也许有人会问,不是已经有 ListView 了吗,为什么还要 RecylerView 呢?这就牵扯到第二个问题了。

2. RecyclerView 的优点是什么?

根据官方的介绍,RecylerView 是 ListView 的升级版,既然如此,那 RecylerView 必然有它的优点,现就 RecyclerView 相对于 ListView 的优点罗列如下:

① RecylerView 封装了 Viewholder 的回收复用,也就是说 RecyclerView 标准化了 ViewHolder,编写 Adapter 面向的是 ViewHolder 而不再是 View 了,复用的逻辑被封装了,写起来更加简单。

② 提供了一种插拔式的体验,高度的解耦,异常的灵活,针对一个 Item 的显示 RecylerView 专门抽取出了相应的类,来控制 Item 的显示,使其的扩展性非常强。例如:想控制横向或者纵向滑动列表效果可以通过 LinearLayoutManager 这个类来进行控制(与 GridView 效果对应的是 GridLayoutManager,与瀑布流对应的还有 StaggeredGridLayoutManager 等),也就是说 RecylerView 不再拘泥于 ListView 的线性展示方式,它也可以实现 GridView 等多种效果。想控制 Item 的分隔线,可以通过继承 RecylerView 的 ItemDecoration 这个类,然后针对自己的业务需求去书写代码。

③ 可以控制 Item 增删的动画,可以通过 ItemAnimator 这个类进行控制,当然,针对增删的动画,RecylerView 有其自己默认的实现。

RecylerView 的设置过程如下:

```
recyclerView = (RecyclerView) findViewById(R.id.recyclerView);
LinearLayoutManager layoutManager = new LinearLayoutManager(this);
//设置布局管理器
recyclerView.setLayoutManager(layoutManager);
//设置为垂直布局,这也是默认的
layoutManager.setOrientation(OrientationHelper.VERTICAL);
//设置 Adapter
recyclerView.setAdapter(recycleAdapter);
//设置分隔线
recyclerView.addItemDecoration( new DividerGridItemDecoration(this));
//设置增加或删除条目的动画
recyclerView.setItemAnimator( new DefaultItemAnimator());
```

可以看到对 RecylerView 的设置过程,比 ListView 要复杂一些,虽然代码书写上有点复杂,但它的扩展性是极高的。在了解了 RecylerView 的一些控制之后,紧接着来看看它的 Adapter 的写法,RecyclerView 的 Adapter 与 ListView 的 Adapter 还是有点区别的,RecyclerView.Adapter,需要实现有 3 个方法:

a) onCreateViewHolder()

这个方法主要为每个 Item inflater 生成出一个 View,但是该方法返回的是一个 ViewHolder。该方法把 View 直接封装在 ViewHolder 中,然后面向的是 ViewHolder 这个实例,当然,这个 ViewHolder 需要自己去编写。直接省去了当初的 convertView.setTag(holder)和 convertView.getTag()这些繁琐的步骤。

b) onBindViewHolder()

这个方法主要用于适配渲染数据到 View 中。方法提供给了一 ViewHolder,而不是原来的 convertView。

c) getItemCount()

这个方法就类似于 BaseAdapter 的 getCount 方法了,即总共有多少个条目。接下来通过一个小的实例帮助大家更深入地了解 RecyclerView 的用法。

例如,用 RecyclerView 实现一个图片滚动的列表,代码如下:

```java
public class MainActivity extends ActionBarActivity {
    private RecyclerView mRecyclerView;
    private List<Integer> mDatas;
    @Override
    protected void onCreate(Bundle savedInstanceState) {
        super.onCreate(savedInstanceState);
        setContentView(R.layout.activity_main);
        initData();
        // 得到控件
        mRecyclerView = (RecyclerView) findViewById(R.id.recyclerview);
        // 设置布局管理器
        LinearLayoutManager layoutManager = new LinearLayoutManager(this);
        layoutManager.setOrientation(LinearLayoutManager.HORIZONTAL);
        mRecyclerView.setLayoutManager(layoutManager);
        // 设置适配器
        mRecyclerView.setAdapter(new MyRecyclerAdapter(this, mDatas));
    }
    private void initData() {
        mDatas = new ArrayList<Integer>(Arrays.asList(R.drawable.kenan1,
                R.drawable.kenan2, R.drawable.kenan3, R.drawable.kenan4,
                R.drawable.kenan5, R.drawable.kenan6, R.drawable.kenan7,
                R.drawable.kenan8));
    }
}
public class DividerItemDecoration extends ItemDecoration {
    public DividerItemDecoration() {super();}
    @Override
    public void getItemOffsets(Rect outRect, View view, RecyclerView parent,
            State state) {
        super.getItemOffsets(outRect, view, parent, state);
    }
    @Override
    @Deprecated
    public void onDraw(Canvas c, RecyclerView parent) {
        super.onDraw(c, parent);
    }
}
public class MyRecyclerAdapter extends Adapter<MyRecyclerAdapter.MyHolder> {
    private Context mContext;
    private List<Integer> mDatas;
```

```java
public MyRecyclerAdapter(Context context, List <Integer >datas) {
    super();
    this.mContext = context;
    this.mDatas = datas;
}
@Override
public int getItemCount() {return mDatas.size();}
@Override
// 填充 onCreateViewHolder 方法返回的 holder 中的控件
public void onBindViewHolder(MyHolder holder, int position) {
    holder.imageView.setImageResource(mDatas.get(position));
}
@Override
// 重写 onCreateViewHolder 方法,返回一个自定义的 ViewHolder
public MyHolder onCreateViewHolder(ViewGroup arg0, int arg1) {
    // 填充布局
    View view = LayoutInflater.from(mContext).inflate(R.layout.item, null);
    MyHolder holder = new MyHolder(view);
    return holder;
}
// 定义内部类继承 ViewHolder
class MyHolder extends ViewHolder {
    private ImageView imageView;
    public MyHolder(View view) {
        super(view);
        imageView = (ImageView) view.findViewById(R.id.iv_item);
    }
}
}
```

运行效果如图 5-13 所示。

图 5-13　RecyclerView 实例运行效果

5.5 列表视图界面的实现

对于物联网智能系统需要在主界面上以列表的形式显示地点（位置）、设备等。这些列表信息，在安卓里以列表视图 ListView 的形式实现。

1. 主布局

根据本章的设计目标，如本章最开始介绍的图 5-2 所示，由于在列表上面还有一个显示图片和文本的小框，因此，需要在主布局中把这些功能加上，地点列表界面的布局代码如下：

```xml
<?xml version="1.0" encoding="utf-8"?>
<LinearLayout xmlns:Android="http://schemas.android.com/apk/res/android"
    Android:orientation="vertical"
    Android:layout_width="fill_parent"
    Android:layout_height="fill_parent">
    <!-- 列表 -->
    <LinearLayout
    Android:gravity="center_horizontal"
    Android:orientation="horizontal"
    Android:layout_width="fill_parent"
    Android:layout_height="40px"
    Android:background="@drawable/bg3"
    >
    <TextView Android:text="列表："
        Android:layout_width="wrap_content"
        Android:layout_height="wrap_content"
        Android:textColor="#FFFFFFFF"
        Android:textSize="20px"/>
    </LinearLayout>
    <!-- 列表 -->
    <ListView Android:id="@+id/list_goods"
        Android:layout_width="fill_parent"
        Android:layout_height="wrap_content"/>
</LinearLayout>
```

2. 列表布局

列表布局每一项列表包括一个图片、两个文本、一个选择框和一个按钮。代码如下：

```xml
<?xml version="1.0" encoding="utf-8"?>
<LinearLayout xmlns:Android="http://schemas.android.com/apk/res/android"
    Android:orientation="horizontal"
    Android:layout_width="fill_parent"
    Android:layout_height="fill_parent"
    Android:background="#eeeeee">
    <!-- 图片 -->
    <ImageView Android:id="@+id/imageItem"
        Android:layout_width="100px"
        Android:layout_height="80px"
        Android:layout_margin="5px"/>
    <!-- 信息 -->
    <LinearLayout Android:orientation="vertical"
        Android:layout_width="wrap_content"
        Android:layout_height="wrap_content"
        >
        <TextView Android:id="@+id/titleItem"
            Android:layout_width="wrap_content"
            Android:layout_height="wrap_content"
            Android:textColor="#FFFF0000"
            Android:textSize="12px"/>
        <TextView Android:id="@+id/infoItem"
            Android:layout_width="wrap_content"
            Android:layout_height="wrap_content"
            Android:textColor="#FFFF0000"
            Android:textSize="23px"/>
    </LinearLayout>
    <!-- 详情 -->
    <LinearLayout Android:gravity="right"
    Android:orientation="horizontal"
    Android:layout_width="fill_parent"
    Android:layout_height="wrap_content">
    <CheckBox Android:id="@+id/checkItem"
        Android:layout_width="wrap_content"
        Android:layout_height="wrap_content"
        Android:layout_margin="5px"/>
    <Button  Android:id="@+id/detailItem"
        Android:layout_width="wrap_content"
        Android:layout_height="wrap_content"
        Android:layout_margin="5px"/>
    </LinearLayout>
</LinearLayout>
```

3. 适配器类

根据本章 5.2.4 小节《有按钮的 ListView》的介绍,在有包含按钮的列表视图时,需要自定义一个适配器类,适配器类文件名 ListViewAdapter.java,实现程序如下:

```java
public class ListViewAdapter extends BaseAdapter {
    private Context context;                              //运行上下文
    private List<Map<String,Object>>listItems;            //信息集合
    private LayoutInflater listContainer;                 //视图容器
    private boolean[] hasChecked;                         //记录选中状态
    public final class ListItemView{                      //自定义控件集合
        public ImageView image;
        public TextView title;
        public TextView info;
        public CheckBox check;
        public Button detail;
        public String id;
    }
    public ListViewAdapter(Context context, List<Map<String,Object>>listItems){
        this.context = context;
        listContainer = LayoutInflater.from(context);     //创建视图容器并设置上下文
        this.listItems = listItems;
        hasChecked = new boolean[getCount()];

    }
    public int getCount() {   return listItems.size();  }
    public Object getItem(int arg0) {return null; }
    public long getItemId(int arg0) { return 0; }
    /**      *记录勾选了哪个物品
     * @param checkedID 选中的物品序号           */
    private void checkedChange(int checkedID) {
        hasChecked[checkedID] = ! hasChecked[checkedID];
    }
    /**      *判断物品是否选择
     * @param checkedID 物品序号
     * @return 返回是否选中状态       */
    public boolean hasChecked(int checkedID) {
        return hasChecked[checkedID];
    }
    /** *显示物品详情      * @param clickID    */
    private void showDetailInfo(int clickID) {
        new AlertDialog.Builder(context)
        .setTitle(listItems.get(clickID).get("title").toString())
```

```java
                .setMessage("详情:" + listItems.get(clickID).get("info").toString())

                .setPositiveButton("确定", null)
                .show();
    }
    /**       * ListView Item 设置      */
    public View getView(int position, View convertView, ViewGroup parent) {
        Log.e("method", "getView");
        final int selectID = position;
        //自定义视图
        if (convertView == null) {
            final ListItemView  listItemView = new ListItemView();
            //获取 list_item 布局文件的视图
            convertView = listContainer.inflate(R.layout.list_item, null);
            //获取控件对象
            listItemView.image = (ImageView)convertView.findViewById(R.id.imageItem);
            listItemView.title = (TextView)convertView.findViewById(R.id.titleItem);
            listItemView.info  = (TextView)convertView.findViewById(R.id.infoItem);
            listItemView.detail = (Button)convertView.findViewById(R.id.detailItem);
            listItemView.check = (CheckBox)convertView.findViewById(R.id.checkItem);
            //设置控件集到 convertView
            convertView.setTag(listItemView);
            //
            listItemView.image.setBackgroundResource((Integer) listItems.get(
                    position).get("image"));
            listItemView.title.setText((String) listItems.get(position)
                    .get("title"));
            listItemView.info.setText((String) listItems.get(position).get("info"));
            listItemView.detail.setText("编辑");
            listItemView.id = ((String) listItems.get(position)
                    .get("id"));
            //注册按钮
            listItemView.detail.setOnClickListener(new View.OnClickListener() {
                    @Override
                    public void onClick(View v) {
                        showDetailInfo(selectID);
                    }
                });
        }else {  }
        return convertView;
    }
}
```

4. 主程序

主程序文件名为 main.java,程序代码如下:

```java
Public class  main extends Activity
{
    private  ListView listView;
    private  ListViewAdapter listViewAdapter;
    private  List <Map <String, Object >>listItems;
    private  Integer[] imgeIDs = {R.drawable.alivingroom, R.drawable.bedroom,
              R.drawable.bedroom, R.drawable.study, R.drawable.kitchen, R.drawable.office};
    private  String[] goodsNames = {"客厅","主卧","客房","书房","厨房","办公室"};
    private  String[] goodsDetails = {
      "客厅:alivingroom",        "主卧:bedroom",
      "客房:bedroom",            "书房:study",
      "厨房:kitchen",            "办公室:office"};
    @Override
    public void onCreate(Bundle savedInstanceState)
    {
        super.onCreate(savedInstanceState);
        setContentView(R.layout.main);
        listView = (ListView) findViewById(R.id.list_goods);
        listItems = getListItems();
        listViewAdapter = new ListViewAdapter(this, listItems); //创建适配器
        listView.setAdapter(listViewAdapter);
    }
    private  List <Map <String, Object >>getListItems() {
        List <Map <String, Object >>listItems = new ArrayList <Map <String, Object >>();
        for(int i = 0; i <goodsNames.length; i ++ ) {
            Map <String, Object >map = new HashMap <String, Object >();
            map.put("image", imgeIDs[i]);              //图片资源
            map.put("title", "信息:");                  //标题
            map.put("info", goodsNames[i]);            //名称
            map.put("detail", goodsDetails[i]);        //详情
            listItems.add(map);
        }
        return listItems;
    }
}
```

运行效果如图 5-14 所示,与本章最开始介绍的图 5-1 物联网智能系统地点列表界面效果图相同。

第 5 章 列表视图界面设计

图 5-14 物联网智能系统列表视图界面运行效果图

第 6 章

导航栏及滑动界面设计

6.1 导航栏及滑动界面设计目标

目前大部分的手机 App,都是带导航栏的滑动主界面,例如 QQ、微信、淘宝等,带导航栏的目的是方便用户进行不同界面的选择,滑动方式是为了提供好的用户体验。物联网智能系统也具有多个界面的选择和切换,例如地点界面、控制界面、设备界面之间的切换,如图 6-1 所示。

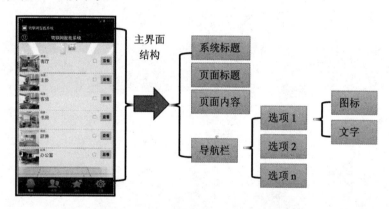

图 6-1 导航栏及滑动界面设计目标

ViewPager 滑动视图页的工作原理如图 6-2 所示。主要包括 PagerAdapter 适配器和 ViewPager 视图页对象两个主要部分,PagerAdapter 适配器是沟通数据与视图页的桥梁。与第 5 章介绍的 ListView 进行比较,可以发现两者实现的功能很相似,一个是左右滑动,一个是上下滑动;一个是视图页面的列表显示,一个是一组数据的列表显示。

导航栏及滑动界面的设计目标包括两部分:

(1) 滑动界面的设计,几个界面可以滑动切换或者通过导航栏上的按钮选择切换,滑动界面的第一页把第 5 章的地点列表内容显示出来。滑动界面采用 ViewPager 实现。

(2) 导航栏的设计,导航栏里包括三个以上的图片按钮,每个图片按钮下显示文字。

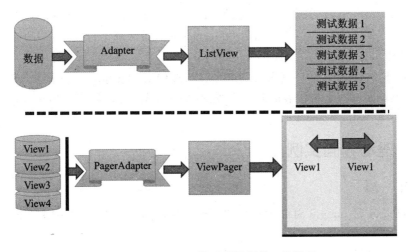

图 6-2　ViewPager 滑动视图页的工作原理

6.2　滑动界面设计

6.2.1　ViewPager 介绍

ViewPager（视图滑动切换工具）是 Android 3.0 后引入的一个 UI 控件，它的大概功能：通过手势滑动可以完成 View 的切换，一般是用来做 App 的引导页或者实现图片轮播，因为是 3.0 后引入的，如果想在低版本下使用，就需要引入 v4 兼容包，也可以看到，ViewPager 在 android.support.v4.view.ViewPager 目录下。

1. ViewPager 的简单介绍

ViewPager 就是一个简单的页面切换组件，可以往里面填充多个 View，然后可以左右滑动，从而切换不同的 View，可以通过 setPageTransformer()方法为 ViewPager 设置切换时的动画效果。和 ListView，GridView 一样，也需要一个 Adapter（适配器）将 View 和 ViewPager 进行绑定，而 ViewPager 则有一个特定的 Adapter——PagerAdapter。

另外，Google 官方是建议使用 Fragment 来填充 ViewPager 的，这样可以更加方便地生成每个 Page，以及管理每个 Page 的生命周期。

2. PagerAdapter 的使用方法

先来介绍最普通的 PagerAdapter，如果想使用这个 PagerAdapter 需要重写下面的四个方法：当然，这只是官方建议，实际上只需重写 getCount()和 isViewFromObject()。

getCount()：获得 Viewpager 中有多少个 View。

destroyItem()：移除一个给定位置的页面。适配器有责任从容器中删除这个视图。这是为了确保在 finishUpdate(viewGroup)返回时视图能够被移除。

instantiateItem()：①将给定位置的 View 添加到 ViewGroup(容器)中，创建并显示出来。②返回一个代表新增页面的 Object(Key)，通常都是直接返回 View 本身就可以了，当然也可以自定义自己的 Key，但是 Key 和每个 View 要一一对应。

isViewFromObject()：判断 instantiateItem(ViewGroup, int)函数所返回来的 Key 与一个页面视图是否代表的同一个视图(即对应的表示同一个 View)，通常直接写 return view==object。

6.2.2 滑动界面实例

1. 布　局

ViewPager 是 google SDk 中自带的一个附加包的一个类，可以用来实现屏幕间的切换。

(1) 在主布局文件里加入如下代码：

```
<RelativeLayout xmlns:android = "http://schemas.android.com/apk/res/android"
    xmlns:tools = "http://schemas.android.com/tools"
    android:layout_width = "fill_parent"
    android:layout_height = "fill_parent"
    tools:context = "com.example.testviewpage_1.MainActivity" >

<android.support.v4.view.ViewPager
    android:id = "@+id/viewpager"
    android:layout_width = "wrap_content"
    android:layout_height = "wrap_content"
    android:layout_gravity = "center" />
</RelativeLayout>
```

其中 <Android.support.v4.view.ViewPager /> 是 ViewPager 对应的组件，要将其放到想要滑动的位置。

(2) 新建三个 Layout，用于滑动切换的视图从效果图中也可以看到，三个视图都非常简单，里面没有任何的控件，大家当然可以往里添加各种控件，但这里是个 DEMO，只详解原理即可，所以，这里仅仅用背景来区别不用 Layout 布局。三个布局代码分别如下：

```
(1)layout1.xml
<? xml version = "1.0" encoding = "utf-8"? >
<LinearLayout xmlns:android = "http://schemas.android.com/apk/res/android"
    android:layout_width = "match_parent"
    android:layout_height = "match_parent"
    android:background = "#ffffff"
```

```
        android:orientation = "vertical" >
</LinearLayout >
(2)layout2.xml
<? xml version = "1.0" encoding = "utf - 8"? >
<LinearLayout xmlns:android = "http://schemas.android.com/apk/res/android"
    android:layout_width = "match_parent"
    android:layout_height = "match_parent"
    android:background = "#ffff00"
    android:orientation = "vertical" >
</LinearLayout >
(3)layout3.xml
<? xml version = "1.0" encoding = "utf - 8"? >
<LinearLayout xmlns:android = "http://schemas.android.com/apk/res/android"
    android:layout_width = "match_parent"
    android:layout_height = "match_parent"
    android:background = "#ff00ff"
    android:orientation = "vertical" >
</LinearLayout > <span style = "color:#660000;">
</span >
```

2. 实现程序

(1) 声明的变量

先看声明的变量的意义：

```
private View view1, view2, view3;
private List <View >viewList;//View 数组
private ViewPager viewPager;   //对应的 ViewPager
```

- viewPager 对应 <android.support.v4.view.ViewPager/>控件。
- view1,view2,view3 对应的三个 Layout,即 layout1.xml,layout2.xml,layout3.xml。
- viewList 是一个 View 数组,盛装上面的三个 View。

(2) 创建视图

接下来是创建视图的过程：

```
viewPager = (ViewPager) findViewById(R.id.viewpager);
LayoutInflater inflater = getLayoutInflater();
view1 = inflater.inflate(R.layout.layout1, null);
view2 = inflater.inflate(R.layout.layout2,null);
view3 = inflater.inflate(R.layout.layout3, null);
viewList = new ArrayList <View >();// 将要分页显示的 View 装入数组中
```

```
viewList.add(view1);
viewList.add(view2);
viewList.add(view3);
```

创建视图过程难度不大，就是将资源与变量联系起来布局，最后将实例化的 view1，view2，view3 添加到 viewList 中。

（3）构建适配器

在 ListView 中也有适配器，listView 通过重写 GetView() 函数来获取当前要加载的 Item。而 PageAdapter 不太相同，毕竟 PageAdapter 是单个 View 的合集。PageAdapter 必须重写的四个函数：

```
boolean isViewFromObject(View arg0, Object arg1)
int getCount()
void destroyItem(ViewGroup container, int position,Object object)
ObjectinstantiateItem(ViewGroup container, int position)
@Override
public int getCount() {
    //getCount():返回要滑动的 View 的个数
    return viewList.size();
}
@Override
public void destroyItem(ViewGroup container, int position,
        Object object) {
        container.removeView(viewList.get(position));
}
destroyItem():从当前 container 中删除指定位置(position)的 View;
@Override
public Object instantiateItem(ViewGroup container, int position) {
        container.addView(viewList.get(position));
        return viewList.get(position);
    }
};
//instantiateItem():做了两件事,第一:将当前视图添加到 container 中,第二:返回当前 View
@Override
public boolean isViewFromObject(View arg0, Object arg1) {
    //TODO Auto - generated method stub
    return arg0 == arg1;
}
```

完整的程序代码如下：

```java
public class MainActivity extends Activity {
    private View view1, view2, view3;
    private ViewPager viewPager;        //对应的 viewPager
    private List <View> viewList;//View 数组
    @Override
    protected void onCreate(Bundle savedInstanceState) {
        super.onCreate(savedInstanceState);
        setContentView(R.layout.activity_main);
        viewPager = (ViewPager) findViewById(R.id.viewpager);
        LayoutInflater inflater = getLayoutInflater();
        view1 = inflater.inflate(R.layout.layout1, null);
        view2 = inflater.inflate(R.layout.layout2, null);
        view3 = inflater.inflate(R.layout.layout3, null);
        viewList = new ArrayList <View>();// 将要分页显示的 View 装入数组中
        viewList.add(view1);
        viewList.add(view2);
        viewList.add(view3);
        PagerAdapter pagerAdapter = new PagerAdapter() {
            @Override
            public boolean isViewFromObject(View arg0, Object arg1) {
                return arg0 == arg1;
            }
            @Override
            public int getCount() { return viewList.size();}
            @Override
            public void destroyItem(ViewGroup container, int position,
                Object object) {
                container.removeView(viewList.get(position));
            }
            @Override
            public Object instantiateItem(ViewGroup container, int position)
            {
                container.addView(viewList.get(position));
                return viewList.get(position);
            }
        };
        viewPager.setAdapter(pagerAdapter);
    }
}
```

上述滑动界面实例程序代码的运行效果如图 6-3 所示。

 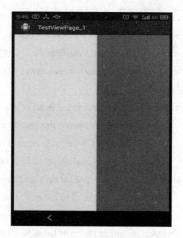

第一个界面向第二个界面滑动　　　　　第二个界面向第三个界面滑动

图 6-3　滑动界面实例运行效果图

6.3　导航栏设计

6.3.1　导航栏设计方法

导航就像书的目录一样,它会讲哪些内容,以及这些内容又是怎样排布的,导航可以让你在页面中不迷路,告诉你当前处于什么位置。底部导航栏是很多 App 都在使用的设计,像 QQ、微信、支付宝等,已经是每个 Android 应用必不可少的。常见的几种安卓导航栏的设计方法有:

(1) 利用 BottomNavigationBar 实现底部导航栏。

(2) FragmentTabHost。

(3) 第三方库:

➢ GitHub - aurelhubert/ahbottomnavigation;

➢ GitHub - roughike/BottomBar;

➢ Ashok - Varma/BottomNavigation。

(4) 用 fragmentTransaction 的 show 和 hide 方法隐藏和显示 Fragment。

(5) ViewPager + List + PagerAdapte(或 FragmentStatePagerAdapter 或 FragmentPagerAdapter)。

(6) 使用 RadioGroup+RadioButton 实现底部导航栏的效果。

(7) 使用 LinearLayout+TextView 实现底部导航栏的效果。

6.3.2　BottomNavigationView 底部导航栏

BottomNavigationView,它是 Android Support Library 25.0.0 版本中,新增加了

一个 API，底部导航视图。

采用 Obtain_Studio 里的"Android 项目\AndroidStudio 项目\AndroidStudio43__简单 BottomNavigationView 模板"，创建一个名为"MyApplication43_new_test_003"的项目。BottomNavigationView 应用程序的结构如图 6-4 所示。

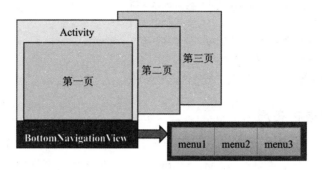

图 6-4　BottomNavigationView 应用程序的结构图

1. 主布局

主布局应用于主 Activety 界面，包括中间内容和底部导航栏。本例子的中间内容很简单，只有一个 TextVew。底部导航栏使用的是 BottomNavigationView 控件。主布局文件名为"activity_main. xml"，程序代码如下：

```
<android.support.constraint.ConstraintLayout xmlns:android = "http://schemas.android.com/apk/res/android"
    xmlns:app = "http://schemas.android.com/apk/res-auto"
    xmlns:tools = "http://schemas.android.com/tools"
    android:id = "@ + id/container"
    android:layout_width = "match_parent"
    android:layout_height = "match_parent"
    tools:context = ".MainActivity">
    <TextView
        android:id = "@ + id/message"
        android:layout_width = "wrap_content"
        android:layout_height = "wrap_content"
        android:layout_marginLeft = "@dimen/activity_horizontal_margin"
        android:layout_marginStart = "@dimen/activity_horizontal_margin"
        android:layout_marginTop = "@dimen/activity_vertical_margin"
        android:text = "@string/title_home"
        app:layout_constraintLeft_toLeftOf = "parent"
        app:layout_constraintTop_toTopOf = "parent" />
    <android.support.design.widget.BottomNavigationView
        android:id = "@ + id/navigation"
        android:layout_width = "0dp"
```

```
        android:layout_height = "wrap_content"
        android:layout_marginEnd = "0dp"
        android:layout_marginStart = "0dp"
        android:background = "? android:attr/windowBackground"
        app:layout_constraintBottom_toBottomOf = "parent"
        app:layout_constraintLeft_toLeftOf = "parent"
        app:layout_constraintRight_toRightOf = "parent"
        app:menu = "@menu/navigation" />
</android.support.constraint.ConstraintLayout>
```

2. 底部导航栏布局

底部导航栏采用 BottomNavigationView 控件，使用的布局是一个菜单布局，在资源目录下有一个 menu 菜单目录，navigation.xml 文件就是底部导航栏布局，程序代码如下：

```
<?xml version = "1.0" encoding = "utf-8"?>
<menu xmlns:android = "http://schemas.android.com/apk/res/android">
    <item android:id = "@+id/navigation_home"
        android:icon = "@drawable/ic_home_black_24dp"
        android:title = "@string/title_home" />
    <item android:id = "@+id/navigation_dashboard"
        android:icon = "@drawable/ic_dashboard_black_24dp"
        android:title = "@string/title_dashboard" />
    <item android:id = "@+id/navigation_notifications"
        android:icon = "@drawable/ic_notifications_black_24dp"
        android:title = "@string/title_notifications" />
</menu>
```

3. Activity 实现

在本例子中，Activity 实现的实现很简单，包括两部分，第一部分是创建窗口，在 onCreate 函数实现；第二部分是导航栏的事件的监听和内容切换。程序代码如下：

```
public class MainActivity extends AppCompatActivity {
    private TextView mTextMessage;
    @Override
    protected void onCreate(Bundle savedInstanceState) {
        super.onCreate(savedInstanceState);
        setContentView(R.layout.activity_main);
        mTextMessage = (TextView) findViewById(R.id.message);
        BottomNavigationView nav = (BottomNavigationView) findViewById(R.id.navigation);
        nav.setOnNavigationItemSelectedListener(mOnNavigationItemSelectedListener);
```

```
    }
    private BottomNavigationView.OnNavigationItemSelectedListenerls =
      new BottomNavigationView.OnNavigationItemSelectedListener() {
    @Override
    public boolean onNavigationItemSelected(@NonNull MenuItem item) {
      switch (item.getItemId()) {
      case R.id.navigation_home:
        mTextMessage.setText(R.string.title_home);
        return true;
      case R.id.navigation_dashboard:
        mTextMessage.setText(R.string.title_dashboard);
        return true;
      case R.id.navigation_notifications:
        mTextMessage.setText(R.string.title_notifications);
        return true;
      }
      return false;
    }};
}
```

简单 BottomNavigationView 模板运行效果如图 6-5 所示。

图 6-5 简单 BottomNavigationView 模板运行效果图

6.4 Fragment

6.4.1 Fragment 简介

自从 Android 3.0 中引入 Fragments 的概念,根据词海的翻译可以译为碎片、片段。其目的是为了解决不同屏幕分辨率的动态和灵活 UI 设计。大屏幕如平板,小屏幕如手机，平板电脑的设计使得其有更多的空间来放更多的UI组件，而多出来的空间存放UI使其会产生更多的交互，从而诞生了Fragments。

Fragments的设计不需要来亲自管理view hierarchy的复杂变化，通过将Activity的布局分散到Frament中，可以在运行时修改Activity的外观，并且由Activity管理的back stack中保存些变化。当一个片段指定了自身的布局时，它能和其他片段配置成不同的组合，在活动中为不同的屏幕尺寸修改布局配置，小屏幕可能每次显示一个片段，而大屏幕则可以显示两个或更多。

Fragment必须被写成可重用的模块。因为Fragment有自己的Layout，自己进行事件响应，拥有自己的生命周期和行为，所以可以在多个Activity中包含同一个Fragment的不同实例，对于让界面在不同的屏幕尺寸下都能给用户完美的体验尤其重要。

Fragment优点：

- Fragment可以使能够将Activity分离成多个可重用的组件，每个都有它自己的生命周期和UI。
- Fragment可以轻松地创建动态灵活的UI设计，可以适应于不同的屏幕尺寸。从手机到平板电脑。
- Fragment是一个独立的模块，紧紧地与Activity绑定在一起。可以在运行中动态地移除、加入、交换等。
- Fragment提供一个新的方式让在不同的安卓设备上统一UI。
- Fragment解决Activity间的切换不流畅，轻量切换。
- Fragment替代TabActivity做导航，性能更好。
- Fragment在4.2.版本中新增嵌套Fragment使用方法，能够生成更好的效果。
- Fragment做局部内容更新更方便，原来为了到达这一点要把多个布局放到一个Activity里面，现在可以用多Fragment来代替，只有在需要的时候才加载Fragment，提高性能。可以从startActivityForResult中接收到返回结果，但是View不能。

6.4.2　Fragment和View的比较

Fragment和View都有助于界面组件的复用，这在大型工程里边是特别重要的，但是二者又有所区别。

（1）Fragment的复用程度更大。Fragment有完整的生命周期，从代码设计角度讲可以提高内聚性，不同情况下还可以设计不同的Fragment，比如横屏和竖屏情况下View的显示不一样，那么可以建立2个不同的Fragment去处理，代码上面可以有效地扩展。从形态上讲和Activity更为接近，当然，从编程角度上看也比View更为复杂。但是Fragment可以组装更多的View同一展示，而且生命周期有助于资源的管理。

（2）简单的直接用View，复杂的才用Fragment，Fragment资源消耗比较大。

（3）一个Fragment必须总是绑定到一个Activity中，虽然Fragment有自己的生命周期,但同时也被它的宿主 Activity 的生命周期直接影响。

大部分情况下,Fragment 用来封装 UI 的模块化组件;但是也可以创建没有 UI 的 Fragment 来提供后台行为,该行为会一直持续到 Activity 重新启动。这特别适合于定期和 UI 交互的后台任务或者当因配置改变而导致 Activity 重新启动时,保存状态变得特别重要的场合。

注意:当 Activity 因为配置发生改变(屏幕旋转)或者内存不足被系统杀死,造成重新创建时,的 Fragment 会被保存下来,但是会创建新的 FragmentManager,新的 FragmentManager 会首先会去获取保存下来的 Fragment 队列,重建 Fragment 队列,从而恢复之前的状态。

6.4.3 Fragment 应用

下面实现一个简单的 Fragment 应用例子,流程如下:
(1) 添加两个类继承 Fragment,代码如下:

```java
import android.app.Fragment;
import android.os.Bundle;
import android.view.LayoutInflater;
import android.view.View;
import android.view.ViewGroup;
public class Fragment01 extends Fragment{
    @Override
    public View onCreateView(LayoutInflater inflater, ViewGroup container, Bundle savedInstanceState) {
        //引用创建好的 XML 布局
        View view = inflater.inflate(R.layout.fragment01,container,false);
        return view;
    }
}
```

(2) 为对应的 Fragment 创建对应的 XML 布局,代码如下:

```xml
<?xml version="1.0" encoding="utf-8"?>
<LinearLayout xmlns:android="http://schemas.android.com/apk/res/android"
    android:layout_width="match_parent"
    android:layout_height="match_parent"
    >
    <TextView
        android:layout_width="match_parent"
        android:layout_height="match_parent"
        android:textSize="50dp"
        android:text="是第一页">
    </TextView>
</LinearLayout>
```

（3）添加到 Activity 中，代码如下：

```java
import android.app.Activity;
import android.app.Fragment;
import android.app.FragmentManager;
import android.app.FragmentTransaction;
import android.os.Bundle;
import android.view.View;
import android.widget.Button;
public class MainActivity extends Activity {
    private Button button01,button02;
    private Fragment fragment01;
    @Override
    protected void onCreate(Bundle savedInstanceState) {
        super.onCreate(savedInstanceState);
        setContentView(R.layout.activity_main);
        setview();
    }
    private void setview() {
        button01 = (Button)findViewById(R.id.button01);
        button01.setOnClickListener(new View.OnClickListener() {
            @Override
            public void onClick(View v) {
                FragmentManager FM = getFragmentManager();
                FragmentTransaction MfragmentTransaction = FM.beginTransaction();
                Fragment01  f1 = new Fragment01();
                MfragmentTransaction.add(R.id.fragment_buju,f1);
                MfragmentTransaction.commit();
            }
        });
        button02 = (Button)findViewById(R.id.button02);
        button02.setOnClickListener(new View.OnClickListener() {
            @Override
            public void onClick(View v) {
                FragmentManager FMs = getFragmentManager();
                FragmentTransaction MfragmentTransactions = FMs.beginTransaction();
                Fragment02 f2 = new Fragment02();
                MfragmentTransactions.replace(R.id.fragment_buju,f2);
                MfragmentTransactions.commit();
            }
        });
    }
}
```

添加到 Activity 中的过程是,获取到 FragmentManager,在 V4 包中通过 getSupportFragmentManager 获得,在系统中原生的 Fragment 是通过 getFragmentManager 获得。开启一个事务,通过调用 beginTransaction 方法开启,把自己创建好的 Fragment 创建一个对象。向容器内加入 Fragment,一般使用 add 或者 replace 方法实现,需要传入容器的 id 和 Fragment 的实例。提交事务,调用 commit 方法提交。运行效果如图 6-6 所示。

图 6-6　Fragment 运行效果图

6.5　SurfaceView 与 TextureView

6.5.1　SurfaceView

应用程序的视频或者 OpenGL 内容往往是显示在一个特别的 UI 控件中:SurfaceView。SurfaceView 的工作方式是创建一个置于应用窗口之后的新窗口。这种方式的效率非常高,因为 SurfaceView 窗口刷新的时候不需要重绘应用程序的窗口(Android 普通窗口的视图绘制机制是一层一层的,任何一个子元素或者是局部的刷新都会导致整个视图结构全部重绘一次,因此效率非常低下,不过满足普通应用界面的需求还是绰绰有余),但是 SurfaceView 也有一些非常不便的限制。

因为 SurfaceView 的内容不在应用窗口上,所以不能使用变换(平移、缩放、旋转等)。也难以放在 ListView 或者 ScrollView 中,不能使用 UI 控件的一些特性,比如 View.setAlpha()。

1. 定　义

可以直接从内存或者 DMA 等硬件接口取得图像数据,是个非常重要的绘图容器。它的特性是:可以在主线程之外的线程中向屏幕绘图。这样可以避免画图任务繁重的时候造成主线程阻塞,从而提高了程序的反应速度。在游戏开发中多用到 SurfaceView,游戏中的背景、人物、动画等尽量在画布 Canvas 中画出。

2. 实　现

首先继承 SurfaceView 并实现 SurfaceHolder.Callback 接口。使用接口的原因:因为使用 SurfaceView 有一个原则,所有的绘图工作必须得在 Surface 被创建之后才能开始(Surface—表面,这个概念在图形编程中常常被提到。基本上可以把它当作显存的一个映射,写入到 Surface 的内容可以被直接复制到显存从而显示出来,这使得显示速度会非常快),而在 Surface 被销毁之前必须结束。所以 Callback 中的 surfaceCreated 和 surfaceDestroyed 就成了绘图处理代码的边界。需要重写的方法如下:

(1) public void surfaceChanged(SurfaceHolder holder,int format,int width,int

height){}

//在Surface的大小发生改变时激发。

（2）public void surfaceCreated(SurfaceHolder holder){}

//在创建时激发,一般在这里调用画图的线程。

（3）public void surfaceDestroyed(SurfaceHolder holder) {}

//销毁时激发,一般在这里将画图的线程停止、释放。

整个过程：继承 SurfaceView 并实现 SurfaceHolder.Callback 接口→SurfaceView.getHolder()获得 SurfaceHolder 对象→SurfaceHolder.addCallback(callback)添加回调函数→SurfaceHolder.lockCanvas()获得 Canvas 对象并锁定画布→Canvas 绘画→SurfaceHolder.unlockCanvasAndPost(Canvas canvas)结束锁定画图,并提交改变,将图形显示。

3. SurfaceHolder

这里用到了一个类 SurfaceHolder,可以把它当成 Surface 的控制器,用来操纵 Surface。处理它的 Canvas 上画的效果和动画,控制表面、大小、像素等。

几个需要注意的方法如下：

（1）abstract void addCallback(SurfaceHolder.Callback callback)：

给 SurfaceView 当前的持有者一个回调对象。

（2）abstract Canvas lockCanvas()：

锁定画布,一般在锁定后就可以通过其返回的画布对象 Canvas,在其上面画图等操作了。

（3）abstract Canvas lockCanvas(Rect dirty)：

锁定画布的某个区域进行画图等,因为画完图后,会调用下面的 unlockCanvasAndPost 来改变显示内容。

相对部分内存要求比较高的游戏来说,可以不用重画 Dirty 外的其他区域的像素,可以提高速度。

（4）abstract void unlockCanvasAndPost(Canvas canvas)：

结束锁定画图,并提交改变。

SurfaceView 使用步骤：获取到 SurfaceView 对应的 SurfaceHolder,给 SurfaceHolder 添加一个 SurfaceHolder.callback 对象。在子线程中开始在 Surface 上面绘制图形,因为 SurfaceView 没有对暴露 Surface,而只是暴露了 Surface 的包装器 SurfaceHolder,所以使用 SurfaceHolder 的 lockCanvas()获取 Surface 上面指定区域的 Canvas,在该 Canvas 上绘制图形,绘制结束后,使用 SurfaceHolder 的 unlockCanvasAndPost()方法解锁 Canvas,并且让 UI 线程把 Surface 上面的东西绘制到 View 的 Canvas 上面。下面是 SurfaceView 使用例子：

```java
public class GameUI extends SurfaceView implements SurfaceHolder.Callback {
    private SurfaceHolder holder;
    private RenderThread renderThread;
    private boolean isDraw = false;// 控制绘制的开关
    public GameUI(Context context) {
        super(context);
        holder = this.getHolder();
        holder.addCallback(this);
        renderThread = new RenderThread();
    }
    @Override
    public void surfaceChanged(SurfaceHolder holder, int format, int width,
        int height) {}
    @Override
    public void surfaceCreated(SurfaceHolder holder) {
        isDraw = true;
        renderThread.start();
    }
    @Override
    public void surfaceDestroyed(SurfaceHolder holder) {isDraw = false;}
    /** * 绘制界面的线程 * @author Administrator     */
    private class RenderThread extends Thread {
        @Override
        public void run() {
            // 不停绘制界面
            while (isDraw) {drawUI();}
            super.run();
        }
    }
    /** * 界面绘制 */
    public void drawUI() {
        Canvas canvas = holder.lockCanvas();
        try {
            drawCanvas(canvas);
        } catch (Exception e) {e.printStackTrace();} finally {
            holder.unlockCanvasAndPost(canvas);
        }
    }
    private void drawCanvas(Canvas canvas) {
        // 在 canvas 上绘制需要的图形
    }
}
```

6.5.2 TextureView

1. TextureView 介绍

TextureView 在 4.0(API level 14)中引入,与 SurfaceView 一样继承自 View,它可以将内容流直接投影到 View 中,可以用于实现 Live preview 等功能。和 SurfaceView 不同,它不会在 WMS 中单独创建窗口,而是作为 View hierachy 中的一个普通 View,因此,可以和其它普通 View 一样进行移动,旋转,缩放,动画等变化。值得注意的是 TextureView 必须在硬件加速的窗口中。它显示的内容流数据可以来自 App 进程或是远端进程。从类图中可以看到,TextureView 继承自 View,它与其他的 View 一样在 View hierachy 中管理与绘制。TextureView 重载了 draw()方法,其中主要以 SurfaceTexture 中收到的图像数据作为纹理更新到对应的 HardwareLayer 中。SurfaceTexture.OnFrameAvailableListener 用于通知 TextureView 内容流有新图像到来。SurfaceTextureListener 接口用于让 TextureView 的使用者知道 SurfaceTexture 已准备好,这样就可以把 SurfaceTexture 交给相应的内容流。Surface 为 BufferQueue 的 Producer 接口实现类,使生产者可以通过它的软件或硬件渲染接口为 SurfaceTexture 内部的 BufferQueue 提供 graphic buffer。

与 SurfaceView 相比,TextureView 并没有创建一个单独的 Surface 用来绘制,这使得它可以像一般的 View 一样执行一些变换操作,设置透明度等。另外,Textureview 必须在硬件加速开启的窗口中。

TextureView 的使用非常简单,唯一要做的就是获取用于渲染内容的 SurfaceTexture。具体做法是先创建 TextureView 对象,然后实现 SurfaceTextureListener 接口,代码如下:

```
private TextureView myTexture;
public class MainActivity extends Activity implements
SurfaceTextureListener{
protected void onCreate(Bundle savedInstanceState) {
    myTexture = new TextureView(this);
    myTexture.setSurfaceTextureListener(this);
    setContentView(myTexture);
    }
}
```

Activity 实现了 SurfaceTextureListener 接口,因此 Activity 中需要重写如下方法:

```
@Override
public void onSurfaceTextureAvailable(SurfaceTexture arg0,int arg1,int arg2){}
@Override
public boolean onSurfaceTextureDestroyed(SurfaceTexture arg0){}
@Override
```

```
public void onSurfaceTextureSizeChanged(SurfaceTexture arg0, int arg1,int arg2) {}
@Override
public void onSurfaceTextureUpdated(SurfaceTexture arg0){}
```

2. TextureView 应用

TextureView 可以使用 setAlpha 和 setRotation 方法达到改变透明度和旋转的效果。下面的例子演示了如何使用 TextureView 类,创建了一个可以在 TextureView 中预览 Camera 的 Demo,可以改变它的角度以及方向。当然,程序需要运行在有摄像头的设备上。下面是 MainActivity.java 中的代码:

```
public class MainActivity extends Activity implements SurfaceTextureListener {
    private TextureView myTexture;
    private Camera mCamera;
    @SuppressLint("NewApi")
    @Override
    protected void onCreate(Bundle savedInstanceState) {
        super.onCreate(savedInstanceState);
        setContentView(R.layout.activity_main);
        myTexture = new TextureView(this);
        myTexture.setSurfaceTextureListener(this);
        setContentView(myTexture);
    }
    @Override
    public boolean onCreateOptionsMenu(Menu menu) {
            getMenuInflater().inflate(R.menu.main, menu);
        return true;
    }
    @SuppressLint("NewApi")
    @Override
    public void onSurfaceTextureAvailable(SurfaceTexture arg0, int arg1,
    int arg2) {
        mCamera = Camera.open();
        Camera.Size previewSize = mCamera.getParameters().getPreviewSize();
        myTexture.setLayoutParams(new FrameLayout.LayoutParams(
        previewSize.width, previewSize.height, Gravity.CENTER));
        try {
           mCamera.setPreviewTexture(arg0);
           } catch (IOException t) {
            }
        mCamera.startPreview();
        myTexture.setAlpha(1.0f);
        myTexture.setRotation(90.0f);
```

```java
    }
    @Override
    public boolean onSurfaceTextureDestroyed(SurfaceTexture arg0) {
        mCamera.stopPreview();
        mCamera.release();
        return true;
    }
    @Override
    public void onSurfaceTextureSizeChanged(SurfaceTexture arg0, int arg1,
    int arg2) { }
    @Override
    public void onSurfaceTextureUpdated(SurfaceTexture arg0) {}
}
```

activity_main.xml

```xml
<RelativeLayout xmlns:android = "http://schemas.android.com/apk/res/android"
    xmlns:tools = "http://schemas.android.com/tools"
    android:layout_width = "match_parent"
    android:layout_height = "match_parent"
    android:paddingBottom = "@dimen/activity_vertical_margin"
    android:paddingLeft = "@dimen/activity_horizontal_margin"
    android:paddingRight = "@dimen/activity_horizontal_margin"
    android:paddingTop = "@dimen/activity_vertical_margin"
    tools:context = ".MainActivity" >
    <TextureView
        android:id = "@+id/textureView1"
        android:layout_width = "wrap_content"
        android:layout_height = "wrap_content"
        android:layout_alignParentTop = "true"
        android:layout_centerHorizontal = "true" />
</RelativeLayout>
```

AndroidManifest.xml

```xml
<?xml version = "1.0" encoding = "utf-8"?>
<manifest xmlns:android = "http://schemas.android.com/apk/res/android"
    package = "com.example.textureview"
    android:versionCode = "1"
    android:versionName = "1.0" >
    <uses-sdk
        android:minSdkVersion = "8"
        android:targetSdkVersion = "17" />
    <uses-permission android:name = "android.permission.CAMERA"/>
    <application
        android:allowBackup = "true"
        android:icon = "@drawable/ic_launcher"
        android:label = "@string/app_name"
        android:theme = "@style/AppTheme" >
```

```
        <activity
            android:name = "com.example.textureview.MainActivity"
            android:label = "@string/app_name" >
            < intent - filter >
                <action android:name = "android.intent.action.MAIN" />
                <category android:name = "android.intent.category.LAUNCHER" />
            </ intent - filter >
        </activity >
    </application >
</manifest >
```

3. SurfaceView 与 TextureView 对比

SurfaceView 继承自类 View,因此它本质上是一个 View。但与普通 View 不同的是,它有自己的 Surface。有自己的 Surface,在 WMS 中有对应的 WindowState,在 SurfaceFlinger 中有 Layer。一般的 Activity 包含的多个 View 会组成 View hierachy 的树形结构,只有最顶层的 DecorView,也就是根结点视图,才是对 WMS 可见的。这个 DecorView 在 WMS 中有一个对应的 WindowState。相应地,在 SF 中有对应的 Layer。而 SurfaceView 自带一个 Surface,这个 Surface 在 WMS 中有自己对应的 WindowState,在 SF 中也会有自己的 Layer。虽然在 App 端它仍在 View hierachy 中,但在 Server 端(WMS 和 SF)中,它与宿主窗口是分离的。这样的好处是对这个 Surface 的渲染可以放到单独线程去做,渲染时可以有自己的 GL context。这对于一些游戏、视频等性能相关的应用非常有益,因为它不会影响主线程对事件的响应。但它也有缺点,因为这个 Surface 不在 View hierachy 中,它的显示也不受 View 的属性控制,所以不能进行平移,缩放等变换,也不能放在其他 ViewGroup 中,一些 View 中的特性也无法使用。

(1) SurfaceView 优点及缺点

优点:可以在一个独立的线程中进行绘制,不会影响主线程;使用双缓冲机制,播放视频时画面更流畅。

缺点:Surface 不在 View hierachy 中,它的显示也不受 View 的属性控制,所以不能进行平移,缩放等变换,也不能放在其他 ViewGroup 中。SurfaceView 不能嵌套使用。TextureView 优点及缺点比较如表 6 - 1 所列。

表 6 - 1 SurfaceView 与 TextureView 优点及缺点比较

	SurfaceView	TextureView
内存	低	高
绘制	及时	1 - 3 帧的延迟
耗电	低	高
动画和截图	不支持	支持

(2) SurfaceView 中双缓冲

双缓冲：SurfaceView 在更新视图时用到了两张 Canvas，一张 frontCanvas 和一张 backCanvas，每次实际显示的是 frontCanvas，backCanvas 存储的是上一次更改前的视图，当使用 lockCanvas() 获取画布时，得到的实际上是 backCanvas，而不是正在显示的 frontCanvas，之后在获取到的 backCanvas 上绘制新视图，再 unlockCanvasAndPost(canvas) 此视图，那么上传的这张 Canvas 将替换原来的 frontCanvas 作为新的 frontCanvas，原来的 frontCanvas 将切换到后台作为 backCanvas。例如，如果已经先后两次绘制了视图 A 和 B，那么再调用 lockCanvas() 获取视图，获得的将是 A 而不是正在显示的 B，之后将重绘的 C 视图上传，那么 C 将取代 B 作为新的 frontCanvas 显示在 SurfaceView 上，原来的 B 则转换为 backCanvas。

(3) 播放器的选择

从性能和安全性角度出发，使用播放器优先选 SurfaceView。

1）在 Android 7.0 上系统 Surfaceview 的性能比 TextureView 更有优势，支持对象的内容位置和包含的应用内容同步更新，平移、缩放不会产生黑边。在 Android 7.0 以下系统如果使用场景有动画效果，可以选择性使用 TextureView。

2）由于失效（invalidation）和缓冲的特性，TextureView 增加了额外 1~3 帧的延迟显示画面更新。

3）TextureView 总是使用 GL 合成，而 SurfaceView 可以使用硬件 Overlay 后端，可以占用更少的内存带宽，消耗更少的能量。

4）TextureView 的内部缓冲队列导致比 SurfaceView 使用更多的内存。

5）SurfaceView：内部自己持有 Surface，Surface 创建、销毁、大小改变时系统来处理的，通过 SurfaceHolder 的 Callback 回调通知。当画布创建好时，可以将 Surface 绑定到 MediaPlayer 中。SurfaceView 如果为用户可见的时候，创建 SurfaceView 的 SurfaceHolder 用于显示视频流解析的帧图片，如果发现 SurfaceView 变为用户不可见的时候，则立即销毁 SurfaceView 的 SurfaceHolder，以达到节约系统资源的目。

6.6 导航栏及滑动界面设计实例

在本章 6.3.2 小节 "BottomNavigationView 底部导航栏" 的基础上完成导航栏及滑动界面设计实例，采用 BottomNavigationView＋ViewPager＋Fragment 的结构。

1. 创建 Fragment

增加四个 Fragment 布局，第一个 Fragment 布局文件名为 page01.xml，程序代码如下：

```
<? xml version = "1.0" encoding = "utf-8"? >
<LinearLayout xmlns:android = "http://schemas.android.com/apk/res/android"
    android:orientation = "vertical" android:layout_width = "match_parent"
```

```
        android:layout_height = "match_parent">
    <TextView
        android:text = "第 1 页"
        android:gravity = "center"
        android:layout_width = "match_parent"
        android:layout_height = "match_parent" />
</LinearLayout>
```

给四个 Fragment 布局编写实现类,第一个 Fragment 布局对应的实现类文件名为 Fragment1.java,程序代码如下:

```
public class Fragment1 extends Fragment {
    @Override
    public View onCreateView(LayoutInflater inflater, ViewGroup container,
            Bundle savedInstanceState){
        View view = inflater.inflate(R.layout.page01, container, false);
        return view;
    }
}
```

其他三个 Fragment 布局的实现类程序代码与第一个类似,直接拷贝过来修改即可。

2. 导航栏布局

导航栏布局在 6.3.2 小节《BottomNavigationVIew 底部导航栏》的基础上修改,程序代码如下:

```
<? xml version = "1.0" encoding = "utf - 8"? >
<menu xmlns:android = "http://schemas.android.com/apk/res/android">
    <item
        android:id = "@ + id/navigation_home"
        android:icon = "@drawable/message_selected"
        android:title = "消息" />
    <item
        android:id = "@ + id/navigation_dashboard"
        android:icon = "@drawable/contacts_selected"
        android:title = "联系人" />
    <item
        android:id = "@ + id/navigation_notifications"
        android:icon = "@drawable/news_selected"
        android:title = "@string/新闻" />
    <item
        android:id = "@ + id/navigation_notifications2"
```

```
            android:icon = "@drawable/setting_selected"
            android:title = "@string/设置" />
    </menu>
```

3. Activity 实现过程

在 Activity 之中,主要进行以下几个步骤:

(1) Fragment 适配器类

FragmentStatePagerAdapter 和前面的 FragmentPagerAdapter 一样,是继承自 PagerAdapter。但是,和 FragmentPagerAdapter 不一样的是,正如其类名中的"State"所表明的含义一样,该 PagerAdapter 的实现将只保留当前页面,当页面离开视线后,就会被消除,释放其资源;而在页面需要显示时,生成新的页面(就像 ListView 的实现一样)。这么实现的好处就是当拥有大量的页面时,不必在内存中占用大量的内存。

FragmentPagerAdapter 继承自 PagerAdapter。相比通用的 PagerAdapter,该类更专注于每一页均为 Fragment 的情况。该类内的每一个生成的 Fragment 都将保存在内存之中,因此适用于那些相对静态的页,数量也比较少的那种;如果需要处理有很多页,并且数据动态性较大、占用内存较多的情况,应该使用 FragmentStatePagerAdapter。FragmentPagerAdapter 重载实现了几个必须的函数,因此来自 PagerAdapter 的函数,只需要实现 getCount() 即可。由于 FragmentPagerAdapter.instantiateItem() 的实现中,调用了一个新增的虚函数 getItem(),因此,还至少需要实现一个 getItem()。总体上来说,相对于继承自 PagerAdapter,更方便一些。

本例中使用的 Fragment 适配器类程序代码如下:

```
class BottomViewAdapter extends FragmentStatePagerAdapter {
    List <Fragment> listFragment;
    public BottomViewAdapter(FragmentManager fm, List <Fragment> listFragment) {
        super(fm);
        this.listFragment = listFragment;
    }
    @Override public Fragment getItem(int position) {return listFragment.get(position);}
    @Override public int getCount() {return listFragment.size();}
}
```

(2) 构建视图

下面是进行 Fragment 的实例化,然后根据 Fragment 列表构建适配器,程序代码如下:

```
private void initView() {
    viewPager = (ViewPager)findViewById(R.id.viewPager);
    //向 ViewPager 添加各页面
    listFragment = new ArrayList <>();
```

```
        listFragment.add(new Fragment1());
        listFragment.add(new Fragment2());
        listFragment.add(new Fragment3());
        listFragment.add(new Fragment4());
        BottomViewAdapter adapter = new BottomViewAdapter(getSupportFragmentManager(),
this,listFragment);
        viewPager.setAdapter(adapter);
    }
```

(3) 事件的监听

需要进行两类事件的监听,一类是监听导航栏的选择,另一类是监听页面的滑动,程序代码如下:

```
    private void initEvent(){
        final BottomNavigationView na = (BottomNavigationView) findViewById(R.id.navigation);
        //监听导航栏的选择
        na.setOnNavigationItemSelectedListener(new BottomNavigationView
            .OnNavigationItemSelectedListener() {
          @Override
          public boolean onNavigationItemSelected(@NonNull MenuItem item) {
            switch (item.getItemId()) {
              case R.id.navigation_home:
                viewPager.setCurrentItem(0);          return true;
              case R.id.navigation_dashboard:
                viewPager.setCurrentItem(1);          return true;
              case R.id.navigation_notifications:
                viewPager.setCurrentItem(2);          return true;
              case R.id.navigation_notifications2:
                viewPager.setCurrentItem(3);          return true;
            }
            return false;
          }
        });
        //监听页面的滑动
        viewPager.addOnPageChangeListener(new ViewPager.OnPageChangeListener() {
          @Override
          public void onPageSelected(int position) {
            na.getMenu().getItem(position).setChecked(true);
            //这里使用 na.setSelectedItemId(position);无效,
          }
          @Override
          public void onPageScrolled(int position,float positionOffset,int positionOffset-
Pixels){}
```

```
        @Override
        public void onPageScrollStateChanged(int state){}
    });
}
```

(4) 构建 Activity 视图

最后一步就是采用上述程序,构建一个完整的 Activity 视图,程序代码如下:

```
protected void onCreate(Bundle savedInstanceState) {
        super.onCreate(savedInstanceState);
        setContentView(R.layout.activity_main);
        initView();//初始化视图
        initEvent();//初始事件处理
}
```

4. 完整的 Activity 实现程序

完整的 Activity 实现程序代码如下:

```
public class MainActivity extends AppCompatActivity {
    ViewPager viewPager;
    List <Fragment>listF;//存储页面对象;
    @Override
    protected void onCreate(Bundle savedInstanceState) {
        super.onCreate(savedInstanceState);
        setContentView(R.layout.activity_main);
        initView();
        initEvent();
    }
    /***实例化 ImageButton 和 ViewPager   */
    private void initView() {
        viewPager = (ViewPager)findViewById(R.id.viewPager);
        listF = new ArrayList<>();//向 ViewPager 添加各页面
        listFragment.add(new Fragment1());
        listFragment.add(new Fragment2());
        listFragment.add(new Fragment3());
        listFragment.add(new Fragment4());
        viewPager.setAdapter(BottomViewAdapter(getSupportFragmentManager(),this,listF));
    }
    /***   *监听事件   */
    private void initEvent(){
        final BottomNavigationViewna = (BottomNavigationView) findViewById(R.id.navigation);
        //监听导航栏的选择
```

```java
            na.setOnNavigationItemSelectedListener(
                new BottomNavigationView.OnNavigationItemSelectedListener() {
                @Override
                public boolean onNavigationItemSelected(@NonNull MenuItem item) {
                switch (item.getItemId()) {
                    case R.id.navigation_home:
                        viewPager.setCurrentItem(0);    return true;
                    case R.id.navigation_dashboard:
                        viewPager.setCurrentItem(1);              return true;
                    case R.id.navigation_notifications:
                        viewPager.setCurrentItem(2);              return true;
                    case R.id.navigation_notifications2:
                        viewPager.setCurrentItem(3);              return true;
                }
                    return false;
                }
            });
            //监听页面的滑动
            viewPager.addOnPageChangeListener(new ViewPager.OnPageChangeListener() {
                @Override public void onPageSelected(int position) {
                    na.getMenu().getItem(position).setChecked(true);
                }
                @Override
                public void onPageScrolled(int position,float positionOffset,int positionOffsetPixels){}
                @Override public void onPageScrollStateChanged(int state){}
            });
        }
        class BottomViewAdapter extends FragmentStatePagerAdapter {
        List<Fragment>listFragment;
        public BottomViewAdapter(FragmentManager fm, List<Fragment>listFragment) {
            super(fm);
            this.listFragment = listFragment;
        }
         @Override public Fragment getItem (int position) { return listFragment.get(position); }
            @Override public int getCount() { return listFragment.size(); }
        }
    }
```

上述导航栏及滑动界面设计实例运行效果如图 6-7 所示。

图6-7 导航栏及滑动界面设计实例运行效果图

6.7 导航栏及滑动界面的实现

导航栏及滑动界面是安卓应用程序,特别是物联网智能系统最常用和最重要的界面,因此,在 Obtain_Studio 中把该界面的基本内容包装成了一个模板,模板名称为 AndroidStudio43__导航与滑动界面模板。

下面将介绍导航栏及滑动界面的实现方法。首先,在 Obtain_Studio 中创建一个 Android 项目,选择与 Android Studio 项目相兼容的模板"\android 项目\AndroidStudio 项目\AndroidStudio43__导航与滑动界面模板"。项目名称为"IoT_IS_App_002",该项目的运行效果如本章最开始介绍的图6-1所示。

主界面带导航栏的滑动界面,布局文件是 mainfragmentpager.xml,内容如下:

```
<LinearLayout xmlns:android = "http://schemas.android.com/apk/res/android"
    android:layout_width = "match_parent"
    android:layout_height = "match_parent"
    android:orientation = "vertical" >
    <include layout = "@layout/top_bar" />
    <android.support.v4.view.ViewPager
        android:id = "@ + id/viewPager"
        android:layout_width = "match_parent"
        android:layout_height = "328dp"
        android:layout_above = "@ + id/ll_Navigation"
        android:layout_weight = "1.34"
        android:background = "@drawable/ay" >
    </android.support.v4.view.ViewPager>
```

第6章 导航栏及滑动界面设计

```xml
<LinearLayout
    android:layout_width = "match_parent"
    android:layout_height = "wrap_content"
    android:background = "@drawable/tab_bg" >
    <RelativeLayout
        android:id = "@+id/message_layout"
        android:layout_width = "0dp"
        android:layout_height = "match_parent"
        android:layout_weight = "1" >
        <LinearLayout
            android:layout_width = "match_parent"
            android:layout_height = "wrap_content"
            android:layout_centerVertical = "true"
            android:orientation = "vertical" >
            <Button
                android:id = "@+id/btnFriendList"
                android:layout_width = "wrap_content"
                android:layout_height = "wrap_content"
                android:background = "@drawable/message_unselected"
                android:layout_gravity = "center_horizontal"
                android:text = "" />
            <TextView
                android:id = "@+id/message_text"
                android:layout_width = "wrap_content"
                android:layout_height = "wrap_content"
                android:layout_gravity = "center_horizontal"
                android:text = "地点"
                android:textColor = "#82858b" />
        </LinearLayout>
    </RelativeLayout>
    <RelativeLayout
        android:id = "@+id/news_layout"
        android:layout_width = "0dp"
        android:layout_height = "match_parent"
        android:layout_weight = "1" >
        <LinearLayout
            android:layout_width = "match_parent"
            android:layout_height = "wrap_content"
            android:layout_centerVertical = "true"
            android:orientation = "vertical" >
            <Button
                android:id = "@+id/btnGroupChat"
```

```xml
            android:layout_width = "wrap_content"
            android:layout_height = "wrap_content"
            android:background = "@drawable/contacts_unselected"
            android:layout_gravity = "center_horizontal"
            android:text = "" />
        <TextView
            android:id = "@ + id/contacts_text"
            android:layout_width = "wrap_content"
            android:layout_height = "wrap_content"
            android:layout_gravity = "center_horizontal"
            android:text = "控制"
            android:textColor = "#82858b" />
    </LinearLayout>
</RelativeLayout>
<RelativeLayout
    android:id = "@ + id/news_layout"
    android:layout_width = "0dp"
    android:layout_height = "match_parent"
    android:layout_weight = "1" >
    <LinearLayout
        android:layout_width = "match_parent"
        android:layout_height = "wrap_content"
        android:layout_centerVertical = "true"
        android:orientation = "vertical" >
        <Button
            android:id = "@ + id/btnTopicList"
            android:layout_width = "wrap_content"
            android:layout_height = "wrap_content"
            android:background = "@drawable/news_unselected"
            android:layout_gravity = "center_horizontal"
            android:text = "" />
        <TextView
            android:id = "@ + id/news_text"
            android:layout_width = "wrap_content"
            android:layout_height = "wrap_content"
            android:layout_gravity = "center_horizontal"
            android:text = "动态"
            android:textColor = "#82858b" />
    </LinearLayout>
</RelativeLayout>
<RelativeLayout
    android:id = "@ + id/setting_layout"
```

```xml
            android:layout_width = "0dp"
            android:layout_height = "match_parent"
            android:layout_weight = "1" >
            <LinearLayout
                android:layout_width = "match_parent"
                android:layout_height = "wrap_content"
                android:layout_centerVertical = "true"
                android:orientation = "vertical" >
                <Button
                    android:id = "@ + id/btnSetList"
                    android:layout_width = "wrap_content"
                    android:layout_height = "wrap_content"
                    android:background = "@drawable/setting_unselected"
                    android:layout_gravity = "center_horizontal"
                    android:text = "" />
                <TextView
                    android:id = "@ + id/setting_text"
                    android:layout_width = "wrap_content"
                    android:layout_height = "wrap_content"
                    android:layout_gravity = "center_horizontal"
                    android:text = "设置"
                    android:textColor = "#82858b" />
            </LinearLayout>
        </RelativeLayout>
    </LinearLayout>
</LinearLayout>
```

主界面外层采用线性布局,内容采用相对布局。主界面的实现采用FragmentActivity类。FragmentActivity继承自Activity,用来解决Android 4.0之前没有Fragment的API,所以,在使用的时候需要导入Support包,同时继承FragmentActivity,这样在Activity中就能嵌入Fragment来实现想要的布局效果。Android 4.0之后就可以直接继承自Activity,可以在其中嵌入使用Fragment。主界面的实现程序如下:

```java
public class MainFragmentPagerActivity extends FragmentActivity
            implements View.OnClickListener {
    private boolean is_closed = false;
    private long mExitTime;
    ViewPager viewpager;
    Fragment1 f1 = null;
    Fragment2 f2 = null;
    Fragment3 f3 = null;
    Fragment4 f4 = null;
```

```java
Button[] btnArray;
TextView[] TextViewArray;
int[] BackgroundResource_unselected;
int[] BackgroundResource_selected;
MainFragmentPagerAdapter adapter;
private int currentPageIndex = 0;
private ArrayList<Fragment> datas;
@Override
protected void onCreate(Bundle arg0) {
    super.onCreate(arg0);
    setContentView(R.layout.mainfragmentpager);
    viewpager = (ViewPager)this.findViewById(R.id.viewPager);
    adapter = new MainFragmentPagerAdapter(this.getSupportFragmentManager(), null);
    viewpager.setAdapter(adapter);
    setupView();
    setListener();
    setButtonColor();
    viewpager.setCurrentItem(currentPageIndex);
}
private void setButtonColor() {
    for (int i = 0; i < btnArray.length; i++) {
        if (i == currentPageIndex) {
            btnArray[i].setBackgroundResource(BackgroundResource_selected[i]);
            TextViewArray[i].setTextColor(Color.WHITE);
        } else {
            btnArray[i].setBackgroundResource(BackgroundResource_unselected[i]);
            TextViewArray[i].setTextColor(Color.parseColor("#82858b"));
        }
    }
}
private void setListener() {
    viewpager.setOnPageChangeListener(new ViewPager.OnPageChangeListener() {
        @Override
        public void onPageScrollStateChanged(int index) {}
        @Override
        public void onPageScrolled(int arg0, float arg1, int arg2) {}
        @Override
        public void onPageSelected(int pageIndex) {
            currentPageIndex = pageIndex;
            setButtonColor();
        }
    });
```

```java
        for (Button btn : btnArray) {btn.setOnClickListener(this);}
    }
    private void setupView() {
    datas = new ArrayList <Fragment>();
    datas.add(f1 = new Fragment1());
    datas.add(f2 = new Fragment2());
    datas.add(f3 = new Fragment3());
    datas.add(f4 = new Fragment4());
    adapter.setDatas(datas);
    adapter.notifyDataSetChanged();
    btnArray = new Button[]{
        (Button)findViewById(R.id.btnFriendList),
        (Button)findViewById(R.id.btnGroupChat),
        (Button)findViewById(R.id.btnTopicList),
        (Button)findViewById(R.id.btnSetList),
    };
    BackgroundResource_unselected = new int[]{R.drawable.message_unselected,R.drawable.contacts_unselected,R.drawable.news_unselected,R.drawable.setting_unselected};
    BackgroundResource_selected = new int[]{R.drawable.message_selected,R.drawable.contacts_selected,R.drawable.news_selected,R.drawable.setting_selected};
    TextViewArray = new TextView[]{
        (TextView)findViewById(R.id.message_text),
        (TextView)findViewById(R.id.contacts_text),
        (TextView)findViewById(R.id.news_text),
        (TextView)findViewById(R.id.setting_text),
    };
    }
    @Override
    public void onClick(View v) {
      try {
        switch (v.getId()) {
          case R.id.btnFriendList: currentPageIndex = 0; break;
          case R.id.btnGroupChat: currentPageIndex = 1; break;
          case R.id.btnTopicList: currentPageIndex = 2; break;
          case R.id.btnSetList: currentPageIndex = 3; break;
        }
        viewpager.setCurrentItem(currentPageIndex);
      } catch (Exception e) {
        e.printStackTrace();
      }
    }
}
```

运行效果如图6-8所示,可以自由左右滑动切换界面,也可以通过导航栏切换界面。

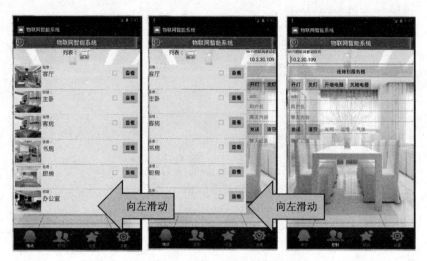

图6-8 导航栏及滑动界面运行效果图

第 7 章

Wi-Fi 物联网移动软件设计

7.1 Wi-Fi 物联网移动软件设计目标

1. Wi-Fi 物联网移动软件设计目标

本章将介绍一个基于 Wi-Fi 通信方式的物联网智能系统的实现方案与设计,主要通过手机 App 实现对物联网智能设备的控制与监测,Wi-Fi 物联网移动软件界面效果如图 7-1 所示。

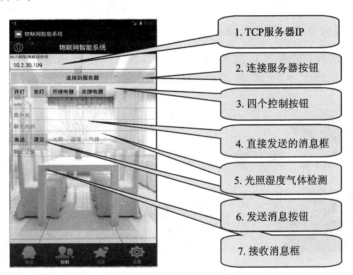

图 7-1 Wi-Fi 物联网移动软件界面效果

本章的设计目标是以 STM32F103 开发板作为物联网智能设备,STM32F103 开发板通过 ESP8266 Wi-Fi 模块连接网络。通过本章所设计的手机 App,可以控制板上的四个 LED 灯亮灭,也可以监测和显示 STM32 板上发送过来的温度、光照、气敏数据,如图 7-2 所示。

本章设计目标如下:

(1) 设计一个手机 App,通过该 App 连接 TCP 服务器,再通过 TCP 服务器连接到物联网智能设备(本章主要是针对 Wi-Fi 通信形式的智能设备)。

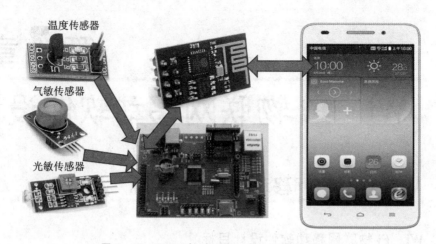

图 7-2　STM32 与 ESP8266Wi-Fi 模块的连接

（2）手机 App 可以实现 STM32 板上 LED 灯的控制，以及实现对温度、湿度和光照度的监测与显示。

（3）手机 App 控制界面建立在第 6 章"导航栏及滑动界面设计"的基础上。

2. 物联网移动软件组网结构

Wi-Fi 物联网移动软件组网结构有很多种，如果采用本地电脑作为 TCP 服务器组网模式，其结构如图 7-3 所示，可以在本地局域网内实现连接与控制。

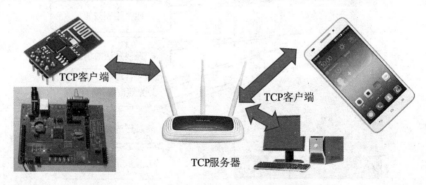

图 7-3　采用本地 TCP 服务器的物联网移动软件组网模式

如果采用 TCP 云服务器的方式，手机可以实现对物联网智能设备的远程监测和控制，Wi-Fi 物联网移动软件组网结构如图 7-4 所示。

物联网感知层常用的通信方式有 Wi-Fi、蓝牙、Zigbee、NB-IoT 和 LoRa 等，如果底层智能设备采用的是蓝牙、Zigbee、NB-IoT 和 LoRa 等通信方式，则需要专门的 Zigbee、LoRa、蓝牙网关，其组网模式如图 7-5 所示。

目前应用最广泛的物联网感知层的通信方式依然是 Wi-Fi，因为手机直接支持与 Wi-Fi 设备的通信，同时也可以直接接入 Internet 网络，是成本最低、最成熟、应用最

第 7 章 Wi-Fi 物联网移动软件设计

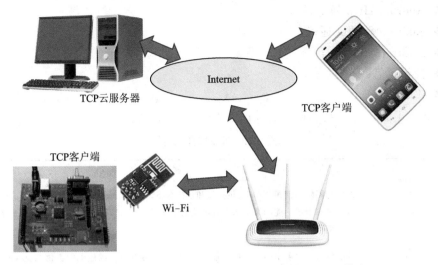

图 7-4 采用 TCP 云服务器的物联网移动软件组网模式

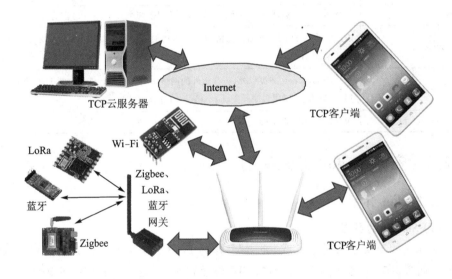

图 7-5 Wi-Fi 物联网移动软件组网结构图

广泛的物联网智能设备无线通信方式。因此,本章重点介绍 Wi-Fi 物联网移动软件设计方法,也是本书介绍的第一个物联网智能系统实际应用案例。

7.2 安卓通信程序设计

7.2.1 物联网 App 安卓端网络编程基础

安卓网络编程主要包括以下内容:

- Android Http 通信；
- Android Socket 通信；
- Android SSL 通信；
- 蓝牙通信；
- WIFI 通信。

Android SDK 中一些与网络有关的包如表 7-1 所列。

表 7-1 网络有关的包

包	描　述
java.net	提供与网络通信相关的类，包括流和数据包 Socket、Internet 协议和常见 HTTP 处理。该包是一个多功能网络资源。有经验的 Java 开发人员可以立即使用这个熟悉的包创建应用程序
java.io	虽然没有提供现实网络通信功能，但是仍然非常重要。该包中的类由其他 Java 包中提供的 Socket 和链接使用。它们还用于与本地文件的交互
java.nio	包含表示特定数据类型的缓冲区的类。适用于两个基于 Java 语言的端点之间的通信
org.apache.*	表示许多为 HTTP 通信提供精确控制和功能的包。可以将 Apache 视为流行的开源 Web 服务器
android.net	除核心 java.net.* 类以外，包含额外的网络访问 Socket。该包包括 URI 类，后者频繁用于 Android 应用程序开发，而不仅仅是传统的联网
android.net.http	包含处理 SSL 证书的类

7.2.2 安卓 Socket 通信基础

1. Socket 介绍

(1) Socket 的意思

Socket 的英文原义是"孔"或"插座"。作为 BSD UNIX 的进程通信机制，取后一种意思。通常也称作"套接字"，用于描述 IP 地址和端口，是一个通信链的句柄，可以用来实现不同虚拟机或不同计算机之间的通信。

(2) Socket 一词的起源

在组网领域的首次使用是在 1970 年 2 月 12 日发布的文献 IETF RFC33 中发现的，撰写者为 Stephen Carr、Steve Crocker 和 Vint Cerf。根据美国计算机历史博物馆的记载，Croker 写道："命名空间的元素都可称为套接字接口。一个套接字接口构成一个连接的一端，而一个连接可完全由一对套接字接口规定。"计算机历史博物馆补充道："这比 BSD 的套接字接口定义早了大约 12 年。"

(3) Socket 工作模式

Socket 起源于 Unix，而 Unix/Linux 基本哲学之一就是"一切皆文件"，都可以用"打开 Open→读写 Write/Read→关闭 Close"模式来操作。其工作模式如图 7-6

所示。

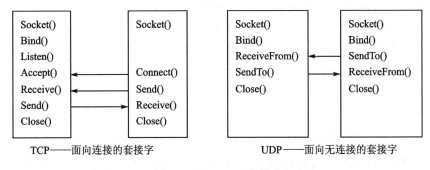

图 7-6 Socket 工作模式

2. Android Socket 通信

Socket(套接字)用于描述 IP 地址和端口。App 常常通过 Socket 向网络发出请求或者应答网络请求。Socket 是支持 TCP/IP 协议的网络通信的基本操作单元,是网络通信过程中端点的抽象表示,包含进行网络通信必需的 5 种信息:网络协议、本地 IP 地址、本地端口、远程 IP 地址、远程端口。

java.net 包中提供 Socket 和 ServerSocket 表示双向连接的 Client 和 Server。

在选择端口时必须小心。每个端口提供一种特定的服务,只有给出正确的端口,才能获得相应的服务。端口号 0~1 023 为系统保留。例如,80 是 Http 服务的,21 是 Telnet 服务的,23 是 Ftp 服务的。我们在选择端口号时,最好选择一个大于 1 023 的数,防止发生冲突。在创建 Socket 或 ServerSocket 时,如果产生错误,将会抛出 IOException。

要想在 Client 使用 Socket 来与一个 Server 通信,就必须在 Client 创建一个 Socket,指定 ServerIP 地址和端口。例如:

Socket socket = newSocket("192.164.1.110",5555);

在 Server 创建 ServerSocket,指定监听的端口。例如:

ServerSocket serverSocket = newServerSocket(5555);

实际应用中 ServerSocket 总是不停地循环调用 accept()方法,一旦收到请求就会创建线程来处理和响应。accept()是一个阻塞方法,接收到请求后会返回一个 Socket 来与 Client 进行通信。

Socket 提供了 getInputStream()和 getOutputStream()用来得到输入流和输出流进行读写操作,这两个方法分别返回 InputStream 和 OutputStream。为了方便读写,我们常常在 InputStream 和 OutputStream 基础上进行包装得到 DataInputStream、DataOutputStream、PrintStream、InputStreamReader、OutputStreamWriter、printWriter 等。示例代码如下:

```
PrintStream printStream = new PrintStream(
new BufferedOutputStream(socket.getOutputStream()));
PrintWriter printWriter = new PrintWriter(new BufferedWriter(
new OutputStreamWriter(socket.getOutputStream(), true)));
printWriter.println(String msg);
DataInputStream dis = new DataInputStream(socket.getInputStream());
BufferedReader br =   new BufferedReader(
new InputStreamReader(socket.getInputStream()));
String line = br.readLine();
```

在关闭 Socket 之前,应将与其有关的 Stream 全部关闭,以释放所有的资源。

3. Android 套接字通信实例

(1) 服务器程序

服务器程序需要在 PC 上运行,该程序比较简单,因此,不需要建立 Android 项目,直接定义一个 Java 类,并且运行该类即可。它仅仅建立 ServerSocket 监听,并使用 Socket 获取输入输出流。服务器程序代码如下,这里的服务器程序,也可以用 Windows 下的 VC++写的通信服务器程序代替。

```java
import java.io.IOException;
import java.io.OutputStream;
import java.net.ServerSocket;
import java.net.Socket;
public class SimpleServer {
  public static void main(String[] args) throws IOException {
    //创建一个 ServerSocket,用于监听客户端 Socket 的连接请求
    ServerSocket ss = new ServerSocket(30000);
    //采用循环不断接收来自客户端的请求,服务器端也对应产生一个 Socket
    while(true){
      Socket s = ss.accept();
      OutputStream os = s.getOutputStream();
      os.write("收到了服务器消息! n".getBytes("utf-8"));
      os.close();
      s.close();
    }
}}
```

(2) 客户端程序

客户端程序运行在 Android 操作系统上。对于复杂的安卓应用程序,只要把下面代码中的通信部分拿出来,放到相应的通信功能实现之处即可。客户端程序代码如下:

```java
package my.learn.tcp;
import java.io.BufferedReader;
import java.io.IOException;
import java.io.InputStreamReader;
import java.net.Socket;
import java.net.UnknownHostException;
import android.app.Activity;
import android.os.Bundle;
import android.util.Log;
import android.widget.EditText;
public class SimpleClient extends Activity {
    private EditText show;
    @Override
    protected void onCreate(Bundle savedInstanceState) {
        super.onCreate(savedInstanceState);
        setContentView(R.layout.main);
        show = (EditText) findViewById(R.id.show);
        try {
            Socket socket = new Socket("自己计算机的 IP 地址", 30000);
            //设置 10 s 之后即认为是超时
            socket.setSoTimeout(10000);
            BufferedReader br = new BufferedReader(new InputStreamReader(
                socket.getInputStream()));
            String line = br.readLine();
            show.setText("来自服务器的数据:" + line);
            br.close();
            socket.close();
        } catch (UnknownHostException e) {
            // TODO Auto-generated catch block
            Log.e("UnknownHost","来自服务器的数据");
            e.printStackTrace();
        } catch (IOException e) {
            Log.e("IOException","来自服务器的数据");
            // TODO Auto-generated catch block
            e.printStackTrace();
        }
    }
}
```

另外,在 Manifest.xml 文件当中,需要对互联网的访问进行授权:
< uses - permission android:name="android.permission.INTERNET"/>

7.3 Wi-Fi 通信概要

7.3.1 WLAN 通信

WLAN（Wireless Local Area Networks）意思是无线局域网络，是一种数据传输系统。它利用射频（RF）技术进行数据的传输，实现无网线、无距离限制的通畅网络。WLAN 使用 ISM（Industrial、Scientific、Medical）无线电广播频段通信。WLAN 的 802.11a 标准使用 5 GHz 频段，支持的最大速度为 54 Mbps，而 802.11b 和 802.11g 标准使用 2.4 GHz 频段，分别支持最大 11 Mbps 和 54 Mbps 的速度。目前 WLAN 常用的协议标准有：IEEE 802.11b 协议、IEEE 802.11a 协议、IEEE 802.11g 协议、IEEE 802.11E 协议、IEEE 802.11i 协议、无线应用协议（WAP）。

无线局域网第一个版本发表于 1997 年，其中定义了介质访问接入控制层（MAC 层）和物理层。物理层定义了工作在 2.4 GHz 的 ISM 频段上的两种无线调频方式和一种红外传输的方式，总数据传输速率设计为 2 Mbps。两个设备之间的通信可以自由直接（ad hoc）的方式进行，也可以在基站（Base Station，BS）或者访问点（Access Point，AP）的协调下进行。

1999 年，加上了两个补充版本：802.11a 定义了一个在 5 GHz 的 ISM 频段上的数据传输速率可达 54 Mbps 的物理层，802.11b 定义了一个在 2.4 GHz 的 ISM 频段上但数据传输速率高达 11 Mbps 的物理层。2.4 GHz 的 ISM 频段为世界上绝大多数国家通用，因此，802.11b 得到了最为广泛的应用。苹果公司把自己开发的 802.11 标准起名叫 AirPort。1999 年工业界成立了 Wi-Fi 联盟，致力解决符合 802.11 标准的产品的生产和设备兼容性问题。802.11 标准和补充标准如下：

802.11，1997 年，原始标准（2 Mbps 工作在 2.4 GHz）。

802.11a，1999 年，物理层补充（54 Mbps 工作在 5 GHz）。

802.11b，1999 年，物理层补充（11 Mbps 工作在 2.4 GHz）。

802.11c，符合 802.1D 的媒体接入控制层（MAC）桥接（MAC Layer Bridging）。

802.11d，根据各国无线电规定做的调整。

802.11e，对服务等级（Quality of Service，QoS）的支持。

802.11f，基站的互连性（Interoperability）。

802.11g，物理层补充（54 Mbps 工作在 2.4 GHz）。

802.11h，无线覆盖半径的调整，室内（Indoor）和室外（Outdoor）信道（5 GHz 频段）。

802.11i，安全和鉴权（Authentification）方面的补充。

802.11n，导入多重输入输出（MIMO）和 40 Mbps 通道宽度（HT40）技术，基本上是 802.11a/g 的延伸版。

802.11n 主要是结合物理层和 MAC 层的优化来充分提高 WLAN 技术的吞吐。主要的物理层技术涉及了 MIMO、MIMO-OFDM、40 MHz、Short GI 等技术，从而将

第7章 Wi-Fi物联网移动软件设计

物理层吞吐提高到600 Mbps。如果仅仅提高物理层的速率，而没有对空口访问等MAC协议层的优化，802.11n的物理层优化将无从发挥，就好比即使建了很宽的马路，但是车流的调度管理如果跟不上，仍然会出现拥堵和低效。所以，802.11n对MAC采用了Block确认、帧聚合等技术，大大提高MAC层的效率。

IEEE 802.11ac，俗称5GWi-Fi，是一个802.11无线局域网（WLAN）通信标准，它通过5 GHz频带（也是其得名原因）进行通信。理论上，它能够提供最少1 Gbps带宽进行多站式无线局域网通信，或是最少500 Mbps的单一连接传输带宽。802.11ac是802.11n的继承者。它采用并扩展了源自802.11n的空中接口（Air Interface）概念，包括：更宽的RF带宽（提升至160 MHz），更多的MIMO空间流（Spatial Streams）（增加到8），多用户的MIMO，以及高密度的调变（Modulation）（达到256 QAM）。

802.11ad主要用于实现家庭内部无线高清音视频信号的传输，为家庭多媒体应用带来更完备的高清视频解决方案。802.11ad抛弃了拥挤的2.4 GHz和5 GHz频段，而是使用高频载波的60 GHz频谱。由于60 GHz频谱在大多数国家有大段的频率可供使用，因此，802.11ad可以在MIMO技术的支持下实现多信道的同时传输，而每个信道的传输带宽都将超过1 Gbps。据了解802.11ad，载频60 GHz，速率是7 Gbps。同时，802.11ad也面临技术上的限制。比如：60 GHz载波的穿透力很差，而且在空气中信号衰减很厉害，其传输距离、信号覆盖范围都大受到影响，这使得它的有效连接只能局限在一个很小的范围内。在理想的状态下，802.11ad最适合被用来作为房间内各个设备之间高速无线传输的通道。

7.3.2 Wi-Fi通信

Wi-Fi是IEEE定义的无线网技术，在1999年IEEE官方定义802.11标准的时候，IEEE选择并认定了CSIRO发明的无线网技术是世界上最好的无线网技术，因此，CSIRO的无线网技术标准，就成为现在Wi-Fi的核心技术标准。

Wi-Fi技术由澳洲政府的研究机构CSIRO在20世纪90年代发明并于1996年在美国成功申请了无线网技术专利。发明人是悉尼大学工程系毕业生Dr John O'Sullivan领导的一群由悉尼大学工程系毕业生组成的研究小组。IEEE曾请求澳洲政府放弃其Wi-Fi专利让世界免费使用Wi-Fi技术，但遭到拒绝。

澳洲政府随后在美国通过官司胜诉或庭外和解，收取了世界上几乎所有电器电信公司（包括苹果、英特尔、联想、戴尔、AT&T、索尼、东芝、微软、宏碁、华硕等）的专利使用费。现在我们每购买一台含有Wi-Fi技术的电子设备的时候，我们所付的价钱就包含了交给澳洲政府的Wi-Fi专利使用费。现在全球每天估计会有30亿台电子设备使用Wi-Fi技术，而到2013年底CSIRO的无线网专利过期之后，这个数字预计会增加到50亿。

Wi-Fi被澳洲媒体誉为澳洲有史以来最重要的科技发明，其发明人John O'Sullivan被澳洲媒体称为"Wi-Fi之父"，并获得了澳洲的国家最高科学奖和全世界的众多赞誉，其中包括最近欧盟机构——欧洲专利局（EPO, European Patent Office），颁发的

European Inventor Award 2012(2012年欧洲发明者大奖)。

7.3.3 ESP8266模块的应用

ESP8266是一个完整且自成体系的Wi-Fi网络解决方案,能够搭载软件应用,或通过另一个应用处理器卸载所有Wi-Fi网络功能。

ESP8266在搭载应用并作为设备中唯一的应用处理器时,能够直接从外接闪存中启动。内置的高速缓冲存储器有利于提高系统性能,并减少内存需求。

另外一种情况是,无线上网接入承担Wi-Fi适配器的任务时,可以将其添加到任何基于微控制器的设计中,连接简单易行,只需通过SPI/SDIO接口或中央处理器AHB桥接口即可。

ESP8266强大的片上处理和存储能力,使其可通过GPIO口集成传感器及其他应用的特定设备,实现了最低前期的开发和运行中最少地占用系统资源。ESP8266高度片内集成,包括天线开关balun、电源管理转换器,因此,仅需极少的外部电路,且包括前端模块在内的整个解决方案在设计时将所占PCB空间降到最低。

装有ESP8266的系统表现出来的领先特征有:节能VoIP在睡眠/唤醒模式之间的快速切换、配合低功率操作的自适应无线电偏置、前端信号的处理功能、故障排除和无线电系统共存特性为消除蜂窝/蓝牙/DDR/LVDS/LCD干扰。

ESP8266模块特征如下:
- 802.11b/g/n;
- Wi-FiDirect(P2P)、soft-AP;
- 内置TCP/IP协议栈;
- 内置TR开关、balun、LNA、功率放大器和匹配网络;
- 内置PLL、稳压器和电源管理组件;
- 802.11b模式下+17.5 dBm的输出功率;
- 内置温度传感器;
- 支持天线分集;
- 断电泄露电流小于10 μA;
- 内置低功率32位CPU:可以兼作应用处理器;
- SDIO2.0、SPI、UART;
- STBC、1x1MIMO、2x1MIMO;
- MPDU、A-MSDU的聚合和0.4 s的保护间隔;
- 2 ms之内唤醒、连接并传递数据包;
- 待机状态消耗功率小于1.0 mW(DTIM3)。

ESP8266常用AT指令:

AT+GMR	查看版本号
AT+RST	重启
AT+CWMODE=3	设置工作模式(1Station模式,2AP模式,3AP+Station模式)

AT+CWMODE?　　　　查询工作模式
AT+CWLIF　　　　　查看接入的客户端 IP
AT+CWLAP　　　　 返回 AP 列表
AT+CWJAP="ssid","87654321" 加入 AP(第一个参数是用户名,第二个参数是密码)
AT+CIFSR　　　　　获取本模块 IP
AT+CIPSTART="TCP","121.7.25.1",9800 　连接服务器
AT+CIPMODE?　　　查询模块传输模式　(1 透传,0 不透传)
AT+CIPMODE=1　　 设置模块传输模式
AT+CIPSEND 进入透传模式,每次重启后需要重新执行一条连接服务器指令

//服务器模式
AT+CIPMUX=1　　　　　　　0 单路连接模式,1 多路连接模式
AT+CIPMUX?　　　　　　　 查询连接模式
AT+CIPSERVER=1,8888　　　配置为服务器(第二个参数是服务器端口号)
AT+ CIPSTATUS:返回当前模块的连接状态和连接参数
STATUS：<stat>＋ CIPSTATUS：<id>，<type>，<addr>，<port>，<tetype>OK
参数说明：
> <id>连接的 id 号 0～4　<type>字符串参数,类型 TCP 或 UDP；
> <addr>字符串参数,IP 地址；
> <port>端口号；
> <tetype>0:本模块做 Client 的连接,1:本模块做 Server 的连接。

7.3.4　Smartconfig

App 需要配置 Wi-Fi 用户名和密码进入智能硬件,目前各个 Wi-Fi 芯片厂家基本采用以下几种方式。通常情况下采用 App 配置,手机连接到智能硬件(Wi-Fi 芯片的 AP),构建成一个局域网,当然该局域网是不能上网的。此时,该局域网内一般有三个设备手机——智能硬件 STATION 模式——智能硬件的 AP。

Smartconfig 也就是所谓的一键配置,速度比 AP 模式快。Smartconfig 就是手机 App 端发送包含 Wi-Fi 用户名和密码的 UDP 广播包或者组播包,智能终端的 Wi-Fi 芯片可以接收到该 UDP 包,只要知道 UDP 的组织形式,就可以通过接收到的 UDP 包解密出 Wi-Fi 用户名和密码,然后智能硬件配置收到的 Wi-Fi 用户名和密码到指定的 Wi-Fi AP 上。

AP 模式比较不好的就是配置时间比 Smartconfig 要长,手机连接的 Wi-Fi 接入点会变化,如果配置不成功,手机就无法上网,需要在手动配置 Wi-Fi,手机才能上网。

QCA4004 芯片和 ESP8266 芯片 SDK 提供了 Smartconfig 和 AP 接入两种连接方式。

1. Smartconfig 方式

Smartconfig 采用 UDP 广播模式（UDP 接收 IP 地址是 255.255.255.255）。ESP8266 先扫描 AP，得到 AP 的相关信息，如工作的 Channel13，然后配置 Wi-Fi 芯片工作在刚才扫描到的 Channel13 上去接收 UDP 包，如果没有接收到，继续配置 ESP8266 工作在另外的 Channel13 上，如此循环，直到收到 UDP 包为止，为什么要提前扫描 Wi-Fi AP 呢？就是为了提高配置效率。假设当前网络中只有 AP1、AP2 这两个 AP，AP1 工作在 Channel1 上，AP2 工作在 Channel13 上，现在需要配置智能硬件连接到 AP2，也就是要让智能硬件工作在 Channel13 上，如果不提前进行扫描，就需要从 Channel1-13 逐个进行扫描，非常浪费时间。如果提前扫描了 AP，芯片将记录提前扫描的结果，在使用的过程中就可以快速从 AP1 和 AP2 获取 UDP 包。

2. AP 接入方式

QCA4004 芯片：Kuaifi 连接也就是 Smartconfig 采用 UDP 组播模式（IP 地址是组播即可，如 239.0.0.254）。

UDP 广播模式跟组播的差异就不用说了，这是基本的 TCP/IP 常识，如果从安全角度看 UDP 组播会安全点，当然目前这方面还不需要考虑。

QCA 4004 跟 ESP8266 不同，它不进行前期 AP 扫描，直接从 1～2～13 循环配置去接收 UDP 包，并且 UDP 包采用组播。

7.4 安卓 TCP 客户端程序实例

1. 安卓 TCP 客户端程序界面

安卓 TCP 客户端设计一个简单的界面，该界面只包含一个发送数据按钮，完整的程序代码如下：

```xml
<?xml version="1.0" encoding="utf-8"?>
<LinearLayout xmlns:android="http://schemas.android.com/apk/res/android"
    android:orientation="vertical"
    android:layout_width="fill_parent"
    android:layout_height="fill_parent"
    >
    <EditText android:id="@+id/showall"
        android:layout_width="fill_parent"
        android:layout_height="380px"
        android:background="#FFFFFF"
        android:paddingTop="0px"
        android:gravity="top"
        android:textColor="#000000"/>
```

```
    <LinearLayout
  android:orientation = "horizontal"
  android:gravity = "center_vertical"
  android:layout_width = "fill_parent"
  android:layout_height = "fill_parent"
  >
    <EditText android:id = "@ + id/input"
      android:layout_width = "260px"
      android:layout_height = "wrap_content"
      android:textColor = "#000000"/>
    <Button android:id = "@ + id/send"
      android:layout_width = "fill_parent"
      android:layout_height = "wrap_content"
      android:text = "发送"/>
    </LinearLayout>
</LinearLayout>
```

2. 安卓 TCP 客户端 Java 程序

安卓 TCP 客户端 Java 程序中创建 Socket 对象进行数据的发送,完整的程序代码如下:

```java
public class ChatClientActivity extends Activity {
EditText showAll,input;
Button send;
String serverIp = "192.167.2.104";
int port = 6806;
Socket socket = null;
OutputStream os;
BufferedWriter bw;
InputStream is;
BufferedReader br;
String Msg = "";
public void onCreate(Bundle savedInstanceState) {
  super.onCreate(savedInstanceState);
  setContentView(R.layout.tcpactivity);
  showAll = (EditText) findViewById(R.id.showall);
  showAll.setCursorVisible(false);
  showAll.setFocusable(false);
  input = (EditText) findViewById(R.id.input);
  send = (Button) findViewById(R.id.send);
  send.setOnClickListener(new OnClickListener() {
   public void onClick(View v) {
```

```java
        sendMsg(input.getText().toString() + "\n");
      }
    });
    showAlert();
}
MyHandler handler = new MyHandler();
class MyHandler extends Handler {
    public void handleMessage(Message msg) {
     super.handleMessage(msg);
     showAll.append(Msg);
    }
}
public void showAlert() {
    LinearLayout ll = new LinearLayout(ChatClientActivity.this);
    ll.setOrientation(LinearLayout.HORIZONTAL);
    TextView tv = new TextView(ChatClientActivity.this);
    tv.setText("请输入昵称:");
    final EditText et = new EditText(ChatClientActivity.this);
    et.setWidth(160);
    ll.addView(tv);
    ll.addView(et);
    AlertDialog alertDialog = new AlertDialog.Builder(
       ChatClientActivity.this)
    // 设置标题
      .setTitle("聊天设置")
      // 设置按钮监听器
      .setPositiveButton("确定", new DialogInterface.OnClickListener() {
       public void onClick(DialogInterface dialogInterface, int i) {
         sendMsg("start:" + et.getText().toString() + "\n");
       }
       // 创建对话框
      }).create();
    alertDialog.setView(ll);
    alertDialog.show();
}

public void sendMsg(String msg) {
    try {
     if (socket == null) {
       socket = new Socket(serverIp, port);
       is = socket.getInputStream();
       os = socket.getOutputStream();
```

```
        bw = new BufferedWriter(new OutputStreamWriter(os));
        new Thread() {
         public void run() {
           while (true) {
            try {
              byte[] b = new byte[is.available()];
              if (is.read(b) != -1) {
                if (b.length != 0) {
                  Msg = new String(b, "utf-8") + "\n";
                  handler.sendEmptyMessage(0);
                }
              }
            } catch (IOException e) {
              e.printStackTrace();
            }
           }
         }
        }.start();
      }
    } catch (UnknownHostException e) {
      e.printStackTrace();
    } catch (IOException e) {
      e.printStackTrace();
    }
    try {
      bw.write(msg);
      bw.flush();
    } catch (IOException e) {
      e.printStackTrace();
    }
  }
}
```

7.5 Wi-Fi 物联网移动软件的实现

  Wi-Fi 物联网移动软件设计目标是设计一个手机 App,通过该 App 连接 TCP 服务器,再通过 TCP 服务器连接到物联网智能设备,本章主要是针对 Wi-Fi 通信形式的智能设备。

  本小节介绍的实例,物联网智能设备采用 STM32F103 开发板,通过 ESP8266 Wi-Fi 模块连接网络。通过本章所设计的手机 App,可以控制板上的四个 LED 灯亮灭,也可以监测和显示 STM32 板上送过来的温度、光照、气敏数据。STM32 与 ESP8266 模块的连接如图 7-7 所示,可以选择 STM32 板上的串口 1/2/3 等进行连接。

图 7-7　STM32 与 ESP8266 模块的连接

STM32F103 开发板的程序采用 Obtain_Studio 自带的"STM32F103ZET6_OS 模板",该模板里已经包含本实例的所有程序代码。采用 main9.cpp 程序,程序代码如下:

```
#include "main.h"
void task0(void);//任务 0
void task1(void);//任务 1
void task2(void);//任务 0
void task3(void);//任务 0
void processing(string str);
void setup()
{
    add(task0);
    add(task1);
    add(task2);
    add(task3);
}
void  task0(void )//任务 0
{
    while(1)
    {
        delay(900);
    }
}
void  task1(void ) //任务 1
```

```c
{
    CKey key1(KEY1);
    if(key1.isDown())
    {
        led1 = 0;
        wifi.TCP_Server_init("Obtain_Net_Init","012345678","5000",
            "1","0001","123456");
        wifi.TCP_Server_start(processing);
    }
    else
    {
        led1 = 1;
        wifi.TCP_Client_init("9999","9999","103.204.177.76","5000",
            "1","0001","123456");
        wifi.TCP_Client_start(processing);
    }
    while(1)
    {
        wifi.send("adc1",adc1.getValue());
        sleep(55000);
    }
}
#include "remote_control.h"
void  task2(void ) //任务2
{
    remote_control_init();//红外遥控学习初始化
    while(1)
    {
        wifi.Processing();
        sleep(100);
    }
}
void  task3(void ) //任务2
{
    while(1)
    {
        led4.isOn()? led4.Off():led4.On();
        sleep(1600);
    }
}
const char * text =
    #include "webview.html.txt"
```

```
    ;
    string m_str = string(text);
    void getControlFile()
    {
        wifi.send(m_str);
    }
    void processing(string str)
    {
        string name = wifi.get_name(str);
        string idata = wifi.get_idata(str);
        unsigned int index = str2int(idata.c_str());
        if(name == "led1"&&idata == "0"){led1.On(); }//wifi.send("led1_s",1);
        if(name == "led1"&&idata == "1"){led1.Off();} //wifi.send("led1_s",0);
        if(name == "led2"&&idata == "0"){led2.On(); } //wifi.send("led2_s",1);
        if(name == "led2"&&idata == "1"){led2.Off();}//wifi.send("led2_s",0);
        if(name == "remote_control"){remote_control(index);}//红外遥控
        if(name == "remote_control_study"){remote_control_study(index);}//红外遥控学习

        if(name == "getControlFile"){getControlFile();}//读取控制界面文件
    }
    }
```

本小节介绍的实例，TCP 服务器采用 Obtain_Studio 自带的"wxWidgets_TCP 服务器模板"，该模板的 TCP 服务器程序采用 wxWidgets 编写，可以在 Windows、Linux 等平台上运行。对于运行的硬件环境，可以运行于本地电脑上，也可以运行于云服务器上，还可以运行在树莓派、香蕉派、香橙派等各种开发板上。TCP 服务器程序运行界面如图 7-8 所示。

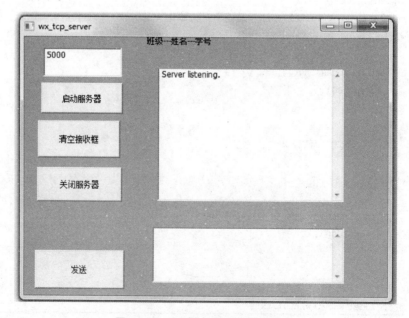

图 7-8　TCP 服务器程序运行效果

第7章 Wi-Fi物联网移动软件设计

手机App控制界面建立在第6章"导航栏及滑动界面设计"的基础上，App可以实现灯的控制，以及实现对温度、湿度和光照度的监测与显示。运行效果如本章开始所述的图7-1所示。在第6章"导航栏及滑动界面设计"里，在第二个界面"控制界面"里已经包括这些功能，对应的布局文件是f2.xml，对应的实现程序是Fragment2.java。

f2.xml布局代码如下：

```xml
<?xml version="1.0" encoding="utf-8"?>
<LinearLayout xmlns:android="http://schemas.android.com/apk/res/android"
    android:orientation="vertical"
    android:layout_width="fill_parent"
    android:layout_height="fill_parent"
    >
<TextView
    android:layout_width="fill_parent"
    android:layout_height="wrap_content"
    android:text="Wi-Fi物联网移动软件"
    />
 <EditText
    android:layout_width="fill_parent"
    android:layout_height="wrap_content"
    android:text="10.2.30.109"
    android:id="@+id/edit0"
    />
 <Button
    android:layout_width="fill_parent"
    android:layout_height="wrap_content"
    android:text="连接到服务器"
    android:id="@+id/button1"
    />
    <LinearLayout
    android:layout_width="fill_parent"
    android:layout_height="wrap_content"
    android:orientation="horizontal"
    >
        <Button android:id="@+id/Led1On"
            android:layout_height="wrap_content"
            android:layout_width="wrap_content"
            android:text="开灯">
        </Button>
        <Button android:id="@+id/Led1Off"
            android:layout_height="wrap_content"
            android:layout_width="wrap_content"
```

```xml
            android:text = "关灯">
        </Button>
        <Button android:id = "@ + id/Led2On"
            android:layout_height = "wrap_content"
            android:layout_width = "wrap_content"
            android:text = "开继电器">
        </Button>
        <Button android:id = "@ + id/Led2Off"
            android:layout_height = "wrap_content"
            android:layout_width = "wrap_content"
            android:text = "关继电器">
        </Button>
</LinearLayout>
<EditText
android:layout_width = "fill_parent"
android:layout_height = "wrap_content"
android:hint = "adc"
android:id = "@ + id/edit1"
/>
<EditText
android:layout_width = "fill_parent"
android:layout_height = "wrap_content"
android:hint = "用户名"
android:id = "@ + id/edit1"
/>
    <EditText
android:layout_width = "fill_parent"
android:layout_height = "wrap_content"
android:hint = "聊天内容"
android:id = "@ + id/edit2"
/>
    <LinearLayout
        android:layout_width = "fill_parent"
        android:layout_height = "wrap_content"
        android:orientation = "horizontal"
        >
    <Button
android:layout_width = "wrap_content"
android:layout_height = "wrap_content"
android:text = "发送"
android:id = "@ + id/button2"
/>
```

```xml
    <Button
    android:layout_width = "wrap_content"
    android:layout_height = "wrap_content"
    android:text = "清空"
    android:id = "@ + id/button5_1"
    />
    <EditText
    android:layout_width = "wrap_content"
    android:layout_height = "wrap_content"
    android:hint = "光照"
    android:id = "@ + id/edit5_1"
    />
    <EditText
    android:layout_width = "wrap_content"
    android:layout_height = "wrap_content"
    android:hint = "温度"
    android:id = "@ + id/edit5_2"
    />
    <EditText
    android:layout_width = "wrap_content"
    android:layout_height = "wrap_content"
    android:hint = "气体"
    android:id = "@ + id/edit5_3"
    />
    </LinearLayout>
    <EditText
    android:layout_width = "fill_parent"
    android:layout_height = "wrap_content"
    android:hint = "聊天记录"
    android:id = "@ + id/edit3"
    />
</LinearLayout>
```

Fragment2.java 程序代码如下:

```java
public String getLocalIpAddress() {
    try {
        for (Enumeration <NetworkInterface>en = NetworkInterface
        .getNetworkInterfaces(); en.hasMoreElements();) {
            NetworkInterface intf = en.nextElement();
            for (Enumeration <InetAddress>enumIpAddr = intf
            .getInetAddresses(); enumIpAddr.hasMoreElements();) {
                InetAddress inetAddress = enumIpAddr.nextElement();
```

```java
      if (! inetAddress.isLoopbackAddress()) {
        return inetAddress.getHostAddress().toString();
      } } }
    } catch (SocketException ex) {
      Log.e("WifiPreference IpAddress", ex.toString());
    }
    return "null";
  }
    public Socket s;
    public EditText edit0,edit1,edit2,edit3;
    public void tct_close()
    {
      try {
      s.close();
     Toast.makeText(getActivity(),"关闭 TCP", Toast.LENGTH_SHORT).show() ;
        } catch (IOException e) {
     Toast.makeText(getActivity(),"关闭 TCP 不成功", Toast.LENGTH_SHORT).show() ;
      e.printStackTrace();
        }
    }
    public class myThread extends Thread {
       String str1 = "";
       EditText edit3_1;
       public void run(){
    String str;
    try {
        //编写线程的代码
BufferedReader input = new BufferedReader(new
InputStreamReader(s.getInputStream(),"GBK"));
        edit3_1 = (EditText)view.findViewById(R.id.edit3);
        while(true)
        {
          String message = input.readLine();
          Log.d("Tcp Demo","message From Server:" + message);
          str1 = edit3_1.getText().toString() + "\r\n" + message;
          // 必须使用 post 方法更新 UI 组件
          edit3_1.post(new Runnable()
          {
       @Override
       public void run()
       {
          edit3_1.setText(str1);
```

```
            }
        });
        //({success:'[{"id":"1","deviceID":"0001","name":"adc1","idata":
"0","strdata":""}]'})
        str = message;    //str = "({success:'[{\"id\":\"1\",\"deviceID\":\"0001\",\"name\":
\"adc1\",\"idata\":\"099\",\"strdata\":\"\"}]'})";
        int i = str.indexOf("\"name\":\"adc1\"",0);
        if(i! = -1)         {
            i = str.indexOf("idata\":\"",i);
            int j = str.indexOf("\",\"strdata",i);
        if(i! = -1&&j > i + 8)      {
         final String str2 = str.substring(i + 8,j);
        // 必须使用post方法更新UI组件
        final EditText   edit5_1 = (EditText)view.findViewById(R.id.edit5_1);
        edit5_1.post(new Runnable()
        {
            @Override
            public void run()
            {
                edit5_1.setText(str2);
            }
        });
        }  }
        }
    } catch (IOException e) {
    Toast.makeText(getActivity(),"接收 TCP 不成功",Toast.LENGTH_SHORT).show();
        e.printStackTrace();
    } }
}
    public void senddata(String str)
    {
        try {
            //传出流重定向到套接字
            OutputStream out = s.getOutputStream();
            //第二个参数为 true 将会自动 flush,否则需要手动操作 out.flush()
            PrintWriter output = new PrintWriter(out, true);
            out.write(str.getBytes("GBK"));
            out.flush();
        } catch (UnknownHostException e) {
            Toast.makeText(getActivity(),"提交不成功",Toast.LENGTH_SHORT).show();
            e.printStackTrace();
        } catch (IOException e) {
```

```java
        Toast.makeText(getActivity(),"提交不成功",Toast.LENGTH_SHORT).show();
        e.printStackTrace();
    }
}
Handler handler = new Handler();
Runnable runnable = new Runnable() {
    @Override
    public void run() {
        try {
            handler.postDelayed(this, 10000);
            senddata("~");//发送心跳数据
        } catch (Exception e) {
            e.printStackTrace();
            System.out.println("exception...");
        }
    }
};
public void tcp_init()
{
    try {   //创建TCP通信Socket
        InetAddress serverAddr = InetAddress.getByName(edit0.getText().toString());
        s = new Socket(serverAddr, 5000);
        myThread newthread = new myThread();//创建接收线程
        newthread.start();
        handler.postDelayed(runnable, 10000); //定时器
    } catch (IOException e) {
        Toast.makeText(getActivity(),"初始化不成功",Toast.LENGTH_SHORT).show();
        e.printStackTrace();
    }
}
public void init()
{
    edit0 = (EditText)view.findViewById(R.id.edit0);
    edit1 = (EditText)view.findViewById(R.id.edit1);
    edit2 = (EditText)view.findViewById(R.id.edit2);
    //edit3 = (EditText)view.findViewById(R.id.edit3);
    Button button1 = (Button)view.findViewById(R.id.button1);
    button1.setOnClickListener(new View.OnClickListener(){
        // @Override
        public void onClick(View v)
        {
            Toast.makeText(getActivity(),"连接到服务器",Toast.LENGTH_SHORT).show();
```

```
        //初始化TCP通信
        tcp_init();
        //tcp_init_timer();
    }
});
    Button button = (Button)view.findViewById(R.id.button2);
    button.setOnClickListener(new View.OnClickListener(){
// @Override
    public void onClick(View v)
    {
      Toast.makeText(getActivity(),"开始提交",Toast.LENGTH_SHORT).show() ;
      String str = "用户";
      str + = edit1.getText() + "发言:\r\n" + edit2.getText() + "\r\n";
      senddata(str);
    }
    });
    Button Led1On = (Button)view.findViewById(R.id.Led1On);
    Led1On.setOnClickListener(new View.OnClickListener(){
// @Override
    public void onClick(View v)
    {
      senddata("({success:'[{\"id\":\"1\",\"deviceID\":\"0001\",\"name\":
\"led1\",\"idata\":\"0\",\"strdata\":\"\"}]'})");
    }
    });
    Button Led1Off = (Button)view.findViewById(R.id.Led1Off);
    Led1Off.setOnClickListener(new View.OnClickListener(){
// @Override
    public void onClick(View v)
    {
      senddata("({success:'[{\"id\":\"1\",\"deviceID\":\"0001\",\"name\":\
"led1\",\"idata\":\"1\",\"strdata\":\"\"}]'})");
    }
    });
    Button Led2On = (Button)view.findViewById(R.id.Led2On);
    Led2On.setOnClickListener(new View.OnClickListener(){
// @Override
    public void onClick(View v)
    {
      senddata("d");
    }
    });
```

```java
      Button Led2Off = (Button)view.findViewById(R.id.Led2Off);
      Led2Off.setOnClickListener(new View.OnClickListener(){
    // @Override
    public void onClick(View v)
    {
       senddata("c");
    }});
      Button edit_clear = (Button)view.findViewById(R.id.button5_1);
       final EditText edit3_1;
       edit3_1 = (EditText)view.findViewById(R.id.edit3);
      edit_clear.setOnClickListener(new View.OnClickListener(){
    // @Override
    public void onClick(View v)
    {
       // 必须使用 post 方法更新 UI 组件
       edit3_1.post(new Runnable()
        {
       @Override
       public void run()
       {
         edit3_1.setText("");
       }
        });
    }});
   }
    String get_data(String src)
  {
    String idata = "";
    String str_f1 = "{{{{<<<<}}}}";
    int a = src.indexOf(str_f1,0);
    if(a>-1)
    {
      a+ = str_f1.length();
      int b = src.indexOf("{{{{>>>>}}}}",a);
      if(b>a) {
          idata = src.substring(a,b);
      }  }
    return idata;
  }
    String get_deviceID(String src)
  {
    String idata = "";
```

```java
    String str_f1 = "\"deviceID\":\"";
    int a = src.indexOf(str_f1,0);
    if(a > -1) {
        a += str_f1.length();
        int b = src.indexOf("\",\"",a);
        if(b > a) {
            idata = src.substring(a,b);
        }
    }
    return idata;
}
String get_name(String src)
{
    String idata = "";
    String str_f1 = "\"name\":\"";
    int a = src.indexOf(str_f1,0);
    if(a > -1) {
        a += str_f1.length();
        int b = src.indexOf("\",\"",a);
        if(b > a) {
            idata = src.substring(a,b);
        }
    }
    return idata;
}
String get_idata(String src)
{
    String idata = "";
    String str_f1 = "\"idata\":\"";
    int a = src.indexOf(str_f1,0);
    if(a > -1) {
        a += str_f1.length();
        int b = src.indexOf("\",\"",a);
        if(b > a) {
            idata = src.substring(a,b);
        }
    }
    return idata;
}
String get_strdata(String src)
{
    String idata = "";
```

```java
    String str_f1 = "\"strdata\":\"";
    int a = src.indexOf(str_f1,0);
    if(a>-1) {
      a+ = str_f1.length();
      int b = src.indexOf("\"",a);
      if(b>a){
         idata = src.substring(a,b);
      }
    }
    return idata;
}
String get_idata(String src,String name)
{
    String idata = "";
    String str_f1 = "\"name\":\"" + name + "\"";
    int i = src.indexOf(str_f1,0);
    if(i>-1)  {
      str_f1 = "\"idata\":\"";
      int a = src.indexOf(str_f1,i);
      if(a>-1)   {
      a+ = str_f1.length();
      int b = src.indexOf("\",\"",a);
      if(b>a)  {
        idata = src.substring(a,b);
      }
    }}  }
    return idata;
}
//初始化定时器
    public void tcp_init_timer()
    {
        final Timer timer = new Timer();
        timer.schedule(new TimerTask() {
            @Override
            public void run() {
              tcp_init();
              if(timer != null){timer.cancel();}
            }
        },1,1);
    }
}
```

需要特别注意的是,由于 Fragment 中不能直接启动 TCP 客户端程序,所以专门设计了一个定时器来启动 TCP 客户端程序,该定时器在 tcp_init_timer() 函数之中实现。

上述程序在模拟器上测试没问题之后,可以把 Android 项目 bin 子目录下的 hello-debug.apk 文件通过 QQ 或者其他方式传到手机上,然后安装和运行。运行界面如图 7-1 所示,第一个界面中的 IP 地址就是 TCP 服务器 IP 地址。单击登录就进入控制界面,在控制界面里,单击"连接到服务器"就可以连接到 TCP 服务器,成功连接之后,可以接收 STM32 开发板送上来的 ADC 数据和温度数据,以及 LED 的状态。单击控制界面上的"开灯"和"关灯",可以控制 STM32 板上的 LED 灯的亮灭。

第 8 章
蓝牙物联网移动软件设计

8.1 蓝牙物联网移动软件设计目标

1. 蓝牙物联网移动软件设计目标

蓝牙物联网移动软件设计目标是设计一个简单的蓝牙手机 App,通过该 App 实现对 STM32 开发板上 LED 灯的控制,以及读取温度、湿度等数据。主要设计目标包括:

(1) 手机 App 可以搜索附近的 BLE 模块;
(2) 连接上 BLE 模块之后,可以对 STM32 板上的两个 LED 灯进行控制;
(3) 可以向 STM32 板发任何的字符数据;
(4) 可以接收 STM32 板上发送上来的数据并显示出来。

手机通过蓝牙物联网移动软件控制 LED 灯的连接和界面如图 8-1 所示。单击左上角的"连接"按钮。正常情况下,可以看到手机上大约每秒接收到一个字符"a",单击界面上的"开 LED1"和"开 LED2",则 STM32 板上的 LED1 和 LED2 亮。

图 8-1 蓝牙物联网移动软件控制界面

2. 蓝牙物联网移动软件硬件连接结构

蓝牙物联网移动软件控制的系统结构如图 8-2 所示,手机 App 与低能耗蓝牙

(BLE)模块进行通信,而蓝牙模块又通过串口与 STM32 开发板和通信,最终实现手机 App 对 STM32 开发板上 LED 灯的控制。

图 8-2　蓝牙物联网移动软件控制系统结构图

8.2　蓝牙通信概要

8.2.1　蓝牙通信介绍

蓝牙通信是一种支持设备短距离通信(一般 10 m 以内)的无线电技术,经过近几年的发展,我们对它已不再陌生,它也是目前数码产品中不可或缺的模块。蓝牙技术的出现让我们在连接各种设备的时候不再被繁多的数据线所束缚,比如音响、电脑、鼠标、键盘,甚至是汽车。这技术是在两个设备间进行无线短距离通信的最简单、最便捷的方法,也能够简化设备与因特网 Internet 之间的通信,从而数据传输变得更加迅速高效。

蓝牙技术最初由电信巨头爱立信公司于 1994 年创制,当时是作为 RS232 数据线的替代方案。蓝牙可连接多个设备,克服了数据同步的难题。如今蓝牙由蓝牙技术联盟(Bluetooth Special Interest Group,简称 SIG)管理。蓝牙技术联盟在全球拥有超过 25 000 家成员公司,它们分布在电信、计算机、网络和消费电子等多重领域。

蓝牙发展历史：

蓝牙 4.2,已经经过 8 个版本的更新,分别为 1.1、1.2、2.0、2.1、3.0、4.0、4.1、4.2。

蓝牙 1.1 标准,1.1 为最早期版本,传输率约在 748～810 kb/s,因是早期设计,容易受到同频率的产品干扰而影响通信质量。

蓝牙 1.2 标准,1.2 同样是只有 748～810 kb/s 的传输率,但加上了(改善 Software)抗干扰跳频功能。

蓝牙 2.0 标准,2.0 是 1.2 的改良提升版,传输率约在 1.8～2.1 M/s,开始支持双工模式——即一面作语音通信,同时亦可以传输档案/高质素图片,2.0 版本当然也支持 Stereo 运作。

应用最为广泛的是 Bluetooth 2.0＋EDR 标准,该标准在 2004 年已经推出,支持 Bluetooth 2.0＋EDR 标准的产品也于 2006 年大量出现。

虽然 Bluetooth 2.0+EDR 标准在技术上作了大量的改进,但从 1.X 标准延续下来的配置流程复杂和设备功耗较大的问题依然存在。

蓝牙 2.1 标准,2007 年 8 月 2 日,蓝牙技术联盟正式批准了蓝牙 2.1 版规范,即"蓝牙 2.1+EDR",可供未来的设备自由使用。和 2.0 版本同时代产品,目前仍然占据蓝牙市场较大份额,相对 2.0 版本主要是提高了待机时间 2 倍以上,技术标准没有根本性变化。

蓝牙 3.0 标准,2009 年 4 月 21 日,蓝牙技术联盟(Bluetooth SIG)正式颁布了新一代标准规范"Bluetooth Core Specification Version 3.0 High Speed"(蓝牙核心规范 3.0 版),蓝牙 3.0 的核心是"Generic Alternate MAC/PHY"(AMP),这是一种全新的交替射频技术,允许蓝牙协议栈针对任一任务动态地选择正确射频。

蓝牙 3.0 的数据传输速率提高到了大约 24 Mbps(即可在需要的时候调用 802.11 Wi-Fi 用于实现高速数据传输)。在传输速度上,蓝牙 3.0 是蓝牙 2.0 的八倍,可以轻松用于录像机至高清电视、PC 至 PMP、UMPC 至打印机之间的资料传输,但是需要双方都达到此标准才能实现功能。

蓝牙 4.0 标准,蓝牙 4.0 规范于 2010 年 7 月 7 日正式发布,新版本的最大意义在于低功耗,同时加强不同 OEM 厂商之间的设备兼容性,并且降低延迟,理论最高传输速率依然为 24 Mbps(即 3 MB/s),有效覆盖范围扩大到 100 m(之前的版本为 10 m)。该标准芯片被大量的手机、平板所采用,如苹果 The New iPad 平板电脑,以及苹果 iPhone 5、魅族 MX4、HTC One X 等手机上带有蓝牙 4.0 功能。

蓝牙 4.1 标准,蓝牙 4.1 于 2013 年 12 月 6 日发布,与 LTE 无线电信号之间如果同时传输数据,那么蓝牙 4.1 可以自动协调两者的传输信息,理论上可以减少其他信号对蓝牙 4.1 的干扰。改进是提升了连接速度并且更加智能化,比如减少了设备之间重新连接的时间,意味着用户如果走出了蓝牙 4.1 的信号范围并且断开连接的时间不算很长,当用户再次回到信号范围中之后设备将自动连接,反应时间要比蓝牙 4.0 更短。最后一个改进之处是提高传输效率,如果用户连接的设备非常多,比如连接了多部可穿戴设备,彼此之间的信息都能即时发送到接收设备上。

除此之外,蓝牙 4.1 也为开发人员增加了更多的灵活性,这个改变对普通用户没有很大影响,但是对于软件开发者来说是很重要的,因为为了应对逐渐兴起的可穿戴设备,那么蓝牙必须能够支持同时连接多部设备。

蓝牙 4.2 标准,2014 年 12 月 4 日,最新的蓝牙 4.2 标准颁布,改善了数据传输速度和隐私保护程度,并接入了该设备将可直接通过 IPv6 和 6LoWPAN 接入互联网。在新的标准下,蓝牙信号想要连接或者追踪用户设备必须经过用户许可,否则蓝牙信号将无法连接和追踪用户设备。速度方面变得更加快速,两部蓝牙设备之间的数据传输速率提高了 2.5 倍,因为蓝牙智能(Bluetooth Smart)数据包的容量提高,其可容纳的数据量相当于此前的 10 倍左右。

8.2.2 低能耗蓝牙(BLE)

低能耗蓝牙(BLE)技术是低成本、短距离、可互操作的鲁棒性无线技术,工作在免许可的 2.4 GHz ISM 射频频段。它从一开始就设计为超低功耗(ULP)无线技术。它利用许多智能手段最大限度地降低功耗。低能耗蓝牙技术采用可变连接时间间隔,这个间隔根据具体应用可以设置为几毫秒到几秒不等。另外,因为 BLE 技术采用非常快速的连接方式,因此,平时可以处于"非连接"状态(节省能源),此时链路两端相互间只是知晓对方,只有在必要时才开启链路,然后在尽可能短的时间内关闭链路。

BLE 技术的工作模式非常适合用于从微型无线传感器(每半秒交换一次数据)或使用完全异步通信的遥控器等其他外设传送数据。这些设备发送的数据量非常少(通常几个字节),而且发送次数也很少(例如每秒几次到每分钟一次,甚至更少)。

1. 超低功耗无线技术

低能耗蓝牙技术的三大特性成就了 ULP 性能,这三大特性分别是最大化的待机时间、快速连接和低峰值的发送/接收功耗。

无线"开启"的时间只要不是很短就会令电池寿命急剧降低,因此,任何必需的发送或接收任务需要很快完成。被低能耗蓝牙技术用来最小化无线开启时间的第一个技巧是仅用 3 个"广告"信道搜索其他设备,或向寻求建立连接的设备宣告自身存在。相比之下,标准蓝牙技术使用了 32 个信道。

这意味着低能耗蓝牙技术扫描其他设备只需"开启"0.6~1.2 ms 时间,而标准蓝牙技术需要 22.5 ms 时间来扫描 32 个信道。结果低能耗蓝牙技术定位其他无线设备所需的功耗要比标准蓝牙技术低 10~20 倍。

一旦连接成功后,低能耗蓝牙技术就会切换到 37 个数据信道之一。在短暂的数据传送期间,无线信号将使用标准蓝牙技术倡导的自适应跳频(AFH)技术以伪随机的方式在信道间切换(虽然标准蓝牙技术使用 79 个数据信道)。

要求低能耗蓝牙技术无线开启时间最短的另一个原因是它具有 1 Mbps 的原始数据带宽——更大的带宽允许在更短的时间内发送更多的信息。举例来说,具有 250 kbps 带宽的另一种无线技术发送相同信息需要开启的时间要长 8 倍(消耗更多电池能量)。

低能耗蓝牙技术完成一次连接(即扫描其他设备、建立链路、发送数据、认证和适当地结束)只需 3 ms。而标准蓝牙技术完成相同的连接周期需要数百毫秒。再次提醒,无线开启时间越长,消耗的电池能量就越多。

2. BLE 的两种芯片架构

低能耗蓝牙架构共有两种芯片构成:单模芯片和双模芯片。蓝牙单模器件是蓝牙规范中新出现的一种只支持低能耗蓝牙技术的芯片——是专门针对 ULP 操作优化的技术的一部分。蓝牙单模芯片可以和其他单模芯片及双模芯片通信,此时后者需要使用自身架构中的低能耗蓝牙技术部分进行收发数据(参考图 8-2)。双模芯片也能与标准蓝牙技术及使用传统蓝牙架构的其他双模芯片通信。

双模芯片可以在目前使用标准蓝牙芯片的任何场合使用。这样安装有双模芯片的手机、PC、个人导航设备(PND)或其他应用就可以和市场上已经在用的所有传统标准蓝牙设备以及所有未来的低能耗蓝牙设备通信。然而,由于这些设备要求执行标准蓝牙和低能耗蓝牙任务,因此,双模芯片针对 ULP 操作的优化程度没有像单模芯片那么高。

单模芯片可以用单节纽扣电池(如 3 V、220 mAh 的 CR2032)工作很长时间(几个月甚至几年)。相反,标准蓝牙技术(和低能耗蓝牙双模器件)通常要求使用至少两节 AAA 电池(电量是纽扣电池的 10~12 倍,可以容忍高得多的峰值电流),并且更多情况下最多只能工作几天或几周的时间(取决于具体应用)。注意,也有一些高度专业化的标准蓝牙设备,它们可以使用容量比 AAA 电池低的电池工作。

8.3 CC2541 BLE 蓝牙模块应用

8.3.1 CC2541 BLE 蓝牙模块介绍

1. CC2541 BLE 蓝牙模块介绍

CC2541 是一款针对低能耗以及私有 2.4 GHz 应用的功率优化的真正片载系统(SoC)解决方案。它使得使用低总体物料清单成本建立强健网络节点成为可能。CC2541 将领先 RF 收发器的出色性能和一个业界标准的增强型 8051 MCU、系统内可编程闪存存储器、8 KB RAM 和很多其他功能强大的特性和外设组合在一起。CC2541 非常适合应用于需要超低能耗的系统。这由多种不同的运行模式指定。运行模式间较短的转换时间进一步使低能耗变为可能。

蓝牙低功耗(BLE,Bluetooth low energy)是蓝牙 4.0 的一项关键功能,增加了创新的超低功耗运行模式,重新定义了蓝牙技术的使用方式。CC2540/41 则是 TI 公司推出的低功耗 SoC 解决方案,适合蓝牙低耗能应用。

CC2540 包含一个 8051 内核的 RF 收发器,系统可编程闪存、8 kB RAM 和其他功能强大的配套特征以及外设。CC2540 适用于低功耗系统,超低的睡眠模式,以及运行模式的超低功耗的转换进一步实现了超低功耗。其集成了 AES-128 加密引擎,具有出色的长距离链路预算(高达+97 dB)以及与其他 2.4 GHz 器件的良好兼容性。其采用 40 引脚 6 mm×6 mm×0.85 mm QFN 封装。

2. OSAL 和 HAL 的好处

HAL:硬件抽象层,典型的操作系统硬件封装架构,远点的类似于 Linux 中常见的 BSP,Android 中的 HAL 等,都可以见到。相比于 51 编程,在 HAL 里面各种宏定义,将所有可以操作的寄存器都进行封装,连中断都不放过,让搞裸驱的突然会有点不适应。

OSAL:操作系统抽象层,在 cc2541 的世界里,OSAL 主要用于应用程序的开发,

API 大行其道,看看 API 手册就能完成编码。

整套类操作系统围绕着:Event 任务,Task 事件,Message 消息来进行。事件发生,由该事件对应的任务来处理,任务有时会将消息做为另一个事件从而去触发别的任务进行消息的处理。

8.3.2　Android 蓝牙 BLE 编程

1. 关键概念

(1) Generic Attribute Profile (GATT)

通过 BLE 连接,读写属性类小数据的 Profile 通用规范。现在所有的 BLE 应用 Profile 都是基于 GATT 的。

(2) Attribute Protocol (ATT)

GATT 是基于 ATT Protocol 的。ATT 针对 BLE 设备做了专门的优化,具体就是在传输过程中使用尽量少的数据。每个属性都有一个唯一的 UUID,属性将以 characteristics and services 的形式传输。

(3) Characteristic

Characteristic 可以理解为一个数据类型,它包括一个 value 和 0 至多个对此 value 的描述(Descriptor)。

(4) Descriptor

对 Characteristic 的描述,例如范围、计量单位等。

(5) Service

Characteristic 的集合。例如一个 Service 叫做"Heart Rate Monitor",它可能包含多个 Characteristics,其中,可能包含一个叫做"heart rate measurement"的 Characteristic。

2. 角色和职责

Android 设备与 BLE 设备交互有两组角色——中心设备和外围设备(Central vs. peripheral;GATT server vs. GATT client)

(1) Central vs. peripheral:

中心设备和外围设备的概念针对的是 BLE 连接本身。Central 角色负责 scan advertisement。而 peripheral 角色负责 make advertisement。

(2) GATT server vs. GATT client:

这两种角色取决于 BLE 连接成功后,两个设备间通信的方式。

举例说明:

现有一个活动追踪的 BLE 设备和一个支持 BLE 的 Android 设备。Android 设备支持 Central 角色,而 BLE 设备支持 peripheral 角色。创建一个 BLE 连接需要这两个角色都存在,都仅支持 Central 角色或者都仅支持 peripheral 角色则无法建立连接。

当连接建立后,它们之间就需要传输 GATT 数据。谁做 server,谁做 client,则取决于具体数据传输的情况。例如,如果活动追踪的 BLE 设备需要向 Android 设备传输 sensor 数据,则活动追踪器自然成为了 server 端;而如果活动追踪器需要从 Android 设备获取更新信息,则 Android 设备作为 server 端可能更合适。

3. 权限及特征

和经典蓝牙一样,应用使用蓝牙,需要声明 BLUETOOTH 权限,如果需要扫描设备或者操作蓝牙设置,则还需要 BLUETOOTH_ADMIN 权限:

<uses-permission android:name="android.permission.BLUETOOTH"/>
<uses-permission android:name="android.permission.BLUETOOTH_ADMIN"/>

除了蓝牙权限外,如果需要 BLE 特征,则还需要声明 uses-feature:

<uses-feature android:name="android.hardware.bluetooth_le" android:required="true"/>

required 为 true 时,则应用只能在支持 BLE 的 Android 设备上安装运行;required 为 false 时,Android 设备均可正常安装运行,需要在代码运行时判断设备是否支持 BLE feature:

```
if (! getPackageManager().hasSystemFeature
(PackageManager.FEATURE_BLUETOOTH_LE)) {
Toast.makeText(this, R.string.ble_not_supported,
Toast.LENGTH_SHORT).show();
    finish();
}
```

4. 启动蓝牙

在使用蓝牙 BLE 之前,需要确认 Android 设备是否支持 BLE feature(required 为 false 时),另外,需要确认蓝牙是否打开。

如果发现不支持 BLE,则不能使用 BLE 相关的功能。如果支持 BLE,但是蓝牙没打开,则需要打开蓝牙。

打开蓝牙的步骤:

(1) 获取 BluetoothAdapter

BluetoothAdapter 是 Android 系统中所有蓝牙操作都需要的,它对应本地 Android 设备的蓝牙模块,在整个系统中 BluetoothAdapter 是单例的。当获取到它的示例之后,就能进行相关的蓝牙操作了。获取 BluetoothAdapter 代码示例如下:

```
//Initializes Bluetooth adapter.
final BluetoothManager bluetoothManager = (BluetoothManager) getSystemService(Context.BLUETOOTH_SERVICE);
mBluetoothAdapter = bluetoothManager.getAdapter();
```

注：这里通过 getSystemService 获取 BluetoothManager，再通过 BluetoothManager 获取 BluetoothAdapter。BluetoothManager 在 Android 4.3 以上支持(API level 18)。

(2) 判断是否支持蓝牙，并打开蓝牙

获取到 BluetoothAdapter 之后，还需要判断是否支持蓝牙，以及蓝牙是否打开。如果没打开，需要让用户打开蓝牙：

```
private BluetoothAdapter mBluetoothAdapter;
...
//Ensures Bluetooth is available on the device and it is enabled. If not,
// displays a dialog requesting user permission to enable Bluetooth.
if (mBluetoothAdapter == null || ! mBluetoothAdapter.isEnabled()) {
    Intent enableBtIntent = new Intent(BluetoothAdapter.ACTION_REQUEST_ENABLE);
    startActivityForResult(enableBtIntent, REQUEST_ENABLE_BT);
}
```

5. 搜索 BLE 设备

通过调用 BluetoothAdapter 的 startLeScan()搜索 BLE 设备。调用此方法时需要传入 BluetoothAdapter.LeScanCallback 参数。

因此，需要实现 BluetoothAdapter.LeScanCallback 接口，BLE 设备的搜索结果将通过这个 callback 返回。

由于搜索需要尽量减少功耗，因此，在实际使用时需要注意：

（1）当找到对应的设备后，立即停止扫描；

（2）不要循环搜索设备，为每次搜索设置适合的时间限制。避免设备不在可用范围的时候持续不停扫描，消耗电量。

搜索的示例代码如下：

```
public class DeviceScanActivity extends ListActivity {
    private BluetoothAdapter mBluetoothAdapter;
    private boolean mScanning;
    private Handler mHandler;
    // Stops scanning after 10 seconds.
    private static final long SCAN_PERIOD = 10000;
    ...
    private void scanLeDevice(final boolean enable) {
        if (enable) {
            // Stops scanning after a pre-defined scan period.
            mHandler.postDelayed(new Runnable() {
                @Override
                public void run() {
                    mScanning = false;
```

```
        mBluetoothAdapter.stopLeScan(mLeScanCallback);
      }
    }, SCAN_PERIOD);

    mScanning = true;
    mBluetoothAdapter.startLeScan(mLeScanCallback);
  } else {
    mScanning = false;
    mBluetoothAdapter.stopLeScan(mLeScanCallback);
  }
  ...
  }
...
}
```

如果只需要搜索指定 UUID 的外设,可以调用 startLeScan(UUID[], BluetoothAdapter.LeScanCallback)方法。

其中 UUID 数组指定应用程序所支持的 GATT Services 的 UUID。

BluetoothAdapter.LeScanCallback 的实现示例如下:

```
private LeDeviceListAdapter mLeDeviceListAdapter;
...
// Device scan callback.
private BluetoothAdapter.LeScanCallback mLeScanCallback =
    new BluetoothAdapter.LeScanCallback() {
  @Override
  public void onLeScan(final BluetoothDevice device, int rssi,
      byte[] scanRecord) {
    runOnUiThread(new Runnable() {
      @Override
      public void run() {
        mLeDeviceListAdapter.addDevice(device);
        mLeDeviceListAdapter.notifyDataSetChanged();
      }
    });
  }
};
```

注意:搜索时,只能搜索传统蓝牙设备或者 BLE 设备,两者完全独立,不可同时被搜索。

6. 连接 GATT Server

两个设备通过 BLE 通信,首先需要建立 GATT 连接。这里讲的是 Android 设备作为 client 端,连接 GATT Server。

连接 GATT Server,需要调用 BluetoothDevice 的 connectGatt()方法。此函数带三个参数:Context、autoConnect(boolean)和 BluetoothGattCallback 对象。调用示例:
mBluetoothGatt=device.connectGatt(this, false, mGattCallback);

函数成功,返回 BluetoothGatt 对象,它是 GATT profile 的封装。通过这个对象,我们就能进行 GATT Client 端的相关操作。BluetoothGattCallback 用于传递一些连接状态及结果。

BluetoothGatt 常规用到的几个操作示例:
- connect():连接远程设备。
- discoverServices():搜索连接设备所支持的 service。
- disconnect():断开与远程设备的 GATT 连接。
- close():关闭 GATT Client 端。
- readCharacteristic(characteristic):读取指定的 characteristic。
- setCharacteristicNotification(characteristic, enabled):设置当指定 characteristic 值变化时,发出通知。
- getServices():获取远程设备所支持的 services。

注:

(1)某些函数调用之间存在先后关系。例如,首先需要连接上才能调用 discoverServices。

(2)一些函数调用是异步的,需要得到的值不会立即返回,而会在 BluetoothGattCallback 的回调函数中返回。例如 discoverServices 与 onServicesDiscovered 回调,readCharacteristic 与 onCharacteristicRead 回调,setCharacteristicNotification 与 onCharacteristicChanged 回调等。

8.4 蓝牙物联网移动软件的实现

在 Obtain_Studio 中,采用"傻瓜 STM32 项目\Android_蓝牙 4.0 应用模板"模板,创建一个蓝牙的安卓应用程序,项目名称为"Android_BLE_silly_001"。然后根据需要修改界面布局文件和 Java 程序代码,对于只进行简单的蓝牙通信和简单的控制 LED 亮灭,则不需要修改界面和程序,直接编译即可。

8.4.1 蓝牙物联网移动软件界面设计

1. 菜单设计

在蓝牙物联网移动软件与蓝牙模块通信实例项目 Android_BLE_silly_001 中,通过一个简单的菜单来选择蓝牙设备的扫描、蓝牙设备的连接功能。

(1)蓝牙设备的扫描菜单实现文件 gatt_services.xml 代码如下:

```xml
<menu xmlns:android = "http://schemas.android.com/apk/res/android">
    <item android:id = "@+id/menu_refresh"
        android:checkable = "false"
        android:orderInCategory = "1"
        android:showAsAction = "ifRoom"/>
    <item android:id = "@+id/menu_scan"
        android:title = "@string/menu_scan"
        android:orderInCategory = "100"
        android:showAsAction = "ifRoom|withText"/>
    <item android:id = "@+id/menu_stop"
        android:title = "@string/menu_stop"
        android:orderInCategory = "101"
        android:showAsAction = "ifRoom|withText"/>
</menu>
```

(2) 蓝牙设备的连接菜单实现文件 main.xml 代码如下:

```xml
<menu xmlns:android = "http://schemas.android.com/apk/res/android">
    <item android:id = "@+id/menu_refresh"
        android:checkable = "false"
        android:orderInCategory = "1"
        android:showAsAction = "ifRoom"/>
    <item android:id = "@+id/menu_connect"
        android:title = "@string/menu_connect"
        android:orderInCategory = "100"
        android:showAsAction = "ifRoom|withText"/>
    <item android:id = "@+id/menu_disconnect"
        android:title = "@string/menu_disconnect"
        android:orderInCategory = "101"
        android:showAsAction = "ifRoom|withText"/>
</menu>
```

2. 界面设计

(1) 蓝牙设备扫描界面主要包括设备的列表功能,蓝牙设备扫描界面布局文件 listitem_device.xml 代码如下:

```xml
<?xml version = "1.0" encoding = "utf-8"?>
<LinearLayout xmlns:android = "http://schemas.android.com/apk/res/android"
    android:orientation = "vertical"
    android:layout_width = "match_parent"
    android:layout_height = "wrap_content">
    <TextView android:id = "@+id/device_name"
```

```
        android:layout_width = "match_parent"
        android:layout_height = "wrap_content"
        android:textSize = "24dp"/>
    <TextView android:id = "@ + id/device_address"
        android:layout_width = "match_parent"
        android:layout_height = "wrap_content"
        android:textSize = "12dp"/>
</LinearLayout>
```

蓝牙设备扫描界面中的扫描进度条布局文件 actionbar_indeterminate_ progress. xml 代码如下：

```
<FrameLayout xmlns:android = "http://schemas.android.com/apk/res/android"
    android:layout_height = "wrap_content"
    android:layout_width = "56dp"
    android:minWidth = "56dp">
  <ProgressBar android:layout_width = "32dp"
        android:layout_height = "32dp"
        android:layout_gravity = "center"/>
</FrameLayout>
```

（2）主控制界面，包括 LED 灯控制按钮和数据接收显示等，界面布局文件 main. xml 代码如下：

```
<LinearLayout android:id = "@ + id/tvLinearLayout"
  xmlns:android = "http://schemas.android.com/apk/res/android"
  android:orientation = "vertical"
  android:layout_width = "fill_parent"
  android:layout_height = "fill_parent"
  android:background = "#FFFFFFFF">
 <LinearLayout
    android:layout_width = "fill_parent"
    android:layout_height = "wrap_content"
    android:orientation = "horizontal"
    >
            <Button android:id = "@ + id/Led1On"
                android:layout_height = "wrap_content"
                android:layout_width = "wrap_content"
                android:text = "开 LED1">
            </Button>
            <Button android:id = "@ + id/Led1Off"
                android:layout_height = "wrap_content"
                android:layout_width = "wrap_content"
```

```xml
            android:text = "关 LED1">
        </Button>
</LinearLayout>
<LinearLayout
    android:layout_width = "fill_parent"
    android:layout_height = "wrap_content"
    android:orientation = "horizontal"
    >
        <Button android:id = "@ + id/Led2On"
            android:layout_height = "wrap_content"
            android:layout_width = "wrap_content"
            android:text = "开 LED2">
        </Button>
        <Button android:id = "@ + id/Led2Off"
            android:layout_height = "wrap_content"
            android:layout_width = "wrap_content"
            android:text = "关 LED2">
        </Button>
</LinearLayout>
<TableLayout android:id = "@ + id/TableLayout02"
    android:layout_width = "fill_parent"
    android:layout_height = "wrap_content"
    android:layout_margin = "0px"
    android:stretchColumns = "0"
    android:layout_gravity = "bottom">
    <TableRow>
        <EditText
            android:id = "@ + id/edtSend"
            android:layout_width = "fill_parent"
            android:layout_height = "wrap_content"
            android:layout_span = "4"/>
        <Button android:id = "@ + id/btnSend"
            android:layout_height = "wrap_content"
            android:layout_width = "wrap_content"
            android:text = "@string/send"
            android:layout_gravity = "bottom">
        </Button>
    </TableRow>
</TableLayout>

<ScrollView android:id = "@ + id/svResult"
    android:layout_width = "fill_parent"
```

```
              android:layout_height = "fill_parent"
              android:layout_weight = "1.0"
              android:background = "#FFFFFFFF">
        <TextView
            android:focusable = "true"
                android:focusableInTouchMode = "true"
            android:id = "@ + id/data_value"
            android:layout_width = "fill_parent"
            android:layout_height = "fill_parent"
            android:text = "@string/no_data"
            android:textSize = "16sp"/>
          </ScrollView>
    </LinearLayout>
```

8.4.2 蓝牙物联网移动软件界面程序设计

1. 蓝牙设备扫描程序

蓝牙设备扫描程序主要完成蓝牙设备的扫描、设备列表、设备选中以及界面切换功能，蓝牙设备扫描程序 DeviceScanActivity.java 的程序代码如下：

```java
public class DeviceScanActivity extends ListActivity {
    private LeDeviceListAdapter mLeDeviceListAdapter;
    private BluetoothAdapter mBluetoothAdapter;
    private boolean mScanning;
    private static final int REQUEST_ENABLE_BT = 1;
    // Stops scanning after 10 seconds.
    private static final long SCAN_PERIOD = 10000;
    @Override
    public void onCreate(Bundle savedInstanceState) {
        super.onCreate(savedInstanceState);
        getActionBar().setTitle(R.string.title_devices);
if (! getPackageManager().hasSystemFeature(
PackageManager.FEATURE_BLUETOOTH_LE)) {
        Toast.makeText(this, R.string.ble_not_supported,
Toast.LENGTH_SHORT).show();
        finish();
    }
    final BluetoothManager bluetoothManager =
        (BluetoothManager) getSystemService(Context.BLUETOOTH_SERVICE);
    mBluetoothAdapter = bluetoothManager.getAdapter();
    if (mBluetoothAdapter == null) {
```

```java
            Toast.makeText(this, R.string.error_bluetooth_not_supported, Toast.LENGTH_SHORT).show();
            finish();
            return;
        }
        mLeDeviceListAdapter = new LeDeviceListAdapter();
        setListAdapter(mLeDeviceListAdapter);
    }
    @Override public boolean onCreateOptionsMenu(Menu menu) {
        getMenuInflater().inflate(R.menu.main, menu);
        if (! mScanning) {
            menu.findItem(R.id.menu_stop).setVisible(false);
            menu.findItem(R.id.menu_scan).setVisible(true);
            menu.findItem(R.id.menu_refresh).setActionView(null);
        } else {
            menu.findItem(R.id.menu_stop).setVisible(true);
            menu.findItem(R.id.menu_scan).setVisible(false);
            menu.findItem(R.id.menu_refresh).setActionView(
                R.layout.actionbar_indeterminate_progress);
        }
        return true;
    }
    @Override
    public boolean onOptionsItemSelected(MenuItem item) {
        switch (item.getItemId()) {
            case R.id.menu_scan:
                mLeDeviceListAdapter.clear();
                scanLeDevice(true);
                break;
            case R.id.menu_stop:
                scanLeDevice(false);
                break;
        }
        return true;
    }
    @Override
    protected void onActivityResult(int requestCode, int resultCode, Intent data) {
        if (requestCode == REQUEST_ENABLE_BT && resultCode == Activity.RESULT_CANCELED) {
            finish();
            return;
        }
        super.onActivityResult(requestCode, resultCode, data);
```

```java
}
@Override   protected void onResume() {super.onResume();}
@Override   protected void onPause() {
    super.onPause();
    scanLeDevice(false);
}
@Override
protected void onListItemClick(ListView l, View v, int position, long id) {
    final BluetoothDevice device = mLeDeviceListAdapter.getDevice(position);
    if (device == null) return;
    final Intent intent = new Intent(this, mainActivity.class);
    intent.putExtra(mainActivity.EXTRAS_DEVICE_NAME, device.getName());
    intent.putExtra(mainActivity.EXTRAS_DEVICE_ADDRESS, device.getAddress());
    if (mScanning) {
        mBluetoothAdapter.stopLeScan(mLeScanCallback);
        mScanning = false;
    }
    startActivity(intent);
}
private void scanLeDevice(final boolean enable) {
    if (enable) {
        mHandler.postDelayed(new Runnable() {
            @Override
            public void run() {
                if(mScanning)
                {
                    mScanning = false;
                    mBluetoothAdapter.stopLeScan(mLeScanCallback);
                    invalidateOptionsMenu();
                }
            }
        }, SCAN_PERIOD);
        mScanning = true;
        //F000E0FF-0451-4000-B000-000000000000
        mLeDeviceListAdapter.clear();
        mHandler.sendEmptyMessage(1);
        mBluetoothAdapter.startLeScan(mLeScanCallback);
    } else {
        mScanning = false;
        mBluetoothAdapter.stopLeScan(mLeScanCallback);
    }
    invalidateOptionsMenu();
```

```java
        }
        private static char findHex(byte b) {
            int t = new Byte(b).intValue();
            t = t < 0? t + 16:t;
            if ((0 <= t) &&(t <= 9)) {
            return (char)(t + '0');
            }
            return (char)(t - 10 + 'A');
        }
    public static String ByteToString(byte[] bytes) {
            StringBuffer sb = new StringBuffer();
            for (int i = 0; i < bytes.length; i++) {
             sb.append(findHex((byte)((bytes[i] & 0xf0) >>4)));
             sb.append(findHex((byte)(bytes[i] & 0x0f)));
            }
            return sb.toString();
           }
    private class LeDeviceListAdapter extends BaseAdapter {
       private ArrayList <BluetoothDevice >mLeDevices;
       private LayoutInflater mInflator;
       public LeDeviceListAdapter() {
          super();
          mLeDevices = new ArrayList <BluetoothDevice >();
          mInflator = DeviceScanActivity.this.getLayoutInflater();
       }
       public void addDevice(BluetoothDevice device) {
           if(! mLeDevices.contains(device)) {
              mLeDevices.add(device);
           }
       }
       public BluetoothDevice getDevice(int position) {
          return mLeDevices.get(position);
       }
       public void clear() {
          mLeDevices.clear();
       }
       @Override public int getCount() {
          return mLeDevices.size();
       }
       @Override
       public Object getItem(int i) {
          return mLeDevices.get(i);
```

```java
        }
        @Override
        public long getItemId(int i) {
            return i;
        }
        @Override
        public View getView(int i, View view, ViewGroup viewGroup) {
            ViewHolder viewHolder;
            if (view == null) {
                view = mInflator.inflate(R.layout.listitem_device, null);
                viewHolder = new ViewHolder();
                viewHolder.deviceAddress = (TextView) view.findViewById(R.id.device_address);
                viewHolder.deviceName = (TextView) view.findViewById(R.id.device_name);
                view.setTag(viewHolder);
            } else {
                viewHolder = (ViewHolder) view.getTag();
            }
            BluetoothDevice device = mLeDevices.get(i);
            final String deviceName = device.getName();
            if (deviceName != null && deviceName.length() > 0)
                viewHolder.deviceName.setText(deviceName);
            else
                viewHolder.deviceName.setText(R.string.unknown_device);
            viewHolder.deviceAddress.setText(device.getAddress());
            return view;
        }
    }
    private BluetoothAdapter.LeScanCallback mLeScanCallback =
            new BluetoothAdapter.LeScanCallback() {
        @Override
        public void onLeScan(final BluetoothDevice device, final int rssi, final byte[] scanRecord) {
            runOnUiThread(new Runnable() {
                @Override
                public void run() {
                    mLeDeviceListAdapter.addDevice(device);
                    mHandler.sendEmptyMessage(1);
                }
            });
        }
    };
    static class ViewHolder {
```

```
        TextView deviceName;
        TextView deviceAddress;
    }
    public final Handler mHandler = new Handler() {
        @Override
        public void handleMessage(Message msg) {
            switch (msg.what) {
            case 1: // Notify change
                mLeDeviceListAdapter.notifyDataSetChanged();
                break;
            }
        }
    };
}
```

2. 主控界面程序

主控界面程序主要功能包括连接蓝牙设备、蓝牙数据的收发、以及简单的控制操作界面和数据接收显示界面，主控界面程序 mainActivity.java 代码如下：

```
public class mainActivity extends Activity {
    private final static String TAG = mainActivity.class.getSimpleName();
    public static final String EXTRAS_DEVICE_NAME = "DEVICE_NAME";
    public static final String EXTRAS_DEVICE_ADDRESS = "DEVICE_ADDRESS";
    private TextView mDataField;
    private String mDeviceName;
    private String mDeviceAddress;
    private BluetoothLeService mBluetoothLeService;
    private boolean mConnected = false;
    EditText edtSend;
        ScrollView svResult;
        Button btnSend;
    private final ServiceConnection mServiceConnection =
new ServiceConnection() {
        @Override
    public void onServiceConnected(ComponentName componentName,
IBinder service) {
            mBluetoothLeService = ((BluetoothLeService.LocalBinder)
service).getService();
            if (! mBluetoothLeService.initialize()) {
                Log.e(TAG, "Unable to initialize Bluetooth");
                finish();
            }
```

```java
            Log.e(TAG, "mBluetoothLeService is okay");
        }
        @Override
        public void onServiceDisconnected(ComponentName componentName) {
            mBluetoothLeService = null;
        }
    };
    private final BroadcastReceiver mGattUpdateReceiver = new BroadcastReceiver() {
        @Override
        public void onReceive(Context context, Intent intent) {
            final String action = intent.getAction();
            if (BluetoothLeService.ACTION_GATT_CONNECTED.equals(action)) {
//连接成功
                Log.e(TAG, "Only gatt, just wait");
            } else if (BluetoothLeService.ACTION_GATT_DISCONNECTED.equals(action)) {
//断开连接
                mConnected = false;
                invalidateOptionsMenu();
                btnSend.setEnabled(false);
                clearUI();
            }else
if(BluetoothLeService.ACTION_GATT_SERVICES_DISCOVERED.equals(action))
            {
                mConnected = true;
                mDataField.setText("");
                ShowDialog();
                btnSend.setEnabled(true);
                Log.e(TAG, "In what we need");
                invalidateOptionsMenu();
            }else if (BluetoothLeService.ACTION_DATA_AVAILABLE.equals(action)) { //收到数据
                Log.e(TAG, "RECV DATA");
                String data = intent.getStringExtra(BluetoothLeService.EXTRA_DATA);
                if (data != null) {
                    if (mDataField.length() >500)
                        mDataField.setText("");
                mDataField.append(data);
                svResult.post(new Runnable() {
                        public void run() {
                            svResult.fullScroll(ScrollView.FOCUS_DOWN);
                        }
                    });
                }
```

```java
            }
        }
    };
    private void clearUI() {
        mDataField.setText(R.string.no_data);
    }
    @Override
    public void onCreate(Bundle savedInstanceState) { //初始化
        super.onCreate(savedInstanceState);
        setContentView(R.layout.main);
        final Intent intent = getIntent();
        mDeviceName = intent.getStringExtra(EXTRAS_DEVICE_NAME);
        mDeviceAddress = intent.getStringExtra(EXTRAS_DEVICE_ADDRESS);
        // Sets up UI references.
        mDataField = (TextView) findViewById(R.id.data_value);
        edtSend = (EditText) this.findViewById(R.id.edtSend);
        edtSend.setText("a");
        svResult = (ScrollView) this.findViewById(R.id.svResult);
        btnSend = (Button) this.findViewById(R.id.btnSend);
            btnSend.setOnClickListener(new ClickEvent());
            btnSend.setEnabled(false);
        getActionBar().setTitle(mDeviceName);
        getActionBar().setDisplayHomeAsUpEnabled(true);
        Intent gattServiceIntent = new Intent(this, BluetoothLeService.class);
        Log.d(TAG, "Try to bindService = " + bindService(gattServiceIntent, mServiceConnection, BIND_AUTO_CREATE));
        registerReceiver(mGattUpdateReceiver, makeGattUpdateIntentFilter());
        Button Led1On = (Button)findViewById(R.id.Led1On);
        Led1On.setOnClickListener(new View.OnClickListener(){
            // @Override
            public void onClick(View v)
            {
                mBluetoothLeService.WriteValue("a");
            }
        });
        Button Led1Off = (Button)findViewById(R.id.Led1Off);
        Led1Off.setOnClickListener(new View.OnClickListener(){
            // @Override
            public void onClick(View v)
            {
                mBluetoothLeService.WriteValue("b");
            }
```

```java
        });
        Button Led2On = (Button)findViewById(R.id.Led2On);
        Led2On.setOnClickListener(new View.OnClickListener(){
            // @Override
            public void onClick(View v)
            {
                mBluetoothLeService.WriteValue("c");
            }
        });
        Button Led2Off = (Button)findViewById(R.id.Led2Off);
        Led2Off.setOnClickListener(new View.OnClickListener(){
            // @Override
            public void onClick(View v)
            {
                mBluetoothLeService.WriteValue("d");
            }
        });
    }
    @Override protected void onResume() {super.onResume();}
    @Override protected void onPause() {
        super.onPause();
        unregisterReceiver(mGattUpdateReceiver);
          unbindService(mServiceConnection);
    }
    @Override   protected void onDestroy() {
        super.onDestroy();
        if(mBluetoothLeService != null) {
            mBluetoothLeService.close();
            mBluetoothLeService = null;
        }
        Log.d(TAG, "We are in destroy");
    }
    @Override
    public boolean onCreateOptionsMenu(Menu menu) {
        getMenuInflater().inflate(R.menu.gatt_services, menu);
        if (mConnected) {
            menu.findItem(R.id.menu_connect).setVisible(false);
            menu.findItem(R.id.menu_disconnect).setVisible(true);
        } else {
            menu.findItem(R.id.menu_connect).setVisible(true);
            menu.findItem(R.id.menu_disconnect).setVisible(false);
        }
```

```java
        return true;
    }
    @Override
    public boolean onOptionsItemSelected(MenuItem item) {//点击按钮
        switch(item.getItemId()) {
            case R.id.menu_connect:
                mBluetoothLeService.connect(mDeviceAddress);
                return true;
            case R.id.menu_disconnect:
                mBluetoothLeService.disconnect();
                return true;
            case android.R.id.home:
                if(mConnected) {
                    mBluetoothLeService.disconnect();
                    mConnected = false;
                }
                onBackPressed();
                return true;
        }
        return super.onOptionsItemSelected(item);
    }
    private void ShowDialog()  {
        Toast.makeText(this,"连接成功,现在可以正常通信!",Toast.LENGTH_SHORT).show();
    }
    class ClickEvent implements View.OnClickListener {
        @Override
        public void onClick(View v) {
            if (v == btnSend) {
                if(! mConnected) return;
                if (edtSend.length() <1) {
                    Toast.makeText(mainActivity.this,"请输入要发送的内容",Toast.LENGTH_SHORT).show();
                    return;
                }
                mBluetoothLeService.WriteValue(edtSend.getText().toString());
                InputMethodManager imm = (InputMethodManager) getSystemService(Context.INPUT_METHOD_SERVICE);
                if(imm.isActive())
                    imm.hideSoftInputFromWindow(edtSend.getWindowToken(), 0);
            }
        }
    }
```

```
    private static IntentFilter makeGattUpdateIntentFilter() {
//注册接收的事件
    final IntentFilter intentFilter = new IntentFilter();
    intentFilter.addAction(BluetoothLeService.ACTION_GATT_CONNECTED);
    intentFilter.addAction(BluetoothLeService.ACTION_GATT_DISCONNECTED);
    intentFilter.addAction(BluetoothLeService.ACTION_GATT_SERVICES_DISCOVERED);
    intentFilter.addAction(BluetoothLeService.ACTION_DATA_AVAILABLE);
    intentFilter.addAction(BluetoothDevice.ACTION_UUID);
    return intentFilter;
    }
}
```

8.4.3 STM32 的蓝牙通信程序设计

STM32 的蓝牙通信主要通过串口与蓝牙模块连接和通信，STM32 板与蓝牙模块的连接如图 8-3 所示。

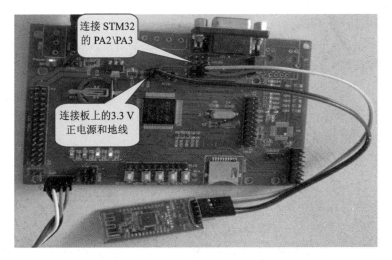

图 8-3 STM32 板与蓝牙模块的连接

STM32 的蓝牙通信程序如下：

```
#include "include/bsp.h"
#include "include/led_key.h"
#include "include/usart.h"

CLed led1(LED1),led2(LED2),led3(LED3),led4(LED4);
CUsart uart1(USART2,9600);

void getChar(int ch)
{
```

```
    if(ch == 'a')led1 = 1;
    if(ch == 'b')led1 = 0;
    if(ch == 'c')led2 = 1;
    if(ch == 'd')led2 = 0;
}
void setup()
{
    uart1.setCallback(getChar);
    uart1.start();
}
void loop()
{
    uart1.Send('a');
    bsp.delay(1000);
}
```

前面的蓝牙安卓项目编译之后，在项目目录的 bin 子目录下，可以找到编译后生成的安卓安装包"hello-debug.apk"。把该安装包发到安卓手机上安装和运行，安卓的版本需要 4.3 以上，并且有蓝牙 BLE 硬件，则可以正常运行。在运行主界面里，单击左上角的"搜索"按钮，可以看到在蓝牙设备的列表，运行效果如图 8-4 所示。

图 8-4　蓝牙设备的列表

单击设备列表中的某一个设备，命名单击名为 HMSoft 的设备，则进入主界面，如图 8-5 所示单击左上角的"连接"按钮。正常情况下，可以看到手机上大约每秒接收到一个字符"a"，单击界面上的"开 LED1"和"开 LED2"，则 STM32 板上的 LED1 和 LED2 亮。

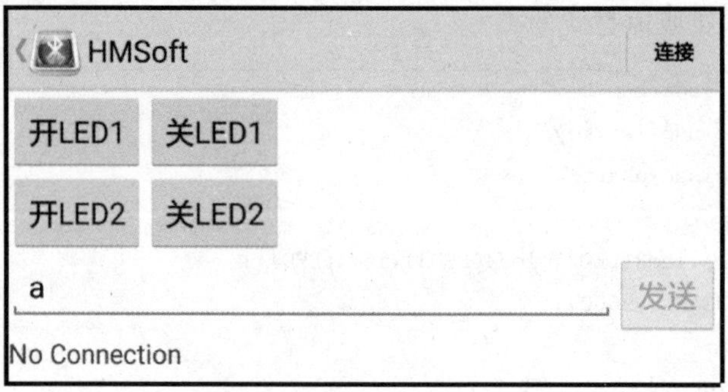

图 8-5　蓝牙控制界面

第 9 章

数据库及动态界面设计

9.1 数据库及动态界面设计目标

在第 6 章"导航栏及滑动界面设计"之中,地点的列表是固定的,这样子做比较简单,但实用性比较差,因为用户在使用的过程中,并不一定正好是这些地点,也不一定是一成变,需要根据实际情况进行增加或者删除地点,这样的界面把它叫做动态界面。

本章的目标是把这些地点列表信息保存在数据库之中,用户可以自行添加、删除、修改每一项列表的内容,实现动态界面显示。在主屏幕右上角添加一个"+"按钮,单击该按钮弹出添加新地点对话框,如图 9-1 所示。

图 9-1 添加新地点功能运行效果图

编辑功能采用类似于添加对话一样的布局。编辑和删除完成之后能更新数据库内容,并且马上更新到地点列表界面中。编辑与删除功能效果如图 9-2 所示。

图 9-2 编辑与删除功能效果图

9.2 物联网 App 安卓端数据存储

Android 提供了 5 种方式来让用户保存持久化应用程序数据。根据自己的需求来做选择，比如数据是否是应用程序私有的，是否能被其他程序访问，需要多少数据存储空间等，分别是：

① 使用 SharedPreferences 存储数据；
② 文件存储数据；
③ SQLite 数据库存储数据；
④ 使用 ContentProvider 存储数据；
⑤ 网络存储数据。

9.2.1 使用 Shared Preferences 存储数据

SharedPreferences，可以把它理解为一种轻量级的 Database，存取形式和 map 一样：<key,value>，以 XML 文件存储。例如，可以用它来存储一下登录信息和登录状态，这样每次登录的时候就可以从本地读取信息。

这种存储方式用于存储原始类型数据，包括 boolean、int、long、float、double、String 等。具体的存储方式是键-值对，若不主动删除，这些数据会一直存在。SharedPreferences 是用 XML 文件存放数据，文件存放在/data/data/<package name>/shared_prefs 目录下。

1. 使用 sharedPreferences 存储数据

使用 sharedPreferences 存储数据的步骤：

(1) 通过 Context 对象创建一个 SharedPreference 对象；
(2) 通过 sharedPreferences 对象获取一个 Editor 对象；
(3) 往 Editor 中添加数据；
(4) 提交 Editor 对象。

使用 sharedPreferences 存储数据的实例：

```
//获取 sharedPreferences 对象
SharedPreferences sharedPreferences = getSharedPreferences("zjl", Context.MODE_PRIVATE);
//获取 editor 对象
SharedPreferences.Editor editor = sharedPreferences.edit();//获取编辑器
//存储键值对
editor.putString("name","周杰伦");
editor.putInt("age", 24);
editor.putBoolean("isMarried", false);
editor.putLong("height", 175L);
editor.putFloat("weight", 60f);
editor.putStringSet("where", set);
//提交
editor.commit();//提交修改
```

2. 使用 sharedPreferences 读取数据

使用 sharedPreferences 读取数据的步骤：

(1) 通过 Context 对象创建一个 SharedPreference 对象；
(2) 通过 sharedPreference 获取存放的数据。

使用 sharedPreferences 读取数据的实例：

```
SharedPreferences sharedPreferences = getSharedPreferences("zjl", Context.MODE_PRIVATE);
//getString()第二个参数为缺省值,如果 preference 中不存在该 key,将返回缺省值
String name = sharedPreferences.getString("name", "");
int age = sharedPreferences.getInt("age", 1);
```

9.2.2 使用文件存储数据

1. 文件存储方式

(1) 内部存储

内部存储位置为/data/data,位置读取方式 context.getFileDir().getPath();是一个应用程序的私有目录,只有当前应用程序有权限访问读写,其他应用无权限访问。一些安全性要求比较高的数据存放在该目录,一般用来存放 size 比较小的数据。

(2) 外部存储(sdcard)

外部存储位置为/sdcard,位置读取方式 Enviroment.getExternalStorage Directory().getPath();是一个外部存储目录,只用应用声明了<uses-permission android:name="android.permission.WRITE_EXTERNAL_STORAGE"/>的一个权限,就可以访问读写 sdcard 目录;所以一般用来存放一些安全性不高的数据,文件 size 比较大的数据。

2. 内部存储

当文件被保存在内部存储中时,默认情况下,文件是应用程序私有的,其他应用不能访问。当用户卸载应用程序时这些文件也跟着被删除。文件默认存储位置:/data/data/包名/files/文件名。

(1) 写入文件使用方法

写入文件使用方法如下:

① 调用 Context 的 openFileOutput()函数,填入文件名和操作模式,它会返回一个 FileOutputStream 对象;

② 通过 FileOutputStream 对象的 write()函数写入数据;

③ FileOutputStream 对象的 close()函数关闭流。

例如:

```
String FILENAME = "a.txt";
String string = "fanrunqi";
try {
    FileOutputStream fos = openFileOutput(FILENAME, Context.MODE_PRIVATE);
    fos.write(string.getBytes());
    fos.close();
} catch (Exception e) {
    e.printStackTrace();
}
```

注解:在 openFileOutput(String name, int mode)方法中:

1) name 参数:用于指定文件名称,不能包含路径分隔符"/",如果文件不存在,Android 会自动创建它。

2) mode 参数:用于指定操作模式,分为四种:

➤ Context.MODE_PRIVATE = 0,为默认操作模式,代表该文件是私有数据,只能被应用本身访问,在该模式下,写入的内容会覆盖原文件的内容;

➤ Context.MODE_APPEND = 32768,该模式会检查文件是否存在,存在就往文件追加内容,否则就创建新文件;

➤ Context.MODE_WORLD_READABLE = 1,表示当前文件可以被其他应用读取;

➤ MODE_WORLD_WRITEABLE,表示当前文件可以被其他应用写入。

(2) 读取一个内部存储的私有文件

读取一个内部存储的私有文件的步骤:

① 调用 openFileInput(),参数中填入文件名,会返回一个 FileInputStream 对象;

② 使用流对象的 read()方法读取字节；
③ 调用流的 close()方法关闭流。
例如：

```
String FILENAME = "a.txt";
try {
    FileInputStream inStream = openFileInput(FILENAME);
    int len = 0;
    byte[] buf = new byte[1024];
    StringBuilder sb = new StringBuilder();
    while ((len = inStream.read(buf)) != -1) {
        sb.append(new String(buf, 0, len));
    }
    inStream.close();
} catch (Exception e) {
    e.printStackTrace();
}
```

其他一些经常用到的方法：
- getFilesDir():得到内部存储文件的绝对路径；
- getDir():在内部存储空间中创建或打开一个已经存在的目录；
- deleteFile():删除保存在内部存储的文件；
- fileList():返回当前应用程序保存的文件的数组,内部存储目录下的全部文件。

(3) 保存编译时的静态文件

如果想在应用编译时保存静态文件,应该把文件保存在项目的 res/raw/目录下,可以通过 openRawResource()方法去打开它(传入参数 R.raw.filename),这个方法返回一个 InputStream 流对象可以读取文件但是不能修改原始文件。

InputStream is =this.getResources().openRawResource(R.raw.filename);

(4) 保存内存缓存文件

有时候只想缓存一些数据而不是持久化保存,可以使用 getCacheDir()去打开一个文件,文件的存储目录(/data/data/包名/cache)是一个应用专门来保存临时缓存文件的内存目录。

当设备的内部存储空间比较低的时候,Android 可能会删除这些缓存文件来恢复空间,但是不应该依赖系统来回收,要自己维护这些缓存文件把它们的大小限制在一个合理的范围内,比如 1 MB,当卸载应用的时候这些缓存文件也会被移除。

3. 外部存储(sdcard)

因为内部存储容量限制,有时候需要存储数据比较大的时候需要用到外部存储,使用外部存储分为以下几个步骤：

(1) 添加外部存储访问限权

首先,要在 AndroidManifest.xml 中加入访问 SD Card 的权限,如下：

```xml
<!-- 在 SD Card 中创建与删除文件权限 -->
<uses-permission android:name=
            "android.permission.MOUNT_UNMOUNT_FILESYSTEMS"/>
<!-- 往 SD Card 写入数据权限 -->
<uses-permission android:name="android.permission.WRITE_EXTERNAL_STORAGE"/>
```

(2) 检测外部存储的可用性

在使用外部存储时需要检测其状态,它可能被连接到计算机、丢失或者只读等。下面代码将说明如何检查状态:

```
//获取外存储的状态 String state = Environment.getExternalStorageState();if (Environment.MEDIA_MOUNTED.equals(state)) {
    // 可读可写
    mExternalStorageAvailable = mExternalStorageWriteable = true;
} else if (Environment.MEDIA_MOUNTED_READ_ONLY.equals(state)) {
    // 可读
} else {
    // 可能有很多其他的状态,但是只需要知道,不能读也不能写}
```

(3) 访问外部存储器中的文件

① 如果 API 版本大于或等于 8,使用 getExternalFilesDir(String type)方法,打开一个外存储目录,此方法需要一个类型,指定想要的子目录,如类型参数 DIRECTORY_MUSIC 和 DIRECTORY_RINGTONES,传 null 就是应用程序的文件目录的根目录。通过指定目录的类型,确保 Android 的媒体扫描仪将扫描分类系统中的文件,例如,铃声被确定为铃声。如果用户卸载应用程序,这个目录及其所有内容将被删除。例如:

```
File file = new File(getExternalFilesDir(null), "fanrunqi.jpg");
```

② 如果 API 版本小于 8(7 或者更低),则用 getExternalStorageDirectory()方法,通过该方法打开外存储的根目录,应该在以下目录下写入应用数据,这样当卸载应用程序时该目录及其所有内容也将被删除。

/Android/data/<package_name>/files/

读写数据方法如下:

```
if(Environment.getExternalStorageState().equals(Environment.MEDIA_MOUNTED))
{
    File sdCardDir = Environment.getExternalStorageDirectory();
    //获取 SD Card 目录"/sdcard"
    File saveFile = new File(sdCardDir,"a.txt");
    //写数据
    try {
```

```
            FileOutputStream fos = new FileOutputStream(saveFile);
            fos.write("fanrunqi".getBytes());
            fos.close();
        } catch (Exception e) {
            e.printStackTrace();
        }
        //读数据
        try {
            FileInputStream fis = new FileInputStream(saveFile);
            int len = 0;
            byte[] buf = new byte[1024];
            StringBuffer sb = new StringBuffer();
            while((len = fis.read(buf)) != -1){
                sb.append(new String(buf, 0, len));
            }
            fis.close();
        } catch (Exception e) {
            e.printStackTrace();
        }
    }
```

也可以在 /Android/data/package_name/cache/目录下做外部缓存。下面是保存文件实例：

```
public void File_Save(String data)
{
    final EditText m_ControlFile = (EditText) findViewById(R.id.id_ControlFile);
    final EditText m_ControlCode = (EditText) findViewById(R.id.id_ControlCode);
    String filename = m_ControlFile.getText().toString();
    if(filename == null||filename.indexOf(".html") < 2)//如果找不到.html
    {
        filename = createNewFileName("html");
    }
    FileService fs = new FileService(mContext);
    fs.save(filename,strUTF8);
    //String data1 = fs.read(filename);
}
```

9.3 安卓端 SQLite 数据库应用设计

9.3.1 安卓端 SQLite 数据库简介

1. SQLite 简介

SQLite 是一款轻型的数据库，是遵守 ACID 的关联式数据库管理系统，它的设计目标是嵌入式,而且目前已经在很多嵌入式产品中得到了应用,它占用资源非常低,在嵌入式设备中,可能只需要几百 KB 的内存就够了。它能够支持 Windows/Linux/Unix 等主流的操作系统,同时能够跟很多程序语言相结合,比如 Tcl、PHP、Java、C++、.Net 等,还有 ODBC 接口,同样比起 Mysql、PostgreSQL 这两款开源世界著名的数据库管理系统来讲,它的处理速度比他们都快。

SQLite 最大的特点是可以把各种类型的数据保存到任何字段中,而不用关心字段声明的数据类型是什么。例如:可以在 Integer 类型的字段中存放字符串,或者在布尔型字段中存放浮点数,或者在字符型字段中存放日期型值。

但有一种情况例外:定义为 INTEGER PRIMARY KEY 的字段只能存储 64 位整数,当向这种字段保存除整数以外的数据时,将会产生错误。

另外,SQLite 在解析 CREATE TABLE 语句时,会忽略 CREATE TABLE 语句中跟在字段名后面的数据类型信息,如下面语句会忽略 name 字段的类型信息:

CREATE TABLE person（personid integer primary key autoincrement,name varchar(20)）

SQLite 可以解析大部分标准 SQL 语句,例如查询语句"select * from 表名 where 条件子句 group by 分组字句 having ... order by 排序子句(顺序一定不能错)"。

(1) 查询语句

select * from person

select * from person order by id desc

select name from person group by name having count(*)>1

分页 SQL 与 mysql 类似,下面 SQL 语句获取 5 条记录,跳过前面 3 条记录

select * from Account limit 5 offset 3

或者

select * from Account limit 3,5

(2) 插入语句

insert into 表名(字段列表) values(值列表)。如:

insert into person(name,age) values('小明',3)

(3) 更新语句

update 表名 set 字段名=值 where 条件子句。如:

update person set name='小明' where id=10

(4) 删除语句

delete from 表名 where 条件子句。如:

delete from person where id=10

2. SQLite 的特点

(1) 轻量级

SQLite 和 C/S 模式的数据库软件不同,它是进程内的数据库引擎,因此,不存在数

据库的客户端和服务器。使用 SQLite 一般只需要带上它的一个动态库,就可以享受它的全部功能。而且那个动态库的尺寸也挺小,以版本 3.6.11 为例,Windows 下 487 KB、Linux 下 347 KB。

(2) 不需要"安装"

SQLite 的核心引擎本身不依赖第三方的软件,使用它也不需要"安装"。有点类似那种绿色软件。

(3) 单一文件

数据库中所有的信息(比如表、视图等)都包含在一个文件内。这个文件可以自由复制到其他目录或其他机器上。

(4) 跨平台/可移植性

除了主流操作系统 Windows,Linux 之后,SQLite 还支持其他一些不常用的操作系统。

(5) 弱类型的字段

同一列中的数据可以是不同类型。

(6) 开　源

3. SQLite 数据类型

一般数据采用的是固定的静态数据类型,而 SQLite 采用的是动态数据类型,会根据存入值自动判断。SQLite 具有以下几种常用的数据类型:

(1) NULL:这个值为空值。

(2) VARCHAR(n):长度不固定且其最大长度为 n 的字串,n 不能超过 4 000。

(3) CHAR(n):长度固定为 n 的字串,n 不能超过 254。

(4) INTEGER:值被标识为整数,依据值的大小可以依次被存储为 1,2,3,4,5,6,7,9。

(5) REAL:所有值都是浮动的数值,被存储为 8 字节的 IEEE 浮动标记序号。

(6) TEXT:值为文本字符串,使用数据库编码存储(TUTF-8、UTF-16BE or UTF-16-LE)。

(7) BLOB:值是 BLOB 数据块,以输入的数据格式进行存储。如何输入就如何存储,不改变格式。

(8) DATA:包含了年份、月份、日期。

(9) TIME:包含了小时、分钟、秒。

9.3.2　SQLiteDatabase 介绍

Android 提供了创建和适用 SQLite 数据库的 API。SQLiteDatabase 代表一个数据库对象,提供了操作数据库的一些方法。在 Android 的 SDK 目录下有 sqlite3 工具,可以利用它创建数据库、创建表和执行一些 SQL 语句。SQLiteDatabase 类封装了一些操作数据库的 API,使用该类可以完成对数据进行添加(Create)、查询(Retrieve)、更新(Update)和删除(Delete)操作,这些操作简称为 CRUD。对 SQLiteDatabase 的学习,应该重点掌握 execSQL()和 rawQuery()方法。

> execSQL()方法可以执行 insert、delete、update 和 CREATE TABLE 之类有更改行为的 SQL 语句;
> rawQuery()方法用于执行 select 语句。

1. execSQL( )方法

首先看一个 execSQL()方法的使用例子:

```
SQLiteDatabase db = ....;
db.execSQL("insert into person(name, age) values('test', 4)");
db.close();
```

执行上面 SQL 语句会往 person 表中添加进一条记录,在实际应用中,语句中的"test"这些参数值会由用户输入界面提供,如果把用户输入的内容原样拼到上面的 insert 语句,当用户输入的内容含有单引号时,拼出来的 SQL 语句就会存在语法错误。要解决这个问题需要对单引号进行转义,也就是把单引号转换成两个单引号。有些时候用户往往还会输入像" & "这些特殊 SQL 符号,为保证组拼好的 SQL 语句语法正确,必须对 SQL 语句中的这些特殊 SQL 符号都进行转义,显然,对每条 SQL 语句都做这样的处理工作是比较烦琐的。SQLiteDatabase 类提供了一个重载后的 execSQL(String sql, Object[] bindArgs)方法,使用这个方法可以解决前面提到的问题,因为这个方法支持使用占位符参数(?)。例子如下:

```
SQLiteDatabase db = ....;
db.execSQL("insert into person(name,age) values(?,?)",new Object[]{"test",4});
db.close();
```

execSQL(String sql, Object[] bindArgs)方法的第一个参数为 SQL 语句,第二个参数为 SQL 语句中占位符参数的值,参数值在数组中的顺序要和占位符的位置对应。

2. rawQuery( )方法

rawQuery()方法用于执行 select 语句,使用例子如下:

```
SQLiteDatabase db = ....;
Cursor cursor = db.rawQuery("select * from person", null);
while (cursor.moveToNext()) {
    int personid = cursor.getInt(0);         //获取第一列的值,第一列的索引从 0 开始
    String name = cursor.getString(1);       //获取第二列的值
    int age = cursor.getInt(2);              //获取第三列的值
}
cursor.close();
db.close();
```

rawQuery()方法的第一个参数为 select 语句;第二个参数为 select 语句中占位符参数的值,如果 select 语句没有使用占位符,该参数可以设置为 null。

3. 游 标

Cursor 是结果集游标,用于对结果集进行随机访问,例如:

```
Cursor cursor = db.rawQuery("select * from person where name like ? and age = ?", new String[]{"%test%", "4"});
```

Cursor 的常用方法:

(1) moveToNext()方法可以将游标从当前行移动到下一行,如果已经移过了结果集的最后一行,返回结果为 false,否则为 true。

(2) moveToPrevious()方法用于将游标从当前行移动到上一行,如果已经移过了结果集的第一行,返回值为 false,否则为 true。

(3) moveToFirst()方法用于将游标移动到结果集的第一行,如果结果集为空,返回值为 false,否则为 true 。

(4) moveToLast()方法用于将游标移动到结果集的最后一行,如果结果集为空,返回值为 false,否则为 true 。

9.3.3 SQLite 数据库编程方法

1. SQLiteDatabase 的常用方法

SQLiteDatabase 的常用方法含义:

(1) 打开或创建数据库

openOrCreateDatabase(String path,SQLiteDatabase.CursorFactory factory)

(2) 插入一条记录

insert(String table,String nullColumnHack,ContentValues values)

(3) 删除一条记录

delete(String table,String whereClause,String[] whereArgs)

(4) 查询一条记录

query(String table,String[] columns,String selection,String[] selectionArgs,String groupBy,String having,String orderBy)

(5) 修改记录

update(String table,ContentValues values,String whereClause,String[] whereArgs)

(6) 执行一条 SQL 语句

execSQL(String sql)

(7) 关闭数据库

close()

2. 打开或者创建数据库

在 Android 中使用 SQLiteDatabase 的静态方法 openOrCreateDatabase(String path,SQLiteDatabae.CursorFactory factory)打开或者创建一个数据库。它会自动去检测是否存在这个数据库,如果存在则打开,不存在则创建一个数据库;创建成功则返回一个 SQLiteDatabase 对象,否则,抛出异常 FileNotFoundException。下面是创建名为"stu.db"数据库的代码:

```
openOrCreateDatabase(String  path,SQLiteDatabae.CursorFactory  factory)
```

参数 1:数据库创建的路径
参数 2:一般设置为 null 就可以了
打开数据库:

```
db = SQLiteDatabase.openOrCreateDatabase("/data/data/com.lingdududu.db/databases/stu.db",null);
```

3. 创建表

创建一张表的步骤很简单,编写创建表的 SQL 语句,调用 SQLiteDatabase 的 execSQL()方法来执行 SQL 语句。下面的代码创建了一张用户表,属性列为:id(主键并且自动增加)、sname(学生姓名)、snumber(学号):

```
private void createTable(SQLiteDatabase db){
//创建表 SQL 语句
String stu_table = "create table usertable(_id integer primary key autoincrement,sname text,snumber text)";
//执行 SQL 语句
db.execSQL(stu_table);
}
```

4. 插入数据

插入数据有两种方法:

(1) SQLiteDatabase 的 insert(String table,String nullColumnHack,ContentValues values)方法,

- 参数 1:表名称;
- 参数 2:空列的默认值;
- 参数 3:ContentValues 类型的一个封装了列名称和列值的 Map。

(2) 编写插入数据的 SQL 语句,直接调用 SQLiteDatabase 的 execSQL()方法来执行

第一种方法的代码:

```
private void insert(SQLiteDatabase db){
    //实例化常量值
    ContentValues cValue = new ContentValues();
    //添加用户名
    cValue.put("sname","xiaoming");
    //添加密码
    cValue.put("snumber","01005");
    //调用 insert()方法插入数据
    db.insert("stu_table",null,cValue);
}
```

第二种方法的代码:

```
private void insert(SQLiteDatabase db){
    //插入数据 SQL 语句
    String stu_sql = "insert into stu_table(sname,snumber)
        values('xiao','01005')";
    //执行 SQL 语句
    db.execSQL(sql);
}
```

5. 删除数据

删除数据也有两种方法:

(1) 调用 SQLiteDatabase 的 delete(String table, String whereClause, String[] whereArgs)方法:

- 参数 1　表名称;
- 参数 2　删除条件;
- 参数 3　删除条件值数组。

(2) 编写删除 SQL 语句,调用 SQLiteDatabase 的 execSQL()方法来执行删除。

第一种方法的代码:

```
private void delete(SQLiteDatabase db) {
    //删除条件
    String whereClause = "id = ?";
    //删除条件参数
    String[] whereArgs = {String.valueOf(2)};
    //执行删除
    db.delete("stu_table",whereClause,whereArgs);
}
```

第二种方法的代码:

```
private void delete(SQLiteDatabase db) {
    //删除 SQL 语句
    String sql = "delete from stu_table where _id = 6";
    //执行 SQL 语句
    db.execSQL(sql);
}
```

6. 修改数据

修改数据有两种方法：

(1) 调用 SQLiteDatabase 的 update(String table, ContentValues values, String whereClause, String[] whereArgs)方法。

- 参数1　表名称；
- 参数2　跟行列 ContentValues 类型的键值对 Key－Value；
- 参数3　更新条件(where 字句)；
- 参数4　更新条件数组。

(2) 编写更新的 SQL 语句，调用 SQLiteDatabase 的 execSQL 执行更新。

第一种方法的代码：

```
private void update(SQLiteDatabase db) {
    //实例化内容值 ContentValues values = new ContentValues();
    //在 values 中添加内容
    values.put("snumber","101003");
    //修改条件
    String whereClause = "id = ?";
    //修改添加参数
    String[] whereArgs = {String.valuesOf(1)};
    //修改
    db.update("usertable",values,whereClause,whereArgs);
}
```

第二种方法的代码：

```
private void update(SQLiteDatabase db){
    //修改 SQL 语句
    String sql = "update stu_table set snumber = 654321 where id = 1";
    //执行 SQL
    db.execSQL(sql);}
```

7. 查询数据

在 Android 中查询数据是通过 Cursor 类来实现，使用 SQLiteDatabase.query()方法时，会得到一个 Cursor 对象，Cursor 指向的就是每一条数据。它提供了很多有关查

询的方法,具体方法如下:

public Cursor query(String table,String[] columns,String selection,String[] selectionArgs,String groupBy,String having,String orderBy,String limit);

各个参数的意义说明:
- 参数 table:表名称;
- 参数 columns:列名称数组;
- 参数 selection:条件字句,相当于 where;
- 参数 selectionArgs:条件字句,参数数组;
- 参数 groupBy:分组列;
- 参数 having:分组条件;
- 参数 orderBy:排序列;
- 参数 limit:分页查询限制;
- 参数 Cursor:返回值,相当于结果集 ResultSet。

Cursor 是一个游标接口,提供了遍历查询结果的方法,如移动指针方法 move(),获得列值方法 getString()等。Cursor 游标常用方法:
- getCount():获得总的数据项数;
- isFirst():判断是否第一条记录;
- isLast():判断是否最后一条记录;
- moveToFirst():移动到第一条记录;
- moveToLast():移动到最后一条记录;
- move(int offset):移动到指定记录;
- moveToNext():移动到下一条记录;
- moveToPrevious():移动到上一条记录;
- getColumnIndexOrThrow(String columnName):根据列名称获得列索引;
- getInt(int columnIndex):获得指定列索引的 int 类型值;
- getString(int columnIndex):获得指定列缩影的 String 类型值。

下面就是用 Cursor 来查询数据库中的数据,具体代码如下:

```
private void query(SQLiteDatabase db) {
    //查询获得游标
    Cursor cursor = db.query ("usertable",null,null,null,null,null,null);
    //判断游标是否为空
    if(cursor.moveToFirst() {
    //遍历游标
    for(int i = 0;i <cursor.getCount();i ++ ){
    cursor.move(i);
    //获得ID
    int id = cursor.getInt(0);
    //获得用户名
```

```
        String username = cursor.getString(1);
        //获得密码
        String password = cursor.getString(2);
        //输出用户信息 System.out.println(id + ":" + sname + ":" + snumber);
        }
    }
}
```

8. 删除指定表

编写插入数据的 SQL 语句,直接调用 SQLiteDatabase 的 execSQL()方法来执行,例如:

```
private void drop(SQLiteDatabase db){
    //删除表的 SQL 语句
    String sql = "DROP TABLE stu_table";
    //执行 SQL
    db.execSQL(sql);
}
```

9.3.4 SQLiteOpenHelper

在开发 SQLite 的过程中为了新建一张数据表,因为数据库不允许同名的数据表存在,所以建表前要先判断该表是否存在,这种处理方式是十分繁琐的。所以 Android 提供了 SQLiteOpenHelper 类来优雅的处理这种问题。

实际项目中很少直接用 SQliteDatabase 的方法操作数据库,都会集成 SQLiteOpen Helper 开发子类,并通过 getReadableDatabase()、getWriteableDatabase()方法打开数据库。

getWritableDatabase()和 getReadableDatabase()方法都可以获取一个用于操作数据库的 SQLiteDatabase 实例。但 getWritableDatabase()方法以读写方式打开数据库,一旦数据库的磁盘空间满了,数据库就只能读而不能写,倘若使用 getWritableDatabase()打开数据库就会出错。getReadableDatabase()方法先以读写方式打开数据库,如果数据库的磁盘空间满了,就会打开失败,当打开失败后会继续尝试以只读方式打开数据库。

```
public class DatabaseHelper extends SQLiteOpenHelper {
    //类没有实例化,是不能用作父类构造器的参数,必须声明为静态
    private static final String name = "count"; //数据库名称
    private static final int version = 1; //数据库版本
    public DatabaseHelper(Context context) {
        //第三个参数 CursorFactory 指定在执行查询时获得一个游标实例的工厂类,
        //设置为 null,代表使用系统默认的工厂类
```

```
            super(context, name, null, version);
        }
        @Override
        public void onCreate(SQLiteDatabase db) {
            db.execSQL("CREATE TABLE IF NOT EXISTS person (personid integer
                primary key autoincrement, name varchar(20), age INTEGER)");
        }
        @Override
        public void onUpgrade(SQLiteDatabase db, int oldVersion, int newVersion){
        db.execSQL("ALTER TABLE person ADD phone VARCHAR(12)");//往表中增加一列
        }
    }
```

在实际项目开发中,当数据库表结构发生更新时,应该避免用户存放于数据库中的数据丢失。该类是 SQLiteDatabase 一个辅助类,这个类主要生成一个数据库,并对数据库的版本进行管理。当在程序当中调用这个类的方法 getWritableDatabase()或者 getReadableDatabase()方法的时候,如果当时没有数据,那么 Android 系统就会自动生成一个数据库。SQLiteOpenHelper 是一个抽象类,通常需要继承它,并且实现里面的3个函数:

(1) onCreate(SQLiteDatabase)

在数据库第一次生成的时候会调用这个方法,也就是说,只有在创建数据库的时候才会调用,当然也有一些其他的情况,一般在这个方法里边生成数据库表。

(2) onUpgrade(SQLiteDatabase, int, int)

当数据库需要升级的时候,Android 系统会主动的调用这个方法。一般在这个方法里边删除数据表,并建立新的数据表,当然是否还需要做其他的操作,完全取决于应用的需求。

(3) onOpen(SQLiteDatabase):

这是当打开数据库时的回调函数,一般在程序中不常使用。

9.4 数据库及动态界面设计目标

数据库及动态界面设计建立在第5章《带导航栏的滑动主界面》的基础上,原来主界面里的第一页是地点列表页,该页采用列表视图来实现,地点的列表是固定的,把所有的地点都固定在界面之中。本章的目标是把这些地点列表信息保存在数据库之中,用户可以自行添加、删除、修改每一项列表的内容,实现动态界面显示。

1. 从数据库中读取界面数据

第5章《带导航栏的滑动主界面》的项目名称是 IoT_IS_App_002,本章将直接利用该项目的代码,方法是整个目录拷贝一份,并且把它命名为 IoT_IS_App_002_test_001

目录。然后在 Obtain_Studio 之中打开该项目。在第 5 章《带导航栏的滑动主界面》之中,列表的信息通过一个 getListItems 函数要获取,回顾一下该函数的内容:

```
private List <Map <String, Object>>getListItems() {
  List <Map <String, Object>>listItems =
           new ArrayList <Map <String, Object>>();
  for(int i = 0; i <goodsNames.length; i++) {
    Map <String, Object>map = new HashMap <String, Object>();
    map.put("image", imgeIDs[i]);        //图片资源
    map.put("title", "信息:");           //标题
    map.put("info", goodsNames[i]);      //名称
    map.put("detail", goodsDetails[i]);  //详情
    listItems.add(map);
  }
  return listItems;
}
```

地点数据保存在 List 列表里,List 列表里的数据又以 map 映射的形式保存,也就是 <key,value> 对的形式。

本章的重点是把该函数修改成数据库的形式保存数据并且从数据库读取数据,返回的内容基本上与 getListItems() 函数相同。下面写一个新的函数 db_test() 来代替原来的 getListItems() 函数,程序代码如下:

```
private SQLiteDatabase db1;
private MyDatabaseUtil myDatabaseUtil;
private List <Map <String, Object>>  db_test()
{
  List <Map <String, Object>>listItems = new ArrayList <Map <String, Object>>();
  //参数二是数据库文件名
  myDatabaseUtil = new MyDatabaseUtil(getActivity(),"obtain_db1.db",
      null,1);
  db1 = getActivity().openOrCreateDatabase("obtain_db1.db",
  Context.MODE_PRIVATE, null);
  /*创建表,并判断是否已经存在此表,没创建,则创建并初始化*/
  if (! myDatabaseUtil.tabIsExist("room")) {
    //创建表 SQL 语句
    String stu_table = "create table room(id integer primary key autoincrement,name
text,ip text,port text,address text,image text)";
    //执行 SQL 语句
    db1.execSQL(stu_table);
    //插入数据 SQL 语句
    String stu_sql = "insert into room(name,ip,port,address,image)
```

第9章 数据库及动态界面设计

```
        values('客厅','192.169.4.100','5000','test1','room.jpg')";
    //执行SQL语句
    db1.execSQL(stu_sql);
    stu_sql = "insert into room(name,ip,port,address,image)
        values('主卧','192.169.4.100','5000','test2','room.jpg')";
    //执行SQL语句
    db1.execSQL(stu_sql);
    stu_sql = "insert into room(name,ip,port,address,image)
        values('客房','192.169.4.100','5000','test2','room.jpg')";
    //执行SQL语句
    db1.execSQL(stu_sql);
}else {
    Log.i("++++++++++","已经创建了,无须再创建");
}
/*查询数据*/
Cursor c = db1.rawQuery("select * from room", null);
c.moveToFirst();
String m_name = "";
while(! c.isAfterLast()){
    String name = c.getString(c.getColumnIndex("name"));
    String ip = c.getString(c.getColumnIndex("ip"));
    String port = c.getString(c.getColumnIndex("port"));
    String address = c.getString(c.getColumnIndex("address"));
    String image = c.getString(c.getColumnIndex("image"));
    String id = c.getString(c.getColumnIndex("id"));
    Map <String, Object >map = new HashMap <String, Object >();
    map.put("image",R.drawable.bedroom);        //图片资源 imgeIDs[i]
    map.put("info", name);                       //物品标题
    map.put("detail", ip);                       //物品名称
    map.put("id", id);                           //物品名称
    listItems.add(map);
    c.moveToNext();
}
return listItems;
}
```

运行效果如图9-3所示。

2. 添加新地点功能

目前还差添加新地点和删除地点的功能。添加新地点功能的基本思路是在主屏幕右上角添加一个"＋"按钮,单击该按钮弹出添加新地点对话框,由于该对话框包含了多个编辑框,结构比较复杂,所以需要创建一个添加和编辑地址对话框布局。在对

图 9-3 采用数据库保存界面信息的运行效果

话框中录入内容之后,首先把该内容保存到数据库之中,然后在第一个页面显示出来。

(1) 添加"+"按钮

在 top_bar.xml 之中添加"+"按钮代码,内容如下:

```
<TextView   android:id = "@ + id/top_bar_add"
            android:layout_width = "wrap_content"
            android:layout_height = "40px"
            android:layout_alignParentRight = "true"
            android:layout_marginRight = "15dp"
            android:text = " + "
            android:textColor = "@color/white"
            android:textSize = "30sp" />
```

(2) 编写添加对话框布局

布局名称为"item_room_dlg.xml",代码如下:

```
<? xml version = "1.0" encoding = "utf - 8"? >
<RelativeLayout xmlns:android = "http://schemas.android.com/apk/res/android"
    android:id = "@ + id/rl_hotelName"
    android:layout_width = "match_parent"
    android:layout_height = "wrap_content"    >
  <LinearLayout android:layout_width = "match_parent"
      android:layout_height = "wrap_content"   android:orientation = "horizontal">
    <ImageView android:id = "@ + id/iv_room_image"
```

```xml
        android:layout_width = "80dp" android:layout_height = "80dp"
        android:src = "@drawable/bedroom" />
<LinearLayout android:layout_width = "match_parent"
        android:layout_height = "wrap_content" android:orientation = "vertical">
    <LinearLayout android:id = "@ + id/rl_addHotel"
        android:layout_width = "match_parent" android:layout_height = "40px"
        android:orientation = "horizontal">
        <TextView android:id = "@ + id/tv_hotelName"
            android:layout_width = "0dp" android:layout_height = "wrap_content"
            android:layout_marginLeft = "5dp" android:text = "地点:" />
        <EditText android:id = "@ + id/ed_hotelName"
            android:layout_width = "0dp" android:layout_height = "wrap_content"
            android:layout_weight = "2" />
        <Button android:id = "@ + id/btn_addHotel"
            android:layout_width = "0dp" android:layout_height = "40dp"
            android:text = "查找" android:textSize = "18sp" />
    </LinearLayout>
    <LinearLayout android:id = "@ + id/ll_addHotelEvaluate"
        android:layout_width = "match_parent" android:layout_height = "wrap_content"
        android:layout_below = "@ + id/rl_addHotel" android:layout_marginTop = "5dp"
        android:orientation = "vertical">
        <RelativeLayout android:id = "@ + id/rl_hotelEvaluate"
            android:layout_width = "match_parent" android:layout_height = "wrap_content"
            android:layout_below = "@ + id/rl_addHotel" android:layout_marginTop = "5dp"
            android:orientation = "horizontal">
            <TextView android:id = "@ + id/tv_hotelServer"
                android:layout_width = "wrap_content" android:layout_height = "wrap_content"
                android:text = "信号:" android:textSize = "18sp" />
            <RatingBar android:id = "@ + id/rb_hotel_evaluate"
                android:layout_width = "wrap_content" android:layout_height = "20dp"
                android:layout_toRightOf = "@ + id/tv_hotelServer"
                android:numStars = "5" />
            <TextView android:id = "@ + id/tv_hotelServer_ID"
                android:layout_width = "wrap_content" android:layout_height = "wrap_content"
                android:layout_toRightOf = "@ + id/rb_hotel_evaluate"
                android:layout_marginLeft = "15dp" android:layout_weight = "1"
                android:text = "ID:" android:layout_centerVertical = "true" />
            <EditText android:layout_toRightOf = "@ + id/tv_hotelServer_ID"
                android:id = "@ + id/ed_hotel_id"
                android:layout_width = "80dp" android:layout_height = "wrap_content"
                android:layout_below = "@ + id/rl_server" />
        </RelativeLayout>
```

```xml
<RelativeLayout android:id = "@ + id/rl_hotelEvaluate"
    android:layout_width = "match_parent" android:layout_height = "wrap_content"
    android:layout_below = "@ + id/rl_addHotel" android:orientation = "horizontal">
    <TextView android:id = "@ + id/ip_hotelServer"
        android:layout_width = "wrap_content" android:layout_height = "wrap_content"
        android:layout_weight = "1" android:text = "IP 地址:"   />
    <EditText android:layout_toRightOf = "@ + id/ip_hotelServer"
        android:id = "@ + id/ed_hotel_ip"
        android:layout_width = "match_parent" android:layout_height = "wrap_content"
        android:layout_below = "@ + id/rl_server" android:singleLine = "true" />
</RelativeLayout> </LinearLayout> </LinearLayout> </LinearLayout> </RelativeLayout>
```

(3) 添加事件响应

在 MainFragmentPagerActivity.java 的 setListener() 函数之中,添加事件响应,程序代码如下:

```java
TextView actionbar_add = (TextView) findViewById(R.id.top_bar_add);
    actionbar_add.setOnClickListener(new View.OnClickListener() {
        @Override
        public void onClick(View v) {
          if(currentPageIndex == 0)
          {
               Fragment1 f1 = (Fragment1)datas.get(currentPageIndex);
               f1.add_new_view("");
}}});
```

(4) 保存到数据库

在 Fragment1 中通过 add_new_view 函数把信息保存到数据库之中,程序代码如下:

```java
public void add_new_view(String str)
{
    LayoutInflater li = getActivity().getLayoutInflater();
    View view = li.inflate(R.layout.item_room_dlg, null);
    final TextView hotelName = (TextView) view.findViewById(R.id.ed_hotelName);
    hotelName.setText("房间");
    final TextView hotel_ip = (TextView) view.findViewById(R.id.ed_hotel_ip);
    hotel_ip.setText("192.169.4.100");
    final TextView hotel_id = (TextView) view.findViewById(R.id.ed_hotel_id);
    hotel_id.setText("null");
    AlertDialog.Builder builder = new AlertDialog.Builder(getActivity());
    builder.setTitle("添加");
    builder.setView(view);
    builder.setPositiveButton("确定", new DialogInterface.OnClickListener()
```

```
        {
            @Override
            public void onClick(DialogInterface dialog, int which)
            {
//show(hotelName.getText().toString());
//插入新记录
        myDatabaseUtil = new MyDatabaseUtil(getActivity(),"obtain_db1.db",null,1);
        db1 = getActivity().openOrCreateDatabase("obtain_db1.db",Context.MODE_PRIVATE,null);
            String sql = "insert into room(name,ip,port,address,image) values('" + hotel-
Name.getText()
            .toString() + "','" + hotel_ip.getText().toString() + "','5000','test1','room.jpg')";
            db1.execSQL(sql);
            Cursor cursor = db1.rawQuery("select last_insert_rowid() from room",null);
            int strid = -1;
            if (cursor.moveToFirst()) strid = cursor.getInt(0);
            String name = hotelName.getText().toString();
            String ip = hotel_ip.getText().toString();
            String port = "";,address = "",image = "";
            String id = Integer.toString(strid);
            addView(name,ip,port,address,image,id);
            } });
        builder.setNegativeButton("取消",null);
        builder.create();
        builder.show();
    }
    public void addView(String name,String ip,String port,String address,String image,String id)
    {
        //addViewItem(this_inflater,null,name,ip,port,address,image,id);
        Map <String, Object>map = new HashMap <String, Object>();
        map.put("image",R.drawable.bedroom );   //图片资源    imgeIDs[i]
        map.put("title", name); //物品标题
        map.put("info", ip); //物品名称
        map.put("id", id); //物品名称
        listItems.add(map);
        listView.setAdapter(listViewAdapter);
        listViewAdapter.notifyDataSetChanged();
    }
}
```

运行效果如本章最开始所介绍的图 9-1 所示。

3. 地点编辑功能

点击第 6 章"带导航栏的滑动主界面"地址列表后面的"查看"按钮,弹出一个对话框,内容如图 9-4 所示。本章的地点编辑功能设计目标将把图 9-4 所示的对话框改

成如图9-5所示的复杂地点编辑对话框。

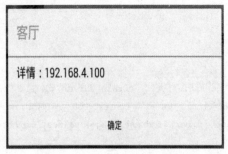

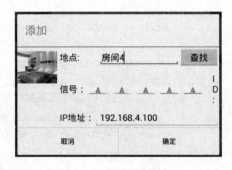

图9-4 查看消息对话框　　　　　图9-5 地点编辑对话框

现在要把该按钮名称修改为"编辑",并且把查看消息对话框的内容修改为地点信息编辑内容。这些内容都在 ListViewAdapter 类的 getView 方法之中。把下面一行程序:

```
listItemView.detail.setText("查看");
```

修改为:

```
listItemView.detail.setText("编辑");
```

原来的查看按钮事件响应程序代码如下:

```
//注册按钮
listItemView.detail.setOnClickListener(new View.OnClickListener() {
    @Override
    public void onClick(View v) {
        //显示物品详情
        private void showDetailInfo(int clickID) {
            new AlertDialog.Builder(context)
            .setTitle("物品详情:" + listItems.get(clickID).get("info"))
            .setMessage(listItems.get(clickID).get("detail").toString())
            .setPositiveButton("确定", null).show();
        }
    }
});
```

把事件响应修改成如下程序:

```
listItemView.detail.setOnClickListener(new View.OnClickListener() {
    @Override
    public void onClick(View v) {
        LayoutInflater li = (LayoutInflater)context.getSystemService(
```

```
            Context.LAYOUT_INFLATER_SERVICE);
        View view = li.inflate(R.layout.item_room_dlg, null);
        AlertDialog.Builder builder = new AlertDialog.Builder(context);
        final TextView m_hotelName = (TextView) view.findViewById(R.id.ed_hotelName);
        m_hotelName.setText(listItemView.title.getText());
        final TextView m_hotel_ip = (TextView) view.findViewById(R.id.ed_hotel_ip);
        m_hotel_ip.setText(listItemView.info.getText());
        final TextView m_hotel_id = (TextView) view.findViewById(R.id.ed_hotel_id);
        m_hotel_id.setText(listItemView.id);
        final ImageView myimageView = (ImageView) view.findViewById(R.id.iv_room_image);
        builder.setTitle("编辑");
        builder.setView(view);
        builder.setPositiveButton("确定", new DialogInterface.OnClickListener()
    {
        @Override
        public void onClick(DialogInterface dialog, int which)
        {
            listItemView.title.setText(m_hotelName.getText());
            listItemView.info.setText(m_hotel_ip.getText());
            //hotel_id.setText(m_hotel_id.getText());
            //更新数据库
            MyDatabaseUtil myDatabaseUtil = new MyDatabaseUtil(context, "obtain_db1.db", null, 1);
            SQLiteDatabase db1 = context.openOrCreateDatabase("obtain_db1.db", Context.MODE
_PRIVATE, null);
            // SQL 修改语句
            String sql = "update room set name = '" + m_hotelName.getText() + "',
                ip = '" + m_hotel_ip.getText() + "' where id = " + m_hotel_id.getText();
            db1.execSQL(sql); //执行 SQL
        }
    });
    builder.setNegativeButton("取消",null);
    builder.create();
    builder.show();
    }
});
```

4. 删除地点功能

删除数据库中的数据和删除 ListViewAdapter 中的数据之后,不能直接改变界面的显示内容,需要进行视图的刷新。直接在 ListViewAdapter 里采 notifyDataSetChanged()函数刷新时,显示不正确,不管删除哪一项都是显示删除最后一项。因为需要采用 Handler 类异步刷新。Handler 的功能主要是接收子线程发送的数据,并用此数据配合主线程更新 UI。删除事件响应程序如下:

```
final AlertDialog dialog1 = builder.show();
final Button m_addHotel = (Button) view.findViewById(R.id.btn_addHotel);
m_addHotel.setText("删除");
m_addHotel.setOnClickListener(new View.OnClickListener()
{
    @Override
    public void onClick(View v)
    {
        AlertDialog dialog2 = new AlertDialog.Builder(context)
        .setTitle("删除信息?")    //创建标题
        .setMessage("您确定要删除" + listItemView.title.getText() + "一项吗?")
        //.setIcon(R.drawable.ic_launcher) //设置LOGO
        .setPositiveButton("删除", new DialogInterface.OnClickListener() {
        public void onClick(DialogInterface dialog, int which) {
        listItems.remove(selectID);
        mHandler.sendEmptyMessage(REFRESH);
        //更新数据库
        MyDatabaseUtil myDatabaseUtil = new MyDatabaseUtil(context, "obtain_db1.db", null,1);
            SQLiteDatabase db1 = context.openOrCreateDatabase("obtain_db1.db",
                Context.MODE_PRIVATE, null);
            String sql = "delete from room where id = " + m_hotel_id.getText();//SQL删除语句
            db1.execSQL(sql); //执行SQL
            dialog1.dismiss();
            return;
        }
    }).setNegativeButton("取消", new DialogInterface.OnClickListener() {
        public void onClick(DialogInterface dialog, int which) {
        }
    }).create();    //创建对话框
        dialog2.show();    //显示对话框
    }});
}});
```

在Fragment1中添加Handler消息接收程序,程序代码如下:

```
Handler mHandler = new Handler(){
    public void handleMessage(android.os.Message msg){
    switch (msg.what) {
      case REFRESH:
            listView.setAdapter(listViewAdapter);
            listViewAdapter.notifyDataSetChanged();
          break;
      default:
          break;
      }
    };
};
```

运行效果如本章最开始所介绍的图9-2所示。

第 10 章
嵌入网页的控制界面设计

10.1 嵌入网页的控制界面设计目标

在第 7 章"Wi-Fi 物联网移动软件设计"中介绍的物联网智能设备控制界面比较容易设计,运行速度也比较快,也可以做到功能很强大。但是如果有很多的地点(应用场景)以及每个地点又有很多不一样的设备,这样子需要使用大量的控制界面,如果继续采用该设计方式则需要大量的 XML 布局文件,并且这些布局文件在软件编译时就必须包括在项目之中,不方便扩展和修改。

为了解决上述难题,本章将介绍采用嵌入网页的形式来实现控制,这样做的优点是:

(1) 界面可以采用 HTML、CCS、JavaScript 编写,可以实现更加复杂和优美的界面;

(2) HTML 控制界面可以在软件编译完成之后,在应用的过程中增加和修改;

(3) 可以从网上加载 HTML 控制界面,不同的设备只要通过扫描该设备的二维码,然后连接到网上的 HTML 控制界面即可实现控制,也可以把该控制界面保存到本地,方便在不连网络的情况下也能进行控制。

嵌入网页的控制界面系统结构如图 10-1 所示,分成两个回路,一个是控制界面下载回路,一个是控制回路。控制界面下载回路通过扫描设备上的二维码获取控制页面

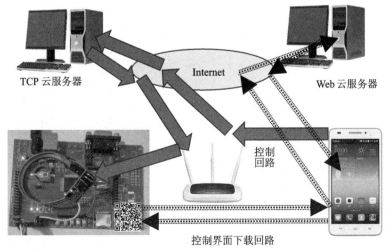

图 10-1 嵌入网页的控制界面系统结构

链接,然后从 Web 服务器上下载 HTML 控制界面。控制回路则启动下载到的 HTML 控制界面,通过底层的 WebView 组件连接到 TCP 云服务器上,然后通过 TCP 云服务器转发控制指令控制物联网智能设备,本章采用 STM32 开发板作为控制物联网智能设备。

本章介绍的嵌入网页控制界面模式,网页只是作为界面显示,真正的网络连接则通过底层的 WebView 组件和 Socket 组件,以通用 TCP 的形式联网和控制,这样做可以避免直接网页控制的时实连接与时实刷新问题。

HTML 界面与 WebView 组件之间的连接,通过 JavaScript 程序来实现,其实现方式如图 10-2 所示。

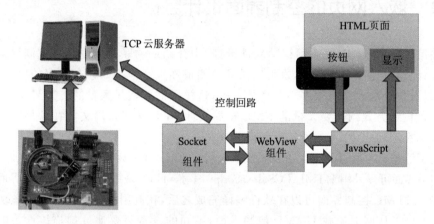

图 10-2　HTML 界面与 WebView 组件之间的连接结构图

本章设计目标是设计一个简单的 HTML 控制界面,界面大致如图 10-3 所示。

图 10-3　简单的 HTML 控制界面

10.2　Android Http

10.2.1　Android Http 通信

　　Android 与服务器通信通常采用 HTTP 通信方式和 Socket 通信方式,而 HTTP 通信方式又分 Get 和 Post 两种方式。

　　Post 请求可以向服务器传送数据,而且数据放在 HTML HEADER 内一起传送到服务端 URL 地址,数据对用户不可见。而 Get 是把参数数据队列加到提交的 URL 中,值和表单内各个字段一一对应,例如:

　　http://www.baidu.com/s?　w=%C4&inputT=2710

　　Get 传送的数据量较小,不能大于 2 KB。Post 传送的数据量较大,一般被默认为不受限制。但理论上,IIS4 中最大量为 80 KB,IIS5 中为 100 KB。Get 安全性非常低,Post 安全性较高。

　　Get 机制用的是在 URL 地址里面通过"?"号间隔,然后以 name=value 的形式给客户端传递参数。所以首先要在 Android 工程下的 AndroidGetTest.java 中 onCreate 方法定义好其 URL 地址以及要传递的参数,然后通过 URL 打开一个 HttpURLConnection 链接,此链接可以获得 InputStream 字节流对象,也是往服务端输出和从服务端返回数据的重要过程,而若服务端 response.getInputStream.write()往 Andorid 返回信息时候,就可以通过 InputStreamReader 作转换,将返回来的数据用 BufferReader 显示出来。

　　Post 传输方式不在 URL 里传递,也正好解决了 Get 传输量小、容易篡改及不安全等一系列不足。主要是通过对 HttpURLConnection 的设置,让其支持 Post 传输方式,然后在通过相关属性传递参数(若需要传递中文字符,则可以通过 URLEncoder 编码,而在获取端采用 URLDecoder 解码即可)。

1. HttpURLConnection 接口

　　首先需要明确的是,Http 通信中的 Post 和 Get 请求方式的不同。Get 可以获得静态页面,也可以把参数放在 URL 字符串后面,传递给服务器。而 Post 方法的参数是放在 Http 请求中。因此,在编程之前,应当首先明确使用的请求方法,然后再根据所使用的方式选择相应的编程方式。

　　HttpURLConnection 是继承自 URLConnection 类,二者都是抽象类。其对象主要通过 URL 的 openConnection 方法获得。创建方法代码如下:

```
URL url = new URL(http://www.51cto.com/index.jsp? par = 123456);
HttpURLConnection urlConn = (HttpURLConnection)url.openConnection();
```

　　通过以下方法可以对请求的属性进行一些设置,代码如下:

```
//设置输入和输出流
urlConn.setDoOutput(true);
urlConn.setDoInput(true);
//设置请求方式为 Post
urlConn.setRequestMethod("POST");
//Post 请求不能使用缓存
urlConn.setUseCaches(false);
urlConn.disConnection();
```

HttpURLConnection 默认使用 Get 方式,代码如下:

```
//使用 HttpURLConnection 打开链接
HttpURLConnection urlConn = (HttpURLConnection) url.openConnection();
//得到读取的内容(流)
InputStreamReader in = new InputStreamReader(urlConn.getInputStream());
// 为输出创建 BufferedReader
BufferedReader buffer = new BufferedReader(in);
String inputLine = null;
//使用循环来读取获得的数据
while (((inputLine = buffer.readLine()) != null))
{
    //在每一行后面加上一个"\n"来换行
    resultData += inputLine + "\n";
}
//关闭 InputStreamReader
in.close();
//关闭 Http 链接
urlConn.disconnect();
```

如果需要使用 Post 方式,则需要 setRequestMethod 设置,代码如下:

```
String httpUrl = "http://192.164.1.110:8080/httpget.jsp";
//获得的数据
String resultData = "";
URL url = null;
try {//构造一个 URL 对象
    url = new URL(httpUrl);
}
catch (MalformedURLException e) {
    Log.e(DEBUG_TAG, "MalformedURLException");
}
if (url != null) {
  try  {// 使用 HttpURLConnection 打开连接
```

```
HttpURLConnection urlConn = (HttpURLConnection) url.openConnection();
//因为这个是Post请求,设立需要设置为true
urlConn.setDoOutput(true);
urlConn.setDoInput(true);
// 设置以Post方式
urlConn.setRequestMethod("POST");
// Post请求不能使用缓存
urlConn.setUseCaches(false);
urlConn.setInstanceFollowRedirects(true);
// 配置本次连接的Content-type,配置为
//application/x-www-form-urlencoded
urlConn.setRequestProperty("Content-Type",
"application/x-www-form-urlencoded");
// 连接,从postUrl.openConnection()至此的配置必须要在connect之前完成,
// 要注意的是connection.getOutputStream会隐含地进行connect
urlConn.connect();
//DataOutputStream流
DataOutputStream out = new DataOutputStream(urlConn.getOutputStream());
//要上传的参数
String content = "par=" + URLEncoder.encode("ABCDEFG", "gb2312");
//将要上传的内容写入流中
out.writeBytes(content);
//刷新、关闭
out.flush();
out.close();
```

2. HttpClient 接口

使用 Apache 提供的 HttpClient 接口同样可以进行 Http 操作。Get 和 Post 请求方法的操作有所不同。Get 方法的操作代码示例如下：

```
// http 地址
String httpUrl =
  "http://192.164.1.110:8080/httpget.jsp?par=HttpClient_android_Get";
//HttpGet 连接对象
HttpGet httpRequest = new HttpGet(httpUrl);
  //取得 HttpClient 对象
  HttpClient httpclient = new DefaultHttpClient();
  //请求 HttpClient,取得 HttpResponse
  HttpResponse httpResponse = httpclient.execute(httpRequest);
  //请求成功
  if (httpResponse.getStatusLine().getStatusCode()
== HttpStatus.SC_OK)
```

```
        {
            //取得返回的字符串
            String strResult = EntityUtils.toString(httpResponse.getEntity());
            mTextView.setText(strResult);
        }
        else
        {
            mTextView.setText("请求错误!");
        }
```

使用 Post 方法进行参数传递时,需要使用 NameValuePair 来保存要传递的参数。另外,还需要设置所使用的字符集,代码如下:

```
// http 地址
String httpUrl = "http://192.164.1.110:8080/httpget.jsp";
//HttpPost 连接对象
HttpPost httpRequest = new HttpPost(httpUrl);
//使用 NameValuePair 来保存要传递的 Post 参数
List <NameValuePair> params = new ArrayList <NameValuePair>();
//添加要传递的参数
params.add(new BasicNameValuePair("par", "HttpClient_android_Post"));
//设置字符集
HttpEntity httpentity = new UrlEncodedFormEntity(params, "gb2312");
//请求 httpRequest
httpRequest.setEntity(httpentity);
//取得默认的 HttpClient
HttpClient httpclient = new DefaultHttpClient();
//取得 HttpResponse
HttpResponse httpResponse = httpclient.execute(httpRequest);
//HttpStatus.SC_OK 表示连接成功
if (httpResponse.getStatusLine().getStatusCode() == HttpStatus.SC_OK)
{
    //取得返回的字符串
    String strResult = EntityUtils.toString(httpResponse.getEntity());
    mTextView.setText(strResult);
}
else
{
    mTextView.setText("请求错误!");
}
```

HttpClient 实际上是对 Java 提供方法的一些封装,在 HttpURLConnection 中的输入输出流操作,在这个接口中被统一封装成了 HttpPost(HttpGet)和 HttpResponse,这样就减少了操作的繁琐性。另外,在使用 Post 方式进行传输时,需要进行字符编码。

3. Android 访问远程网页取回 json 数据

php 代码如下:

```php
$array = array(
  'username' => 'test',
  'password' => '123456',
  'user_id' => );
echo json_encode($array);
```

Java 代码如下:

```java
private void startUrlCheck(String username, String password)
{
    HttpClient client = new DefaultHttpClient();
    StringBuilder builder = new StringBuilder();
    HttpGet myget = new HttpGet("http://10.0.2.2/Android/index.php");
    try {
        HttpResponse response = client.execute(myget);
        BufferedReader reader =
            new BufferedReader(new InputStreamReader(
            response.getEntity().getContent()));
        for (String s = reader.readLine();
            s != null; s = reader.readLine()) {
            builder.append(s);
        }
        JSONObject jsonObject = new JSONObject(builder.toString());
        String re_username = jsonObject.getString("username");
        String re_password = jsonObject.getString("password");
        int re_user_id = jsonObject.getInt("user_id");
        setTitle("用户 id_" + re_user_id);
        Log.v("url response", "true = " + re_username);
        Log.v("url response", "true = " + re_password);
    } catch (Exception e) {
        Log.v("url response", "false");
        e.printStackTrace();
    }
}
```

其中 http://10.0.2.2 为 Android 访问本机 URL 的 IP 地址。对应电脑上测试的 http://127.0.0。另外,执行代码时会抛出异常:

java.net.SocketException:Permission denied

此为应用访问网络的权限不足，在AndroidManifest.xml中，需要进行如下配置：
<uses-permission android:name="android.permission.INTERNET" />

10.2.2　Okhttp

自从 Android 4.4 开始，Google 已经开始将源码中的 HttpURLConnection 替换为 OkHttp，而在 Android 10.0 之后的 SDK 中 Google 更是移除了对于 HttpClient 的支持，而市面上流行的 Retrofit 同样是使用 OkHttp 进行再次封装而来的。由此看见学习 OkHttp 的重要性。OkHttp 是一个精巧的网络请求库，有如下特性：

1) 支持 Http2，对一台机器的所有请求共享同一个 Socket；
2) 内置连接池，支持连接复用，减少延迟；
3) 支持透明的 gzip 压缩响应体；
4) 通过缓存避免重复的请求；
5) 请求失败时自动重试主机的其他 IP，自动重定向；
6) 好用的 API。

OkHttp3 是 Java 和 Android 都能用，Android 还有一个著名网络库叫 Volley，那个只有 Android 能用。自己导入 jar 包，别漏了 okio：

➢ okhttp-3.3.0.jar；
➢ okio-1.8.0.jar。

maven 方式：

```
<dependency>
    <groupId>com.squareup.okhttp3</groupId>
    <artifactId>okhttp</artifactId>
    <version>3.3.0</version>
</dependency>
```

gradle 方式：

```
compile 'com.squareup.okhttp3:okhttp:3.3.0'
Get 请求
Stringurl = "https://www.baidu.com/";
OkHttpClient okHttpClient = newOkHttpClient();
Request request = newRequest.Builder()
    .url(url)
    .build();
Call call = okHttpClient.newCall(request);
try {
    Response response = call.execute();
    System.out.println(response.body().string());
} catch (IOException e) {
    e.printStackTrace();
}
```

如果需要在 request 的的 header 添加参数。例如 Cookie,User－Agent 什么的,就是

```
Request request = newRequest.Builder()
    .url(url)
    .header("键","值")
    .header("键","值")
    ...
    .build();
```

response 的 body 有很多种输出方法,string()只是其中之一,注意是 string()不是 toString()。如果是下载文件就是 response.body().bytes()。

另外,可以根据 response.code()获取返回的状态码。

1. Post 请求

```
Stringurl = "https://www.baidu.com/";
OkHttpClient okHttpClient = newOkHttpClient();

RequestBody body = newFormBody.Builder()
    .add("键", "值")
    .add("键", "值")
    ...
    .build();

Request request = newRequest.Builder()
    .url(url)
    .post(body)
    .build();

Call call = okHttpClient.newCall(request);
try {
    Response response = call.execute();
    System.out.println(response.body().string());
} catch (IOException e) {
    e.printStackTrace();
}
```

Post 请求创建 request 和 Get 是一样的,只是 Post 请求需要提交一个表单,就是 RequestBody。表单的格式有好多种,普通的表单是:

```
RequestBody body = newFormBody.Builder()
    .add("键", "值")
    .add("键", "值")
    ...
    .build();
```

RequestBody 的数据格式都要指定 Content－Type,常见的有三种:

➢ application/x-www-form-urlencoded 数据是个普通表单；
➢ multipart/form-data 数据里有文件；
➢ application/json 数据是个 json。

但是好像以上的普通表单并没有指定 Content-Type，这是因为 FormBody 继承了 RequestBody，它已经指定了数据类型为 application/x-www-form-urlencoded。

private static final MediaType CONTENT_TYPE = MediaType.parse("application/x-www-form-urlencoded");

再看看数据为其他类型的 RequestBody 的创建方式。

如果表单是个 json：

MediaType JSON = MediaType.parse("application/json; charset=utf-8");
RequestBody body = RequestBody.create(JSON, "你的 json");

如果数据包含文件：

RequestBody requestBody = newMultipartBody.Builder()
　　.setType(MultipartBody.FORM)
　　　　.addFormDataPart("file", file.getName(), RequestBody.create(MediaType.parse("image/png"), file))
　　　　.build();

上面的 MultipartBody 也是继承了 RequestBody，看下源码可知它适用于这五种 Content-Type：

public static final MediaType MIXED = MediaType.parse("multipart/mixed");
public static final MediaType ALTERNATIVE = MediaType.parse("multipart/alternative");
public static final MediaType DIGEST = MediaType.parse("multipart/digest");
public static final MediaType PARALLEL = MediaType.parse("multipart/parallel");
public static final MediaType FORM = MediaType.parse("multipart/form-data");

另外，如果上传一个文件不是一张图片，但是 MediaType.parse("image/png")里的"image/png"不知道该填什么，可以参考下这个页面。

2. 同步与异步

从上文已经能知道 call.execute()就是在执行 Http 请求了，但是这是个同步操作，是在主线程运行的。如果在 Android 的 UI 线程直接执行这句话就出异常了。

OkHttp 也实现了异步，写法是：

```
Stringurl = "https://www.baidu.com/";
OkHttpClient okHttpClient = newOkHttpClient();
Request request = newRequest.Builder()
```

```java
            .url(url)
            .build();
Call call = okHttpClient.newCall(request);
call.enqueue(new Callback() {
    @Override
    public void onFailure(Call call, IOException e) {
        e.printStackTrace();
    }
    @Override
    public void onResponse(Call call, Response response) throws IOException {
        System.out.println("我是异步线程,线程 Id 为:" +
            Thread.currentThread().getId());
    }
});
for (int i = 0; i <10; i++) {
    System.out.println("我是主线程,线程 Id 为:"
        + Thread.currentThread().getId());
    try {
        Thread.currentThread().sleep(100);
    } catch (InterruptedException e) {
        e.printStackTrace();
    }
}
```

执行结果是:

我是主线程,线程 Id 为:1

我是主线程,线程 Id 为:1

我是主线程,线程 Id 为:1

我是异步线程,线程 Id 为:11

显然 onFailure()和 onResponse()分别是在请求失败和成功时会调用的方法。这里有个要注意的地方,onFailure()和 onResponse()是在异步线程里执行的,所以,如果你在 Android 把更新 UI 的操作写在这两个方法里面是会报错的,这个时候可以用 runOnUiThread 这个方法。

3. 自动管理 Cookie

Request 经常都要携带 Cookie,上面说过 request 创建时可以通过 header 设置参数,Cookie 也是参数之一。就像下面这样:

```
Request request = newRequest.Builder()
    .url(url)
    .header("Cookie", "xxx")
    .build();
```

然后可以从返回的 response 里得到新的 Cookie，你可能得想办法把 Cookie 保存起来。但是 OkHttp 可以不用我们管理 Cookie，自动携带，保存和更新 Cookie。

方法是在创建 OkHttpClient 设置管理 Cookie 的 CookieJar：

```
private final HashMap <String, List <Cookie>>cookieStore = new HashMap <>();
OkHttpClient okHttpClient = newOkHttpClient.Builder()
    .cookieJar(new CookieJar() {
        @Override
        public void saveFromResponse(HttpUrl httpUrl, List <Cookie>list) {
            cookieStore.put(httpUrl.host(), list);
        }
        @Override
        public List <Cookie>loadForRequest(HttpUrl httpUrl) {
            List <Cookie>cookies = cookieStore.get(httpUrl.host());
            return cookies != null ? cookies : new ArrayList <Cookie>();
        }
    })
    .build();
```

这样以后发送 Request 都不用管 Cookie 这个参数，也不用去 response 获取新 Cookie 了。还能通过 cookieStore 获取当前保存的 Cookie。最后，"new OkHttpClient()" 一行程序只是一种快速创建 OkHttpClient 的方式，更标准的是使用 OkHttpClient.Builder()。后者可以设置一堆参数，例如超时时间。

10.3　WebView 应用

10.3.1　WebView 介绍

1. Android WebView

Android WebView 在 Android 平台上是一个特殊的 View，基于 webkit 引擎、展现 Web 页面的控件，这个类可以被用来在 App 中仅仅显示一张在线的网页，还可以用来开发浏览器。WebView 内部实现是采用渲染引擎来展示 View 的内容，提供网页前进后退，网页放大、缩小，搜索。Android 的 WebView 在低版本中和高版本中采用了不同的 webkit 版本内核，Android4.4 后直接使用了 Chrome。

现在很多 APP 都内置了 Web 网页，比如说很多电商平台，淘宝、京东、聚划算等。WebView 比较灵活，不需要升级客户端，只需要修改网页代码即可。一些经常变化的页面可以用 WebView 这种方式去加载网页。例如，中秋节跟国庆节打开的页面不一样，如果是用 WebView 显示的话，只修改 HTML 页面就行，而不需要升级客户端。

WebView 功能强大，可以直接使用 HTML 文件（本地 sdcard/assets 目录），还可以直接加载 Url,使用 JavaScript 可以 HTML 跟原生 APP 互调。

2. 四种加载 HTML 方式

四种加载 HTML 方式如下：

(1) 方式 1:webView. loadUrl("https://www.duba.com ");//加载 Url

(2) 方式 2:webView. loadUrl("file://android_asset/test. HTML");//加载 asset 文件夹下 HTML

(3) 方式 3:加载手机 SDcard 上的 HTML 页面
webView. loadUrl("content://com. ansen. webview/sdcard/test. HTML");

(4) 方式 4:使用 WebView 显示 HTML 代码
webView. loadDataWithBaseURL(null," < HTML > < head > < title > 欢迎您 </title > </head > < body > < h2 > 使用 WebView 显示 HTML 代码 </h2 > </body > </HTML >", "text/HTML" , "utf-8", null);

10.3.2　WebView 应用

WebView 可以使得网页轻松地内嵌到 App 里,还可以直接跟 JS 相互调用。WebView 有两个方法,setWebChromeClient 和 setWebClient:

(1) setWebClient:主要处理解析,渲染网页等浏览器做的事情;

(2) setWebChromeClient:辅助 WebView 处理 Javascript 的对话框,网站图标,网站 Title,加载进度等。

WebViewClient 就是帮助 WebView 处理各种通知、请求事件的。在 AndroidManifest. xml 设置访问网络权限：

< uses - permission android:name = "android. permission. INTERNET"/>

控件用法如下：

```
<WebView
    android:layout_width = "match_parent"
    android:layout_height = "match_parent"
    android:id = "@ + id/webView"
    />
```

1. 加载本地/Web 资源

example. HTML 存放在 assets 文件夹内。调用 WebView 的 loadUrl()方法,加载本地资源：

webView = (WebView) findViewById(R. id. webView);

webView. loadUrl("file://android_asset/example. HTML");

加载 Web 资源：

webView = (WebView) findViewById(R. id. webView);

```
webView.loadUrl("http://baidu.com");
```

2. 在程序内打开网页

创建一个自己的 WebViewClient，通过 setWebViewClient 关联，例如：

```
package com.example.testopen;
import android.app.Activity;
import android.os.Bundle;
import android.webkit.WebView;
import android.webkit.WebViewClient;
public class MainActivity extends Activity {
private WebView webView;
    @Override
    protected void onCreate(Bundle savedInstanceState) {
        super.onCreate(savedInstanceState);
        setContentView(R.layout.test);
        init();
    }
private void init(){
        webView = (WebView) findViewById(R.id.webView);
        //WebView 加载 Web 资源
        webView.loadUrl("http://baidu.com");
        //覆盖 WebView 默认使用第三方或系统默认浏览器打开网页的行为，使网页用 Web-
View 打开
        webView.setWebViewClient(new WebViewClient(){
            @Override
        public boolean shouldOverrideUrlLoading(WebView view, String url) {
                //返回值是 true 的时候控制去 WebView 打开
                //为 false 调用系统浏览器或第三方浏览器
                view.loadUrl(url);
            return true;
            }
        });
    }
}
```

10.3.3　Android 与 JS 通过 WebView 互相调用方法

Android 与 JS 通过 WebView 互相调用方法，Android 与 JS 二者沟通的桥梁是 WebView，实际上分为以下两种情况：

(1) Android 去调用 JS 的代码

对于 Android 调用 JS 代码的方法有两种：

➢ 通过 WebView 的 loadUrl();

➢ 通过 WebView 的 evaluateJavascript()。

Android 去调用 JS 的代码,例如:

```
/ * * * Android --- 传值 --->JS
    * 调用 JS 中的函数:setData(msg)
    * @param name world 方法中的响应参数 */
public void sendToJs(String name,String data) {
    webView.loadUrl("javascript:setData('" + name + "','" + data + "')");
}
```

(2) JS 去调用 Android 的代码

对于 JS 调用 Android 代码的方法有三种:

➢ 通过 WebView 的 addJavascriptInterface()进行对象映射;

➢ 通过 WebViewClient 的 shouldOverrideUrlLoading()方法回调拦截 Url;

➢ 通过 WebChromeClient 的 onJsAlert()、onJsConfirm()、onJsPrompt()方法回调拦截 JS 对话框 alert()、confirm()、prompt()消息。

JS 去调用 Android 的代码,例如:

```
webView.addJavascriptInterface(new MyJsInterface(),"javaObject");
//要用来被 JS 调用的 Java 对象
private final class MyJsInterface{
//要用来被 JS 调用的 Java 方法
@JavascriptInterface
public void javaDoIt(final String str){
    new AlertDialog.Builder(getActivity())
        .setTitle("Java 对话框")
        .setMessage("来自 Web 的内容:" + str)
        .setPositiveButton("确定", null)
        .show();
}
};
```

10.4　嵌入网页的控制界面的实现

本项目在第 9 章"数据库及动态界面设计"中的项目"IoT_IS_App_002_test_001"的基础之上实现,采用该项目的第三个页面来显示网页。第三个页面的布局文件为 f3.xml,程序代码如下:

```
<? xml version = "1.0" encoding = "utf-8"? >
<LinearLayout xmlns:android = "http://schemas.android.com/apk/res/android"
    android:layout_width = "match_parent"
```

```xml
        android:layout_height = "match_parent"
        android:orientation = "vertical" >
         <EditText
             android:id = "@ + id/input_et"
             android:layout_width = "wrap_content"
             android:layout_height = "wrap_content"
             android:hint = "请输入信息" />
              <Button
             android:id = "@ + id/button1"
             android:text = "刷新"
             android:layout_width = "wrap_content"
             android:layout_height = "wrap_content" />
            <WebView
        android:id = "@ + id/webView"
        android:layout_width = "match_parent"
        android:layout_height = "match_parent" />

</LinearLayout>
```

第三个页面的实现文件为 Fragment3.java，程序代码如下：

```java
public class Fragment3 extends Fragment{
    private WebView webView;
    private View view;
    @SuppressLint("SetJavaScriptEnabled")
    @Override
    public View onCreateView(LayoutInflater inflater, ViewGroup container,
            Bundle savedInstanceState) {
        view = inflater.inflate(R.layout.f3, container, false);
        webView = (WebView)view.findViewById(R.id.webView);
        webView.getSettings().setJavaScriptEnabled(true);
        webView.addJavascriptInterface(new MyJsInterface(),"javaObject");
        webView.setWebChromeClient(new WebChromeClient());
        webView.setWebViewClient(new WebViewClient());
        webView.getSettings().setJavaScriptEnabled(true);
        final EditText edit1 = (EditText)view.findViewById(R.id.edit1);
        view.findViewById(R.id.button1).setOnClickListener(new
          View.OnClickListener() {
            @Override
            public void onClick(View view) {
                //在 Java 中调用 JS 代码
                sendToJs("Java 命令:", edit1.getText().toString());
            }
        });
```

```
            //导入和显示网页
            webView.loadUrl("file:///android_asset/webview.HTML");
            return view;
        }
        /*** Android ---传值--->JS
         * 调用 JS 中的函数:setData(msg)
         * @param name world 方法中的响应参数 */
        public void sendToJs(String name,String data) {
            webView.loadUrl("javascript:setData('" + name + "','" + data + "')");
        }
        //要用来被 JS 调用的 Java 对象
        private final class MyJsInterface{
            //要用来被 JS 调用的 Java 方法
            @JavascriptInterface
            public void javaDoIt(final String str){
                new AlertDialog.Builder(getActivity())
                    .setTitle("Java 对话框")
                    .setMessage("来自 Web 的内容:" + str)
                    .setPositiveButton("确定", null)
                    .show();
            }
            @JavascriptInterface
            public void send3(final String name,final String idata,final String strd){
            String str = "({success:'[{\"id\":\"1\",\"deviceID\":\"01\",\"pwd\":\"123\",";
                str + = "\"name\":\"" + name + "\",\"idata\":\"" + idata + "\",\"strd\":\"" + strd + "\"";
                str + = "}]'})\r\n";
                //senddata(str);
                new AlertDialog.Builder(getActivity())
                    .setTitle("来自 Web 的内容")
                    .setMessage(str)
                    .setPositiveButton("确定", null)
                    .show();
            }
        }
    }
```

网页文件为 webview.html,保存在项目的 asset 目录之中,程序代码如下:

```
<!DOCTYPE HTML>
<HTML>
<head>
    <meta charset = "gbk">
    <title>Android WebView 与 Javascript 交互</title>
```

```
</head>
<body>
    <div>
        <button id = "button1" onclick = "OnClicked('led1','1')">LED1 亮 </button>
        <button id = "button2" onclick = "OnClicked('led1','0')">LED1 灭 </button>
        <br/>
        <button id = "button3" onclick = "OnClicked('led2','1')">LED2 亮 </button>
        <button id = "button4" onclick = "OnClicked('led2','0')">LED2 灭 </button>
    </div>
<br/> <div id = "main"> </div>
</body>
    <script>
        function OnClicked(name,i_data)
        {
            alert(name + ":" + data);
            javaObject.javaDoIt_send3(name,i_data, "");
        }
        function setData(name, data)
        {
            var text = name + ":" + data + " <br/>"
                        + document.getElementById('main').innerHTML;
            document.getElementById('main').innerHTML = text;
        }
    </script>
</HTML>
```

编译之后运行,运行效果如图 10-4 所示,实现了 WebView 组件显示网页并通过 JavaScript 与网页交互。

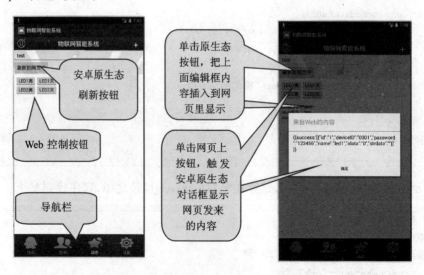

图 10-4 WebView 组件显示网页并通过 JavaScript 与网页交互的效果

第10章 嵌入网页的控制界面设计

为了让上述程序可以通过单击网页上的按钮来控制STM32板上LED灯的亮灭，还需要加入TCP客户端程序，然后通过上述程序中的"senddata(str);"一行程序把控制命令发到STM32板上，实现LED的控制。在Fragment3.java里，添加TCP客户端程序，在初始化程序之中或者在按钮的事件响应程序中调用定时器函数tcp_init_timer()来启动TCP客户端程序。程序代码如下：

```java
////TCP客户端程序
    public Socket s;
    private myThread newthread;
    public boolean tcp_init()
    {
        try {   //创建TCP通信Socket//
            s = new Socket("192.168.4.100",5000);
            newthread = new myThread();
            newthread.start();
        } catch (IOException e) {
                Toast.makeText(getActivity(), "TCP初始化不成功",
                Toast.LENGTH_SHORT).show() ;
                e.printStackTrace();
        }
        return true;
    }
    public class   myThread extends Thread {
      public void run(){
        String str;
        try {
          //编写线程的代码
          BufferedReader input = new BufferedReader(
              new InputStreamReader(s.getInputStream(),"gbk"));
          while(true)
          {
              String message = input.readLine();
              //sendToJs("接收到的数据:", message);
          }
        } catch (IOException e) {
          Toast.makeText(getActivity(), "myThread中接收TCP不成功",
              Toast.LENGTH_SHORT).show() ;
          e.printStackTrace();
        }
      }
    }
    public void senddata(String str)
```

```
{
    try {
        OutputStream out = s.getOutputStream();
        // 注意第二个参数为 true 将会自动 flush,否则,需要手动操作 out.flush()
        PrintWriter output = new PrintWriter(out, true);
        out.write(str.getBytes("GBK"));
        out.flush();
    } catch (UnknownHostException e) {
        Toast.makeText(getActivity(),"提交不成功",Toast.LENGTH_SHORT).show();
        e.printStackTrace();
    } catch (IOException e) {
        Toast.makeText(getActivity(),"提交不成功",Toast.LENGTH_SHORT).show();
        e.printStackTrace();
    }
}
//初始化定时器
public void tcp_init_timer()
{
    final Timer timer = new Timer();
    timer.schedule(new TimerTask() {
        @Override
        public void run() {
            tcp_init();
            if(timer != null){timer.cancel();}
        }
    },1,1);
}
```

第 11 章 传感器应用及拍照更换界面图片设计

11.1 传感器应用及拍照更换界面图片设计目标

1. 本章的设计目标

(1) 安卓设备中常用传感器的程序设计

传感器是物联网应用系统的核心部分之一,以传感器为核心的无线传感网络是物联网的起源之一,也是物联网应用系统重要的组成部分,因此,传感器的应用也是学习物联网移动软件开发的重要内容之一。本章学习的是安卓系统的传感器应用,与普通的单片机连接传感器的程序有所不同,安卓里的传感器一般都是标准配置的传感器,其应用程序也由安卓系统底层提供驱动以及由安卓系统提供应用程序接口,在开发安卓传感器应用程序时,直接应用这些接口即可。

(2) 拍照更换界面图片设计

在很多移动应用程序中,都具有动态更换界面图标图片的功能,例如 QQ、微信中的更换头像。在物联网移动应用程序中,拍照更换界面图片也是特别重要的内容之一,例如前面介绍的地点列表主界面之中,每一项地点列表都带有一个图片,这就需要让用户根据实际的应用场景变换列表中的图片(图标)。

2. 拍照更换界面图片设计目标

动态更换界面图片的方法主要有三种,第一种是选择相册之中已经存在的图片;第二种是拍照然后选择该新拍摄到的图片;第三种是选择网上的图片。本章主要介绍前面两种方法,重点介绍拍照更换界面图片的方法。

拍照更换界面图片设计目标包括一个地点列表、地点编辑界面、拍照界面、图片编辑界面以及图片更换功能,如图 11-1 所示。

在地点编辑界面里,包括了一个当前图片、选择系统自带图标列表、图片文件名等信息,如图 11-2 所示。

图 11-1 拍照更换界面图片设计目标

图 11-2 地点编辑界面

11.2 物联网 App 安卓端传感器编程

11.2.1 安卓传感器(OnSensorChanged)使用介绍

当传感器的值发生变化时,例如磁阻传感器方向改变时会调用 OnSensorChanged()。当传感器的精度发生变化时会调用 OnAccuracyChanged()方法。下面是 API 中定义的几个代表 Sensor 的常量:

- IntTYPE_ACCELEROMETER:加速度传感器;
- intTYPE_ALL:所有类型;
- intTYPE_GRAVITY:恒定的重力传感器类型描述;

- intTYPE_GYROSCOPE:回转仪传感器;
- intTYPE_LIGHT:光线传感器;
- intTYPE_LINEAR_ACCELERATION:恒定线性加速度传感器类型描述;
- intTYPE_MAGNETIC_FIELD:磁场传感器;
- intTYPE_ORIENTATION:方向传感器;
- intTYPE_PRESSURE:压力计传感器;
- intTYPE_PROXIMITY:距离传感器;
- intTYPE_ROTATION_VECTOR:旋转矢量传感器类型描述;
- intTYPE_TEMPERATURE:温度传感器。

在编写传感器相关的代码时可以按照以下步骤:

第一步:获得传感器管理器:

SensorManger sm = (SensorManager).getSystemService(SENSOR_SERVICE);

第二步:为具体的传感器注册监听器,这里使用磁阻传感器 Sensor.TYPE_ORIENTATION:

sm.registerListener(this,sm.getDefaultSensor(Sensor.TYPE_ORIENTATION),SensorManager.SENSOR_DELAY_FASTEST);

这里如果想注册其他的传感器,可以改变第一个参数值的传感器类型属性。应该根据手机中的实际存在的传感器来进行注册。如果手机中不存在注册的传感器,就算注册了也不起什么作用。

第三个参数值表示获得传感器数据的速度,SENSOR_DELAY_FASTEST 表示尽可能快的获取传感器数据,除了该值以外,还可以设置3个获取传感器数据的速度值,这些值如下:

- SENSOR_DELAY_GAME:如果利用传感器开发游戏,建议使用该值。一般大多数实时性较高的游戏使用该级别。
- SENSOR_DELAY_NORMAL:默认的获取传感器数据的速度。标准延迟,对于一般的益智类游戏或者 EASY 界别的游戏可以使用,但过低的采样率可能对一些赛车类游戏有跳帧的现象。
- SENSOR_DELAY_UI:若使用传感器更新 UI,建议使用该值。
- SENSOR_DELAY_FASTEST:最低延迟,一般不是特别灵敏的处理不推荐使用,该模式可能造成手机电力大量消耗,而且由于传递的为大量的原始数据,算法处理不好将会影响游戏逻辑和 UI 的性能。

第三步:既然在第二步已经为传感器设置了监听。就要实现具体的监听方法,在Android 中,应用程序使用传感器主要依赖于 android.hardware.SensorEventListener 接口。该接口可以监听传感器各种事件。SensorEventListener 接口代码如下:

```
public interface SensorEventListener {
    public  void onSensorChanged(SensorEvent event) {
    }
    public void onAccuracyChanged(Sensor sensor , int accuracy ){
    }
}
```

当传感器的值发生变化时,例如磁阻传感器方向改变时会调用 OnSensorChanged()。当传感器的精度发生变化时会调用 OnAccuracyChanged()方法。首先,可以先看一下 Android 开发文档中的事例代码,该事例是一个简单的方向传感器应用例子,是安卓方向传感器编程的一个框架,实际的应用之中可以在该框的基础之上扩展,程序代码如下:

```
public class SensorActivity extends Activity, implements SensorEventListener {
    private final SensorManager mSensorManager;
    private final Sensor mAccelerometer;
    public SensorActivity() {
    mSensorManager = (SensorManager)getSystemService(SENSOR_SERVICE);
    mAccelerometer = mSensorManager.getDefaultSensor(
    Sensor.TYPE_ACCELEROMETER);
        }
    protected void onResume() {
    super.onResume();
    mSensorManager.registerListener(this, mAccelerometer,
    SensorManager.SENSOR_DELAY_NORMAL);
    }
    protected void onPause() {
    super.onPause();
    mSensorManager.unregisterListener(this);
    }
    public void onAccuracyChanged(Sensor sensor, int accuracy) {
    }
    public void onSensorChanged(SensorEvent event) {
    }
}
```

不需要的传感器尽量要解除注册,特别是 Activity 处于失去焦点的状态时,否则,手机电池很快会被用完。还要注意的是当屏幕关闭的时候,传感器也不会自动地解除注册。所以可以利用 Activity 中的 onPause()方法和 onresume()方法。在 onresume 方法中对传感器注册监听器,在 onPause()方法中解除注册。

11.2.2 方向传感器应用编程

下面是完整的方向传感器应用程序,可以在界面之中实时显示当前手机的方向传感器输出的参数。程序代码如下:

```java
package net.blogjava.mobile.sensor;
import android.app.Activity;
import android.hardware.Sensor;
import android.hardware.SensorEvent;
import android.hardware.SensorEventListener;
import android.hardware.SensorManager;
import android.os.Bundle;
import android.widget.TextView;
public class OrientationSensorTest extends Activity implements
    SensorEventListener {
    private SensorManager sensorManager = null;
    private Sensor orientaionSensor = null;
    private TextView textView;
    @Override
    protected void onCreate(Bundle savedInstanceState) {
        super.onCreate(savedInstanceState);
        setContentView(R.layout.main);
        setTitle("方向传感器 DEMO");
        textView = (TextView) findViewById(R.id.textview);
        sensorManager = (SensorManager) getSystemService(SENSOR_SERVICE);
        orientaionSensor = sensorManager
                .getDefaultSensor(Sensor.TYPE_ORIENTATION);
    }
    @Override
    protected void onPause() {
        super.onPause();
        sensorManager.unregisterListener(this);  // 解除监听器注册
    }
    @Override
    protected void onResume() {
        super.onResume();
        sensorManager.registerListener(this, orientaionSensor,
                SensorManager.SENSOR_DELAY_NORMAL);  // 为传感器注册监听器
    }
    @Override
    public void onAccuracyChanged(Sensor sensor, int accuracy) {
    }
    @Override
    public void onSensorChanged(SensorEvent event) {
        float x = event.values[SensorManager.DATA_X];
        float y = event.values[SensorManager.DATA_Y];
        float z = event.values[SensorManager.DATA_Z];
```

```
        textView.setText("x = " + (int) x + "," + "y = " + (int) y + "," + "z = "
        + (int) z);
    }
}
```

11.2.3　安卓坐标系的定义

安卓坐标系的方向包括如下(x,y,z)三轴:

(1) x 轴:x 轴的方向是沿着屏幕的水平方向从左向右,如果手机不是正方形的话,较短的边需要水平放置,较长的边需要垂直放置。

(2) y 轴:y 轴的方向是从屏幕的左下角开始沿着屏幕的的垂直方向指向屏幕的顶端。

(3) z 轴:将手机放在桌子上,z 轴的方向是从手机指向天空。

自从苹果公司在 2007 年发布第一代 iPhone 以来,以前看似和手机挨不着边的传感器也逐渐成为手机硬件的重要组成部分。如果使用过 iPhone、HTC Dream、HTC Magic、HTC Hero 以及其他的 Android 手机,会发现通过将手机横向或纵向放置,屏幕会随着手机位置的不同而改变方向。这种功能就需要通过重力传感器来实现,除了重力传感器,还有很多其他类型的传感器被应用到手机中,例如磁阻传感器就是最重要的一种传感器。虽然手机可以通过 GPS 来判断方向,但在 GPS 信号不好或根本没有 GPS 信号的情况下,GPS 就形同虚设。这时通过磁阻传感器就可以很容易判断方向(东、南、西、北)。有了磁阻传感器,也使罗盘(俗称指向针)的电子化成为可能。

在 Android 应用程序中使用传感器要依赖于 android.hardware.SensorEventListener 接口。通过该接口可以监听传感器的各种事件。SensorEventListener 接口的代码如下:

```
package android.hardware;
public interface SensorEventListener
{
    public void onSensorChanged(SensorEvent event);
    public void onAccuracyChanged(Sensor sensor, int accuracy);
}
```

在 SensorEventListener 接口中定义了 onSensorChanged 和 onAccuracyChanged。当传感器的值发生变化时,例如磁阻传感器的方向改变时会调用 onSensorChanged 方法。当传感器的精度变化时会调用 onAccuracyChanged 方法。

onSensorChanged 方法只有一个 SensorEvent 类型的参数 Event,其中 SensorEvent 类有一个 values 变量非常重要,该变量的类型是 float[]。但该变量最多只有 3 个元素,而且根据传感器的不同,values 变量中元素所代表的含义也不同。

11.2.4 安卓传感器 values 变量的定义

1. 方向传感器

在方向传感器中 values 变量的 3 个值都表示度数,它的含义如下:

(1) values[0]

该值表示方位,也就是手机绕着 Z 轴旋转的角度。0 表示北(North);90 表示东(East);180 表示南(South);270 表示西(West)。如果 values[0]的值正好是这 4 个值,并且手机是水平放置,表示手机的正前方就是这 4 个方向。可以利用这个特性来实现电子罗盘。

(2) values[1]

该值表示倾斜度,或手机翘起的程度。当手机绕着 x 轴倾斜时该值发生变化。values[1]的取值范围是 $-180 \leqslant values[1] \leqslant 180$。假设将手机屏幕朝上水平放在桌子上,这时如果桌子是完全水平的,values[1]的值应该是 0(由于很少有桌子是绝对水平的,因此,该值很可能不为 0,但一般都是-5 和 5 之间的某个值)。这时从手机顶部开始抬起,直到将手机沿 x 轴旋转 180 度(屏幕向下水平放在桌面上)。在这个旋转过程中,values[1]会在 0~-180 之间变化,也就是说,从手机顶部抬起时,values[1]的值会逐渐变小,直到等于-180。如果从手机底部开始抬起,直到将手机沿 x 轴旋转 180 度,这时 values[1]会在 0~180 之间变化。也就是 values[1]的值会逐渐增大,直到等于 180。可以利用 values[1]和下面要介绍的 values[2]来测量桌子等物体的倾斜度。

(3) values[2]

表示手机沿着 y 轴的滚动角度。取值范围是 $-90 \leqslant values[2] \leqslant 90$。假设将手机屏幕朝上水平放在桌面上,这时如果桌面是平的,values[2]的值应为 0。将手机左侧逐渐抬起时,values[2]的值逐渐变小,直到手机垂直于桌面放置,这时 values[2]的值是-90。将手机右侧逐渐抬起时,values[2]的值逐渐增大,直到手机垂直于桌面放置,这时 values[2]的值是 90。在垂直位置时继续向右或向左滚动,values[2]的值会继续在-90 至 90 之间变化。

2. 加速传感器

该传感器 values 变量的 3 个元素值分别表示 x、y、z 轴的加速值。例如,水平放在桌面上的手机从左侧向右侧移动,values[0]为负值;从右向左移动,values[0]为正值。要想使用相应的传感器,仅实现 SensorEventListener 接口是不够的,还需要使用下面的代码来注册相应的传感器。代码如下:

```
//获得传感器管理器
SensorManager sm = (SensorManager)getSystemService(SENSOR_SERVICE);
//注册方向传感器
sm.registerListener(this,
    sm.getDefaultSensor(Sensor.TYPE_ORIENTATION),
    SensorManager.SENSOR_DELAY_FASTEST);
```

如果想注册其他的传感器，可以改变 getDefaultSensor 方法的第 1 个参数值，例如，注册加速传感器可以使用 Sensor.TYPE_ACCELEROMETER。在 Sensor 类中还定义了很多传感器常量，但要根据手机中实际的硬件配置来注册传感器。如果手机中没有相应的传感器硬件，就算注册了相应的传感器也不起任何作用。getDefaultSensor 方法的第 2 个参数表示获得传感器数据的速度。SensorManager.SENSOR_DELAY_FASTEST 表示尽可能快地获得传感器数据。除了该值以外，还可以设置 3 个获得传感器数据的速度值，这些值如下：

SensorManager.SENSOR_DELAY_NORMAL：默认的获得传感器数据的速度。

SensorManager.SENSOR_DELAY_GAME：如果利用传感器开发游戏，建议使用该值。

SensorManager.SENSOR_DELAY_UI：如果使用传感器更新 UI 中的数据，建议使用该值。

3. 重力感应器

加速度传感器的类型常量是 Sensor.TYPE_GRAVITY。重力传感器与加速度传感器使用同一套坐标系。values 数组中三个元素分别表示了 x、y、z 轴的重力大小。Android SDK 定义了一些常量，用于表示星系中行星、卫星和太阳表面的重力。代码如下：

```
public static final float GRAVITY_SUN = 275.0f;
public static final float GRAVITY_MERCURY = 3.70f;
public static final float GRAVITY_VENUS = 11.87f;
public static final float GRAVITY_EARTH = 9.80665f;
public static final float GRAVITY_MOON = 1.6f;
public static final float GRAVITY_MARS = 3.71f;
public static final float GRAVITY_JUPITER = 23.12f;
public static final float GRAVITY_SATURN = 11.96f;
public static final float GRAVITY_URANUS = 11.69f;
public static final float GRAVITY_NEPTUNE = 11.0f;
public static final float GRAVITY_PLUTO = 0.6f;
public static final float GRAVITY_DEATH_STAR_I = 0.000000353036145f;
public static final float GRAVITY_THE_ISLAND = 4.815162342f;
```

4. 光线传感器

光线传感器的类型常量是 Sensor.TYPE_LIGHT。values 数组只有第一个元素（values[0]）有意义，表示光线的强度。最大的值是 120 000.0f。Android SDK 将光线强度分为不同的等级，每一个等级的最大值由一个常量表示，这些常量都定义在 SensorManager 类中，代码如下：

```
public static final float LIGHT_SUNLIGHT_MAX = 120000.0f;
public static final float LIGHT_SUNLIGHT = 110000.0f;
public static final float LIGHT_SHADE = 20000.0f;
public static final float LIGHT_OVERCAST = 10000.0f;
public static final float LIGHT_SUNRISE = 400.0f;
public static final float LIGHT_CLOUDY = 100.0f;
public static final float LIGHT_FULLMOON = 0.25f;
public static final float LIGHT_NO_MOON = 0.001f;
```

上面八个常量只是临界值。在实际使用光线传感器时要根据实际情况确定一个范围。例如,当太阳逐渐升起时,values[0]的值很可能会超过LIGHT_SUNRISE,当values[0]的值逐渐增大时,就会逐渐越过LIGHT_OVERCAST,而达到LIGHT_SHADE,当然,如果天气特别好的话,也可能会达到LIGHT_SUNLIGHT,甚至更高。

5. 陀螺仪传感器

陀螺仪传感器的类型常量是 Sensor.TYPE_GYROSCOPE。values 数组的三个元素表示的含义如下:

values[0]:沿 x 轴旋转的角速度。

values[1]:沿 y 轴旋转的角速度。

values[2]:沿 z 轴旋转的角速度。

当手机逆时针旋转时,角速度为正值,顺时针旋转时,角速度为负值。陀螺仪传感器经常被用来计算手机已转动的角度,代码如下:

```
private static final float NS2S = 1.0f / 1000000000.0f;
private float timestamp;
public void onSensorChanged(SensorEvent event)
{
    if (timestamp != 0)
    { //event.timesamp 表示当前的时间,单位是纳秒(100万分之一毫秒)
        final float dT = (event.timestamp - timestamp) * NS2S;
        angle[0] += event.values[0] * dT;
        angle[1] += event.values[1] * dT;
        angle[2] += event.values[2] * dT;
    }
    timestamp = event.timestamp;
}
```

上面代码中通过陀螺仪传感器相邻两次获得数据的时间差(dT)来分别计算在这段时间内手机延 x、y、z 轴旋转的角度,并将值分别累加到 angle 数组的不同元素上。

6. 其他传感器

在前面几节介绍了加速度传感器、重力传感器、光线传感器、陀螺仪传感器以及方向传感器。除了这些传感器外,Android SDK 还支持如下的几种传感器。关于这些传感器的使用方法以及与这些传感器相关的常量、方法,可以参阅官方文档。
- 近程传感器(Sensor.TYPE_PROXIMITY)
- 线性加速度传感器(Sensor.TYPE_LINEAR_ACCELERATION)
- 旋转向量传感器(Sensor.TYPE_ROTATION_VECTOR)
- 磁场传感器(Sensor.TYPE_MAGNETIC_FIELD)
- 压力传感器(Sensor.TYPE_PRESSURE)
- 温度传感器(Sensor.TYPE_TEMPERATURE)

虽然 AndroidSDK 定义了十多种传感器,但并不是每一部手机都完全支持这些传感器。例如,Google Nexus S 支持其中的 9 种传感器(不支持压力和温度传感器),而 HTC G7 只支持其中的 5 种传感器。如果使用了手机不支持的传感器,一般不会抛出异常,但也无法获得传感器传回的数据。在使用传感器时最好先判断当前的手机是否支持所使用的传感器。

11.3 摄像头及拍照应用

11.3.1 Camera2 应用

API 21 中将原来的 Camera API 弃用,转而推荐使用新增的 Camera2 API,这是一个大的动作,因为新 API 换了架构,让开发者用起来更难了。先来看看 Camera2 包架构示意图,如图 11-3 所示。

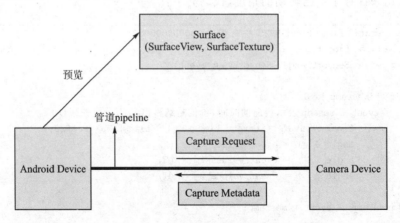

图 11-3 Camera2 包架构示意图

这里引用了管道的概念将安卓设备和摄像头之间联通起来,系统向摄像头发送 Capture 请求,而摄像头会返回 CameraMetadata。这一切建立在一个叫作 CameraCaptureSession 的会话中。Camera2 包中的主要类如图 11-4 所示。

其中 CameraManager 是那个站在高处统管所有摄像投设备(CameraDevice)的管理者,而每个 CameraDevice 自己会负责建立 CameraCaptureSession 以及建立 CaptureRequest。

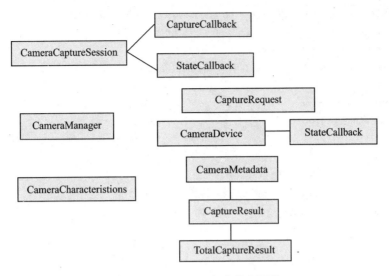

图 11-4 Camera2 包中的主要类

CameraCharacteristics 是 CameraDevice 的属性描述类，非要做个对比的话，那么它与原来的 CameraInfo 有相似性。

类图中有三个重要的 Callback，虽然这增加了阅读代码的难度，但是必须要习惯，因为这是新包的风格。其中 CameraCaptureSession.CaptureCallback 将处理预览和拍照图片的工作，需要重点对待。这些类是如何相互配合的？简单的拍照流程如图 11-5 所示。

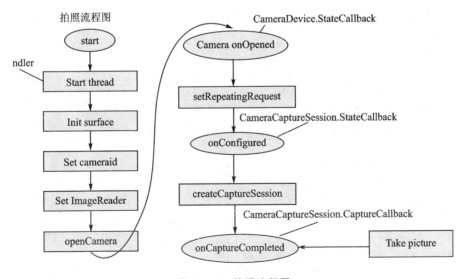

图 11-5 拍照流程图

用 SurfaceView 作为显示对象（当然还可以用 TextureView 去显示），核心代码如下：

```java
mCameraManager = (CameraManager) this.getSystemService(Context.CAMERA_SERVICE);
mSurfaceView = (SurfaceView)findViewById(R.id.surfaceview);
mSurfaceHolder = mSurfaceView.getHolder();
mSurfaceHolder.addCallback(new SurfaceHolder.Callback() {
    @Override
    public void surfaceCreated(SurfaceHolder holder) {
initCameraAndPreview();
    }
});
private void initCameraAndPreview() {
    Log.d("linc","init camera and preview");
    HandlerThread handlerThread = new HandlerThread("Camera2");
    handlerThread.start();
    mHandler = new Handler(handlerThread.getLooper());
    try {
mCameraId = "" + CameraCharacteristics.LENS_FACING_FRONT;
mImageReader = ImageReader.newInstance(mSurfaceView.getWidth(), mSurfaceView.getHeight(),ImageFormat.JPEG,/* maxImages */7);
mImageReader.setOnImageAvailableListener(mOnImageAvailableListener, mHandler);
mCameraManager.openCamera(mCameraId, DeviceStateCallback, mHandler);
    } catch (CameraAccessException e) {
Log.e("linc", "open camera failed." + e.getMessage());
    }
}
private CameraDevice.StateCallback DeviceStateCallback = new CameraDevice.StateCallback() {
    @Override
    public void onOpened(CameraDevice camera) {
Log.d("linc","DeviceStateCallback:camera was opend.");
mCameraOpenCloseLock.release();
mCameraDevice = camera;
try {
    createCameraCaptureSession();
} catch (CameraAccessException e) {
    e.printStackTrace();
}}
};
private void createCameraCaptureSession() throws CameraAccessException {
    Log.d("linc","createCameraCaptureSession");
    mPreviewBuilder = mCameraDevice.createCaptureRequest(CameraDevice.TEMPLATE_PREVIEW);
```

```
        mPreviewBuilder.addTarget(mSurfaceHolder.getSurface());
        mState = STATE_PREVIEW;
        mCameraDevice.createCaptureSession(
        Arrays.asList(mSurfaceHolder.getSurface(), mImageReader.getSurface()),
        mSessionPreviewStateCallback, mHandler);
    }
    private CameraCaptureSession.StateCallback mSessionPreviewStateCallback = new Camera-
CaptureSession.StateCallback() {
        @Override
        public void onConfigured(CameraCaptureSession session) {
Log.d("linc","mSessionPreviewStateCallback onConfigured");
mSession = session;
try {
            mPreviewBuilder.set(CaptureRequest.CONTROL_AF_MODE,
         CaptureRequest.CONTROL_AF_MODE_CONTINUOUS_PICTURE);
            mPreviewBuilder.set(CaptureRequest.CONTROL_AE_MODE,
         CaptureRequest.CONTROL_AE_MODE_ON_AUTO_FLASH);
            session.setRepeatingRequest ( mPreviewBuilder.build ( ), mSessionCaptureCallback,
mHandler);
    } catch (CameraAccessException e) {
        e.printStackTrace();
        Log.e("linc","set preview builder failed." + e.getMessage());
    }
    }
    };
    private CameraCaptureSession.CaptureCallback mSessionCaptureCallback =
    new CameraCaptureSession.CaptureCallback() {
        @Override
        public void onCaptureCompleted(CameraCaptureSession session, CaptureRequest request,
        TotalCaptureResult result) {
        //    Log.d("linc","mSessionCaptureCallback, onCaptureCompleted");
mSession = session;
checkState(result);
        }
        @Override
        public void onCaptureProgressed(CameraCaptureSession session, CaptureRequest request,
        CaptureResult partialResult) {
Log.d("linc","mSessionCaptureCallback,  onCaptureProgressed");
mSession = session;
checkState(partialResult);
        }
    private void checkState(CaptureResult result) {
```

```
switch (mState) {
    case STATE_PREVIEW:
// NOTHING
break;
    case STATE_WAITING_CAPTURE:
int afState = result.get(CaptureResult.CONTROL_AF_STATE);

if (CaptureResult.CONTROL_AF_STATE_FOCUSED_LOCKED == afState ||
    CaptureResult.CONTROL_AF_STATE_NOT_FOCUSED_LOCKED == afState
|| CaptureResult.CONTROL_AF_STATE_PASSIVE_FOCUSED == afState
|| CaptureResult.CONTROL_AF_STATE_PASSIVE_UNFOCUSED == afState) {
    //do something like save picture
}
break;
}}};
```

按下 Capture 按钮的事件响应函数代码如下：

```
public void onCapture(View view) {
    try {
        Log.i("linc", "take picture");
        mState = STATE_WAITING_CAPTURE;
        mSession.setRepeatingRequest(mPreviewBuilder.build(),
        mSessionCaptureCallback, mHandler);
        } catch (CameraAccessException e) {
        e.printStackTrace();
    }}
```

11.3.2 使用 TensorFlow API 构建视频物体识别系统

摄像头的主要应用是拍照，而拍照的目的有很多：

(1) 相机的作用，为了采集保存图片和视频。

(2) 图像的识别，目前摄像头拍照还有一个特别重要的用途，就是为了进行图像的识别，例如目前的扫二维码添加好友、扫二维码支付等。同时，目前的图像识别已经开始大量应用于其他更加广泛的识别之中，例如动物的识别、植物的识别、人脸的识别等。目标识别是物联网应用里一个特别重要的部分，也是最有发展潜力的方向之一。

移动软件的图像识别的实现方案之中，目前应用最为广泛的是 TensorFlow 库。在谷歌 TensorFlow API 推出后，构建属于自己的图像识别系统似乎变成了一件轻松的任务。谷歌开源的 TensorFlow Object Detection API 是其中非常引人注目的一个，任何来自谷歌的产品都是功能强大的，效果如图 11-6 所示。

可以在 GitHub 上找到这个小项目的全部代码：https://github.com/priya-

图 11-6 来自 TensorFlow API 的视频物体检测

dwivedi/Deep-Learning/blob/master/Object_Detection_Tensorflow_API.ipynb

TensorFlow Object Detection API 的代码库是一个建立在 TensorFlow 之上的开源框架,旨在为人构建、训练和部署目标检测模型提供帮助。

该 API 的第一个版本包含:

① 一个可训练性检测模型的集合,包括:

带有 MobileNets 的 SSD(Single Shot Multibox Detector);

带有 Inception V2 的 SSD;

带有 Resnet 101 的 R-FCN(Region-Based Fully Convolutional Networks);

带有 Resnet 101 的 Faster RCNN;

带有 Inception Resnet v2 的 Faster RCNN。

② 上述每一个模型的冻结权重(在 COCO 数据集上训练)可被用于开箱即用推理。

③ 一个 Jupyter notebook 可通过的模型之一执行开箱即用的推理。

④ 借助谷歌云实现便捷的本地训练脚本以及分布式训练和评估管道。

SSD 模型使用了轻量化的 MobileNet,这意味着它可以轻而易举地在移动设备中实时使用。在赢得 2016 年 COCO 挑战的研究中,谷歌使用了 Fast RCNN 模型,它需要更多计算资源,但结果更为准确。如需了解更多细节,请参阅谷歌发表在 CVPR 2017 上的论文:https://arxiv.org/abs/1611.10012。

在 TensorFlow API 的 GitHub 中,已经有经过 COCO 数据集训练过的可用模型了。COCO 数据集包含 30 万张图片,90 种常见事物类别。其中的类别如图 11-7 所示。

COCO 数据集的部分类别:

● TensorFlow Object Detection API 的 GitHub:https://github.com/tensor-

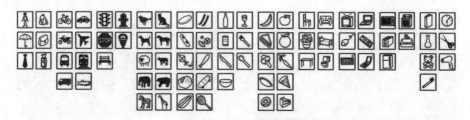

图 11-7 事物类别

flow/models/tree/master/object_detection；
- COCO 数据集：http://mscoco.org/。

如上所述，在 API 中，谷歌提供了 5 种不同的模型，从耗费计算性能最少的 MobileNet 到准确性最高的带有 Inception Resnet v2 的 Faster RCNN，如表 11-1 所列。

表 11-1 5 种不同的 COCO 数据集包模型

Model name	Speed	COCO mAP	Outputs
ssd_mobilenet_v1_coco	fast	21	Boxes
ssd_inception_v2_coco	fast	24	Boxes
rfcn_resnet101_coco	medium	30	Boxes
faster_rcnn_resnet101_coco	medium	32	Boxes
faster_rcnn_inception_resnet_v2_atrous_coco	slow	37	Boxes

在这里 mAP（平均准确率）是精度和检测边界盒的乘积，它是测量网络对目标物体敏感度的一种优秀标准。mAP 值越高就说明神经网络的识别精确度越高，但代价是速度变慢。想要了解这些模型更多的信息，请访问：https://github.com/tensorflow/models/blob/477ed41e7e4e8a8443bc633846eb01e2182dc68a/object_detection/g3doc/detection_model_zoo.md。

使用 API，首先尝试使用了其中最轻量级的模型（ssd_mobilenet）。主要步骤如下：
- 下载封装好的模型（.pb - protobuf），将其载入内存，链接：https://developers.google.com/protocol-buffers/。
- 使用内建帮助代码来载入标签、分类、可视化工具等内容。
- 打开一个新的会话并在一个图像上运行模型。

API 文件还提供了一个 Jupyter 笔记本来帮助记录主要步骤：https://github.com/tensorflow/models/blob/master/object_detection/object_detection_tutorial.ipynb

这个模型在示例图片中的表现非常不错，如图 11-8 所示。

在视频中运行，主要步骤如下：
- 使用 VideoFileClip 函数从视频中抓取图片。
- fl_image 函数非常好用，可以用来将原图片替换为修改后的图片，把它用于传递物体识别的每张抓取图片。

图 11 - 8 模型在示例图

● 最后，所有修改的剪辑图像被组合成为一个新的视频。

这段代码需要一段时间来运行，3 s 到 4 s 的剪辑需要约 1 min 的处理，但鉴于使用的是预制模型内固定的加载内存空间，所有这些都可以在一台普通电脑上完成，甚至无需 GPU 的帮助。只需要几行代码，就可以检测并框住视频中多种不同的事物了，而且准确率很高。当然，它还有一些可以提高的空间，如图 11 - 9 所示，它几乎没有识别出鸭子的存在。

图 11 - 9 在视频中运行

11.4 Android 拍照和选择照片

11.4.1 Android 媒体库 MediaStore

1. MediaStore 媒体库介绍

在物联网移动软件开发之中，拍照不仅仅只是采集图片和保存图片，同时还是刷新相册，而且还需要把拍摄到的图片返回到应用程序之中，这些处理会涉及到大量的文件操作和消息处理过程，为了简化这些开发过程，Android 专门提供了一个媒体库 MediaStore。MediaStore 类是 Android 系统提供的一个多媒体数据库，Android 中多媒体信息都可以从这里提取。这个 MediaStore 包括了多媒体数据库的所有信息，包括音频、视频和图像，Android 把所有的多媒体数据库接口进行了封装，所有的数据库不用自己进行创建，直接调用，利用 ContentResolver 去调用那些封装好的接口就可以进行数据库的操作了。

下面是最简单的打开摄像头和拍照的程序：

```
Intent intent = new Intent(MediaStore.ACTION_IMAGE_CAPTURE);startActivity(intent);
```

上面两行代码可以实现打开摄像头和拍照的功能，但是不能返回拍摄到的图片。如果想要返回图片，则需要改用 startActivityForResult 方法。

2. startActivityForResult 方法

startActivity()仅仅是跳转到目标页面，若是想跳回当前页面，则必须再使用一次 startActivity()。startActivityForResult()可以一次性完成这项任务，当程序执行到这段代码的时候，假如从 T1Activity 跳转到下一个 Text2Activity，而当这个 Text2Activity 调用了 finish()方法以后，程序会自动跳转回 T1Activity，并调用前一个 T1Activity 中的 onActivityResult()方法。

大家都知道 Activity 间的跳转可以使用 startActivity()，然后传入一个 Intent，指定组件即可，然后跳转的那个 Activity 要返回时，需要再使用一次 startActivity()，如果需要传递数据回来，肯定需要用 Intent，但是不断地调用 startActivity()，每次跳转系统都会在 task 中生成一个新的 Activity 实例，并且放于栈结构的顶部，当按下后退键时，才能看到原来的 Activity 实例。当然这是在 standard 启动模式，不管有没有已存在的实例，都生成新的实例。

所以要使用 finish()来返回，但是如果不使用 startActivity()，也就是不能传入 Intent 时，该怎么把 Intent（也就是数据）传递回去呢？

假设有两个 Activity，A 和 B，这时候 Activity A 就该使用 startActivityForResult()了，使用 startActivityForResult()的同时必须使用 onActivityResult()，顾名思义，就是得到 Activity B 返回的结果，也就是通过 Intent 携带的数据。

怎么使用 startActivityForResult()呢？startActivityForResult()需要一个 int 类型的请求码,这个请求码会随着 Activity A 的跳转而带过去,而跳转过去的 Activity B 调用 finish()的时候,需要使用 setResult()来设置一个结果码,这个结果码必须为 RESULT_OK,因为 Activity A 的 onActivityResult()会依次判断结果码和请求码,只有都符合的时候,才可以从 onActivityResult()的参数 Intent data 中获取数据,这个 data 就是 Activity B 中的 Intent。

11.4.2　Android 拍照和返回照片

1. 布局文件

设计一个布局文件,布局里包括一个"打开相机"按钮和一个图片视图,用于显示返回的图片内容,代码如下：

```xml
<?xml version="1.0" encoding="utf-8"?>
<LinearLayout xmlns:android="http://schemas.android.com/apk/res/android"
    android:layout_width="match_parent"
    android:layout_height="match_parent"
    android:orientation="vertical" >
    <Button
        android:id="@+id/photo_btn"
        android:layout_width="wrap_content"
        android:layout_height="wrap_content"
        android:text="拍照"
        />
    <ImageView
        android:layout_width="75dp"
        android:layout_height="75dp"
        android:id="@+id/img"
        />
</LinearLayout>
```

2. 实现文件

拍照和返回照片的 Java 代码如下：

```java
public class hello extends Activity{
    private ImageView img;
    @Override
    protected void onCreate(Bundle savedInstanceState) {
        super.onCreate(savedInstanceState);
        setContentView(R.layout.main);
        img = (ImageView) findViewById(R.id.img);
        Button btn = (Button) findViewById(R.id.photo_btn);
```

```
            btn.setOnClickListener(new View.OnClickListener() {
                @Override
                public void onClick(View v) {
                    Intent i = new Intent(MediaStore.ACTION_IMAGE_CAPTURE);
                    startActivityForResult(i,0);
                }
            });
        }
        @Override
        protected void onActivityResult(int requestCode,int resultCode,
    Intent data){
            if(requestCode == 0 && resultCode == Activity.RESULT_OK)
            {
                Bitmap bitmap = (Bitmap) data.getExtras().get("data");
                img.setImageBitmap(bitmap);
            }
        }
    }
```

3. 权　限

要使用摄像头和保存图片,则需要打开摄像头和保存文件的权限,因此需要在AndroidManifest.xml文件中获取以下权限:

```
<uses-permission android:name = "android.permission.WRITE_EXTERNAL_STORAGE"/>
<uses-permission android:name = "android.permission.READ_EXTERNAL_STORAGE"/>
<uses-permission android:name = "android.permission.MOUNT_UNMOUNT_FILESYSTEMS"/>
```

11.4.3　Android拍照和保存图片

1. 保存图片的方法

如果要保存图片,则需要把使用MediaStore.EXTRA_OUTPUT参数来指定需要保存的文件名和文件路径。下面是一个Android拍照和保存图片函数,该函数直接在上述实现文件的按钮事件响应中调用即可。

```
public void openCamera() {
        String path = Environment.getExternalStorageDirectory() + File.separator + "images"; //获取路径
        String fileName = new Date().getTime() + ".png";//定义文件名
        File file = new File(path, fileName);
        if (! file.getParentFile().exists()) {//文件夹不存在
            file.getParentFile().mkdirs();
```

```
                }
                Uri imageUri = Uri.fromFile(file);// FileProvider.getUriForFile(this, "com.
picker.fileprovider", file);
                Intent intent = new Intent(MediaStore.ACTION_IMAGE_CAPTURE);
                intent.putExtra(MediaStore.EXTRA_OUTPUT, imageUri);
                startActivityForResult(intent, 5223);//takePhotoRequestCode 是自己定义的一个
请求码
    }
```

2. Android 7.0 下调用相机闪退的解决方案

在项目中调用相机拍照和录像的时候,android 4.x、Android 5.x、Android 6.x 均没有问题,但在 Android 7.x 下面直接闪退,原因是 Android 升级到 7.0 后对权限又做了一个更新,即不允许出现以 file:// 的形式调用隐式 App,需要用共享文件的形式: content://URI3。

下面的实例是采用 Obtain_Studio 里的"AndroidStudio43__导航与滑动界面模板",然后修改登录文件 login.java,登录之后直接跳转到 main 界面。

其中 main.java 文件实现 Android 拍照和保存图片的功能,并且增加了 Android 7.0 下调用相机闪退的处理程序。

main.java 文件代码如下:

```
public class main extends Activity implements View.OnClickListener {
  public static final int PHOTO_REQUEST_CAREMA = 1;// 拍照
  public static final int CROP_PHOTO = 2;
  private Button takePhoto;
  private ImageView picture;
  private Uri imageUri;
  public static File tempFile;
  @Override
  protected void onCreate(Bundle savedInstanceState) {
    super.onCreate(savedInstanceState);
    setContentView(R.layout.main);
    takePhoto = (Button) findViewById(R.id.take_photo);
    picture = (ImageView) findViewById(R.id.picture);
    takePhoto.setOnClickListener(this);
  }
  @Override
  public void onClick(View v) {
    switch (v.getId()) {
      case R.id.take_photo:
        openCamera(this);
```

```java
      break; }
    }
    @Override
    protected void onActivityResult(int requestCode, int resultCode, Intent data) {
    switch (requestCode) {
      case PHOTO_REQUEST_CAREMA:
      if (resultCode == RESULT_OK) {
        Intent intent = new Intent("com.android.camera.action.CROP");
        intent.setDataAndType(imageUri, "image/*");
        intent.putExtra("scale", true);
        intent.putExtra(MediaStore.EXTRA_OUTPUT, imageUri);
        startActivityForResult(intent, CROP_PHOTO); // 启动裁剪程序
      }
      break;
      case CROP_PHOTO:
      if (resultCode == RESULT_OK) {
        try {
        Bitmap bitmap = BitmapFactory.decodeStream(getContentResolver()
          .openInputStream(imageUri));
        picture.setImageBitmap(bitmap);
        } catch (FileNotFoundException e) {
        e.printStackTrace();
        }
      }
      break;
    }
    }
    public void openCamera(Activity activity) {
    //获取系统版本
    int currentapiVersion = android.os.Build.VERSION.SDK_INT;
    // 激活相机
    Intent intent = new Intent(MediaStore.ACTION_IMAGE_CAPTURE);
    // 判断存储卡是否可以用,可用进行存储
    if (hasSdcard()) {
      SimpleDateFormat timeStampFormat = new SimpleDateFormat(
        "yyyy_MM_dd_HH_mm_ss");
      String filename = timeStampFormat.format(new Date());
      tempFile = new File(Environment.getExternalStorageDirectory(),
        filename + ".jpg");
      if (currentapiVersion < 24) {
      // 从文件中创建 uri
      imageUri = Uri.fromFile(tempFile);
```

```
        intent.putExtra(MediaStore.EXTRA_OUTPUT,imageUri);
      } else {
      //兼容 android7.0 使用共享文件的形式
      ContentValues contentValues = new ContentValues(1);
      contentValues.put(MediaStore.Images.Media.DATA,tempFile.getAbsolutePath());
      //检查是否有存储权限,以免崩溃
      if(ContextCompat.checkSelfPermission(this,Manifest.permission.WRITE_EXTERNAL_
STORAGE)
         !=PackageManager.PERMISSION_GRANTED){
         //申请 WRITE_EXTERNAL_STORAGE 权限
         Toast.makeText(this,"请开启存储权限",Toast.LENGTH_SHORT).show();
         return;
      }
      imageUri = activity.getContentResolver().insert(MediaStore.Images.Media.EXTERNAL
_CONTENT_URI,contentValues);
      intent.putExtra(MediaStore.EXTRA_OUTPUT,imageUri);
      }
   }
   // 开启一个带有返回值的 Activity,请求码为 PHOTO_REQUEST_CAREMA
   activity.startActivityForResult(intent,PHOTO_REQUEST_CAREMA);
   }
   /** 判断 sdcard 是否被挂载 */
   public static boolean hasSdcard(){
   return Environment.getExternalStorageState().equals(
     Environment.MEDIA_MOUNTED);
   }
}
```

在 build.gradle 中,修改一下目录版本号如下:

```
compileSdkVersion 24
    defaultConfig {
        applicationId "com.example.administrator.myapplication"
      minSdkVersion 18
      targetSdkVersion 18
      versionCode 1
      versionName "1.0"
   testInstrumentationRunner "android.support.test.runner.AndroidJUnitRunner"
}
```

11.5 拍照更换界面图片的实现

本项目在第 10 章"嵌入网页的控制界面设计"中的项目"IoT_IS_App_002_test_001"的基础上实现。实现方案是在第 10 章的基础上增加一个 roomActivity 页面,用于负责 IoT_IS_App_002_test_001 主界面地点列表的编辑功能。为了不影响 IoT_IS_App_002_test_001 项目原来的功能,本章将拷贝一份 IoT_IS_App_002_test_001 项目的整个目录,并且把得到的目录更新为 IoT_IS_App_002_test_002。这个 IoT_IS_App_002_test_002 也就是本章所设计的项目。

因为原来 IoT_IS_App_002_test_001 项目中地点列表的编辑功能是采用一个对话框来实现,但是在对话框中要实现拍照及更换界面图片非常困难或者根本没办法实现,因此,在本章增加一个 Activity,在 Activity 中比较容易实现拍照和更换界面图片功能。

1. 设置跳转到 roomActivity 界面

当用户选择列表中的编辑功能时,跳转到 roomActivity 界面,该功能在 ListViewAdapter 中设置,代码如下:

```
listItemView.detail.setOnClickListener(new View.OnClickListener() {
    @Override
    public void onClick(View v) {
        Intent intent = new Intent(context,roomActivity.class);//要跳转的界面
        intent.putExtra("room_id",(String) listItems.get(selectID).get("id"));
        context.startActivity(intent);
    }
});
```

2. 设计编辑界面布局

根据本章最开始介绍的编辑界面设计目标来设计编辑界面布局,文件名为 item_room_edit.xml,文件代码如下:

```
<?xml version = "1.0" encoding = "utf-8"?>
    <RelativeLayout xmlns:android = "http://schemas.android.com/apk/res/android"
        android:id = "@ + id/rl_hotelName"
        android:layout_width = "match_parent"
        android:layout_height = "wrap_content"  >
        <LinearLayout android:layout_width = "match_parent"
            android:layout_height = "wrap_content"
            android:orientation = "horizontal">
            <LinearLayout android:layout_width = "120dp"
```

```xml
        android:layout_height = "wrap_content"
        android:orientation = "vertical"
        android:background = "#eeeeff"              >
    <TextView  android:layout_width = "wrap_content"
            android:layout_height = "wrap_content"
            android:layout_marginLeft = "15dp"
            android:text = "选择拍照或相册图片:"
            android:textSize = "18sp" />
    <ImageView android:id = "@ + id/iv_room_image"
                android:layout_width = "80dp"
                android:layout_height = "80dp"
                android:layout_marginLeft = "15dp"
                android:src = "@drawable/add" />
        <TextView   android:layout_width = "wrap_content"
            android:layout_height = "wrap_content"
            android:layout_marginLeft = "15dp"
            android:text =  "选择系统自带图标:"
            android:textSize = "18sp" />
    <Spinner   android:layout_width = "100dp"
    android:layout_height = "wrap_content"
    android:id = "@ + id/spinner"          />
</LinearLayout>
<LinearLayout  android:layout_width = "match_parent"
        android:layout_height = "wrap_content"
        android:orientation = "vertical">
    <LinearLayout android:id = "@ + id/rl_addHotel"
        android:layout_width = "match_parent"
        android:layout_height = "40dp"
        android:orientation = "horizontal">
        <TextView    android:layout_width = "0dp"
            android:layout_height = "40dp"
            android:text = "地点:" />
        <EditText android:id = "@ + id/ed_hotelName"
            android:layout_width = "0dp"
            android:layout_height = "40dp"
            android:layout_weight = "2" />
        <Button android:id = "@ + id/id_room_delete"
            android:layout_width = "0dp"
            android:layout_height = "40dp"
            android:text = "删除"/>
    </LinearLayout>
    <LinearLayout android:id = "@ + id/ll_addHotelEvaluate"
```

```xml
        android:layout_width = "match_parent"
        android:layout_height = "wrap_content"
        android:layout_below = "@ + id/rl_addHotel"
        android:orientation = "vertical">
    <RelativeLayout android:id = "@ + id/rl_hotelEvaluate"
            android:layout_width = "match_parent"
            android:layout_height = "wrap_content"
            android:layout_below = "@ + id/rl_addHotel"
            android:orientation = "horizontal">
        <TextView android:id = "@ + id/tv_hotelServer"
            android:layout_width = "wrap_content"
            android:layout_height = "wrap_content"
            android:text = "信号:" />
        <RatingBar android:id = "@ + id/rb_hotel_evaluate"
            android:layout_width = "wrap_content"
            android:layout_height = "20dp"
            android:layout_toRightOf = "@ + id/tv_hotelServer" />
        <TextView android:id = "@ + id/tv_hotelServer_ID"
            android:layout_width = "wrap_content"
            android:layout_height = "wrap_content"
            android:layout_toRightOf = "@ + id/rb_hotel_evaluate"/>
        <EditText android:layout_toRightOf = "@ + id/tv_hotelServer_ID"
            android:id = "@ + id/ed_hotel_id"
            android:layout_width = "80dp"
            android:layout_height = "wrap_content"
            android:layout_below = "@ + id/rl_server" />
    </RelativeLayout>
    <RelativeLayout android:id = "@ + id/rl_hotelEvaluate"
            android:layout_width = "match_parent"
            android:layout_height = "wrap_content"
            android:layout_below = "@ + id/rl_addHotel"
            android:orientation = "horizontal">
        <TextView android:id = "@ + id/ip_hotelServer"
            android:layout_width = "wrap_content"
            android:layout_height = "wrap_content"
            android:text = "IP 地址:"/>
    <EditText android:layout_toRightOf = "@ + id/ip_hotelServer"
            android:id = "@ + id/ed_hotel_ip"
            android:layout_width = "match_parent"
            android:layout_height = "wrap_content"
            android:layout_below = "@ + id/rl_server"/>
</RelativeLayout>
```

```
        </LinearLayout>
        <RelativeLayout android:id = "@ + id/rl_hotelEvaluate"
            android:layout_width = "match_parent"
            android:layout_height = "wrap_content"
            android:layout_below = "@ + id/rl_addHotel"
            android:layout_marginTop = "5dp"
            android:orientation = "horizontal">
            <TextView android:id = "@ + id/ip_hotel_img"
                android:layout_width = "wrap_content"
                android:layout_height = "wrap_content"
                android:text = "图标:" />
            <TextView android:layout_toRightOf = "@ + id/ip_hotel_img"
                android:id = "@ + id/ip_room_img_file"
                android:layout_width = "match_parent"
                android:layout_height = "wrap_content"
                android:layout_below = "@ + id/rl_server" />
        </RelativeLayout>
    </LinearLayout>
</LinearLayout>
</RelativeLayout>
```

3. 图片编辑的实现文件

编辑的实现文件名为 roomActivity.java,由于程序代码比较长,下面只列出了拍照的核心程序代码,其他部分代码与上述介绍的两个拍照例子基本相同,可以直接参考上述程序代码即可。

在下面介绍的程序中,用到了 ImageUtil 类,该类在开源项目 CropDemo 中提供。roomActivity.java 中还调用了 CropDemo 项目部分有关拍照和相片保存的方法。CropDemo 是一个证件拍照裁剪框架,下载地址为:

https://github.com/chengzichen/CropDemo。

(1) 选择图片编辑的功能,可以选择相机或者相册两种,程序代码如下:

```
@Override
  public void onClick(View v) {
  final AlertDialog.Builder builder = new AlertDialog.Builder(this);
   builder.setTitle("选择照片");
   builder.setPositiveButton("相机",
     new DialogInterface.OnClickListener() {
     @Override
     public void onClick(DialogInterface dialog, int which) {
       ImageUtil.takePhotoForResult(myactivity, PICK_FROM_CAMERA);
```

```
            }
        });
    builder.setNegativeButton("相册",
        new DialogInterface.OnClickListener() {
        @Override
        public void onClick(DialogInterface dialog, int which) {
            ImageUtil.openAbleForResult(myactivity, SELECT_FROM_AMBL);
        }
        });
    AlertDialog alert = builder.create();
    alert.show();
}
```

(2)返回图片处理:在相机或者相册界面返回之后,读取返回的图片,并且显示到编辑界面之中,同时更新数据库中该项地点的图片文件名。返回图片处理程序如下:

```
@Override
    public void onActivityResult(int requestCode, int resultCode, Intent data) {
        if (resultCode != Activity.RESULT_OK) {
            return;
        }
        switch (requestCode) {
        case PICK_FROM_CAMERA:
            ImageUtil.cropForReslt(this, CROP_FROM_CAMERA, mCropParams);
            break;
        case CROP_FROM_CAMERA:
            if (null != data) {
                setCropImg(data);
            }
            break;
        case SELECT_FROM_AMBL:
            ImageUtil.cropForReslt(this,data.getData(),CROP_FROM_CAMERA, mCropParams);
            break;
        }
    }
```

第 12 章 苹果手机移动软件设计

12.1 苹果手机移动软件设计目标

1. 学习和设计目标

本章学习和设计目标很简单,一是初步了解苹果手机 iOS 操作系统的特点,二是初步了解应用程序的开发过程,并设计一个最简单的 HelloWrold 程序。HelloWrold 程序运行效果如图 12-1 所示。

图 12-1 HelloWrold 程序运行效果

2. iOS 操作系统的特点

2004 年,苹果公司召集了 1 000 多名内部员工组成研发 iPhone 团队,开始了被列为高度机密的项目——Project Purple,当中包括 iPhone 的幕后设计师 Jonathan Ive。当时苹果公司的首席执行官史蒂夫·乔布斯把研发重点从 iPad 平板、电脑转向手机。2007 年 6 月 29 日下午 6 时正在美国正式发售,苹果公司销售商店外有数百名苹果粉丝为了抢购而提早排队购买。由于刚推出的 iPhone 上市后反应热烈,被部分媒体誉为"上帝手机"。

iOS 是由苹果公司为 iPhone、iPad 开发的操作系统。它主要是给 iPhone、iPod touch 以及 iPad 使用。就像其基于 Mac OS X 操作系统一样,它也是以 Darwin(Unix-like,类 UNIX 系统)为基础。原本这个系统名为 iPhone OS,直到 2010 年 6 月 7 日

WWDC 大会上宣布改名为 iOS。

目前 iOS 系统的应用开发人才还是很紧缺。随着互联网的不断发展和进步，各大企业对 iPhone 开发人才的需求量越来越多。目前我国的 iOS 开发人才缺口大，培养 iOS 开发与互联网络和软件技术相互融合的人才已经成为当务之急。

负责开发苹果 iOS 系统应用的开发人才价格仍然在上涨，目前 iOS 开发人才的价格要远远高于 Android 开发人才。例如，腾讯在上海招聘 iPhone 高级研发工程师，有 3 年工作经验者年薪能达到 35 万，远远高于 Android 的开发人才。

这是因为大多数公司在进军移动互联网时都将较为高端的 iOS 平台作为重点。此外，苹果营造的生态系统让应用开发者们能够在 iOS 上赚到较多的收入，即使是一些较小的团队，也多少有一笔收入让团队先活下来。此外，由于开发语言的关系，iOS 的技术人才比较难像 Android 那样批量培养，市场供给较少。

12.2　iOS 开发环境搭建

1. 安装 Mac OS

安装 VMware Workstation，最好选择 10.0 或以上版本。然后安装补丁软件 "VMware Unlocker for mac"。

VMwareUnlocker for mac 是 VMware 下安装 Mac OS X 所需要的安装补丁，是针对 VMware 的 Mac OS X 解锁。

补丁代码进行了以下修改：
- 修复 vmware-VMX 及衍生工具，允许 Mac OS X 启动；
- 修正 vmwarebase 或 dll。让创建过程中苹果可以被选中；
- 复制 darwin.iso 到 VMware 文件夹。

在 Windows 上，将需要以管理员身份运行 cmd.exe 或使用资源管理器的命令文件上单击右键，选择"以管理员身份运行"。下载 VM 的 MAC OS 补丁，解压后以管理员身份运行"install"。

下载 Mac.OS，选择的版本一般为 Mac.OS.X.Lion，版本号为 10.8 或者以上版本，文件名一般为 Mac.OS.X.Lion.iso。

运行 VMware，建立虚拟机，客户机操作系统选择"Apple Mac OS X"，CD/DVD 指向下载的"Mac.OS.X.Lion.iso"文件。

在 VM 管理界面运行虚拟机，进入启动界面，稍等片刻之后开始安装，进入 Mac OS X 实用工具界面，选择磁盘工具，如图 12-2 所示。

选择重新安装 OS X，如图 12-3 所示。

其他步骤按安装提示进行选择，完成整个系统的安装即可。

2. 安装 Xcode 12.0

Xcode 12.0 版本是 iPhone 5s 和 iOS 7.0 开发必须的工具。另外也支持 64 bit

第 12 章　苹果手机移动软件设计

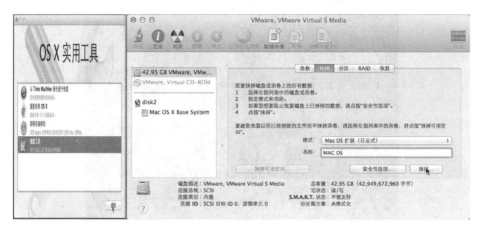

图 12-2　Mac OS X 实用工具界面

图 12-3　重新安装 OS X

App 开发，还支持 OS X Mavericks 开发，以及 TDD、Continuous Integration、自动化配置等。改进方面有界面扁平风格、可视化调试、静态分析、代码管理等。

(1) 下载 Xcode 12.0

Xcode 12.0 正式版发布于 2013 年 9 月 18 日（美国当地时间），可以从 Mac Store 上免费下载安装。前提条件是系统必须是 Mountain Lion 10.8.5 及以上版本。

(2) 安装 Xcode 12.0

下载好 DMG 安装包后，打开会发现 Xcode 以 App 的形式发布，直接复制到本地磁盘即可，比如/Development 下，与之前的 Xcode 版本路径不同即可。

12.3　iOS 入门实例

12.3.1　创建 iOS 项目

选择 Mac 桌面工具条上的应用程序图标，然后选择 Xcode，如图 12-4 所示。

图 12-4 应用程序界面

单击 Xcode 欢迎屏幕上的 Create A New Xcode Project,如图 12-5 所示。

图 12-5 Xcode 启动界面

选择 Single View Application,相当于以前版本的 View-Application,如图 12-6 所示。

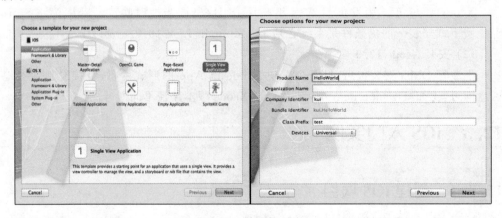

图 12-6 创建新项目

创建新项目完成之后,生成的项目文件结构如图 12-7 所示。这个时候可以直接编译和运行,在模拟器中可以看到一个空的窗口,说明项目基本功能正常。

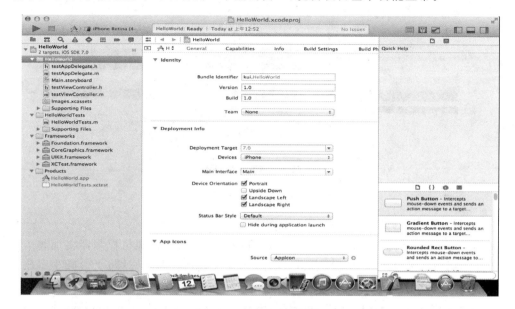

图 12-7 项目文件结构

这个新项目包含两个类:AppDelegate 和 ViewController,以及这个例子中的新型界面文件 Main.storyboard。请注意,这个项目中并没有 xib 文件。

这是一个只支持竖屏的应用,所以在编译项目之前,钩掉 Deployment Info——Device Orientation 下面的 Landscape Left 和 Landscape Right 选项。

12.3.2 编辑 main.storyboard 文件

1. Storyboard 介绍

Storyboard 在 iOS 5 中首次推出,在开发 App 界面时可以极大地节省时间。如图 12-8 所示,这是一个应用项目的 Storyboard,可以很清晰标识这些视图之间的关系,这就是使用 Storyboard 的强大之处。

当应用程序中有许多不同页面时,使用 Storyboard,可以大大减少页面之间跳转的代码。过去需要为每个视图控制器创建一个 nib 文件,现在只需要使用一个 Storyboard,它包含了应用中所有的视图控制器以及它们之间的关系。

相比传统的 nib 文件,storyboard 有以下优点:

- 使用 Storyboard,可以更好地理解应用中所有视图在概念上的概览,以及它们之间的关系。掌控所有的视图变得很容易,因为所有的设计都是在一个文件中,而不是在很多单独的 nib 文件中;
- Storyboard 描述了视图之间的动画,这些动画叫做"segues",可以很容易地从一个视图控制器(点 ctrl-dragging)拖拽到另一个视图控制器来实现,"segues"

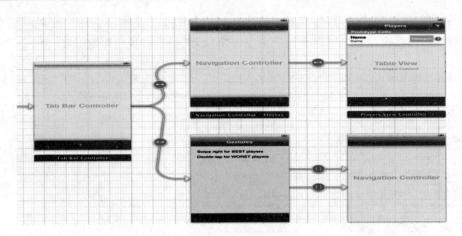

图 12-8 Storyboard 结构图

不需要写代码即可实现页面的跳转控制；
- Storyboard 通过新的 cell 原型,以及静态 cell 特性,让表格控制器实现起来更容易;几乎可以完全通过 Storyboard 来设计表格控制器,这也大大减少了软件开发的代码量。

当然使用 Storyboard 也有一些限制,Storyboard 的 Interface builder 远没有旧版 nib 编辑器那么强大,并且有一些东西只能在 nib 中做而不能在 Storyboard 编辑器中做。

2. 编辑 Storyboard

单击 Main.storyboard,在 interface builder 中打开它,如图 12-9 所示。

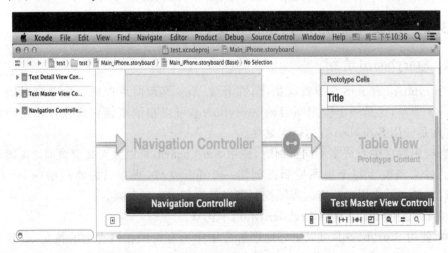

图 12-9 Storyboard 视图

在 Interface builder 中编辑 Storyboard 就跟编辑 nib 文件差不多,可以从 Object Library 中拖拽新的元素到视图控制器中,并且可以编辑它的布局。区别在于 Storyboard 不仅仅有一个视图控制器，而且它把应用中的所有视图控制器全都包含了。

　　Storyboard 的一个视图控制器就是一个场景。可以交替地使用这个模式，场景是呈现在 Storyboard 中的视图控制器，以往需要为每一个场景/视图控制器创建一个 nib 文件。现在只需要把他们都集中在一个 Storyboard 里。在 iPhone 中，一次只能看到一个场景，而 iPad 应用中，一次可能会看到多个场景。例如 master/detail 或者 popover。

　　注意：Xcode 5 默认 Storyboard 以及 nib 中的 Auto Layout 属性是打开的。Auto Layout 是一个新的可以自动调整控件大小的技术，在应用适配 iPad 以及 iPhone 5 的时候很有用，它只能在 iOS 6 以上运行。

　　要在 Storyboard 中的 File inspector 禁用 Auto Layout 选项，如图 12 - 10 所示。

图 12 - 10　File inspector 禁用

可以拖拽一些控件到空的视图控制器上，如图 12 - 11 所示。

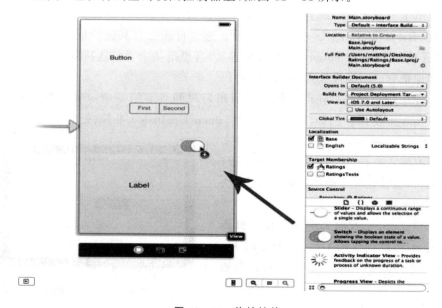

图 12 - 11　拖拽控件

在 Storyboard 编辑页面中找到图 12-12 中这个标上红色左箭头的按钮。单击它打开左侧的 Document Outline 视图，如图 12-12 所示。

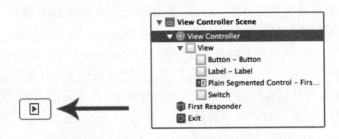

图 12-12　Document Outline 视图

在编辑 nib 的时候，列表中显示的是这个 nib 中所有的控件。对于 Storyboard，它会显示应用中的所有视图控制器的内容。

3. Dock 功能

在场景（scene）下面有个微型版的 Document Outline，叫做 Dock，Dock 原意是停靠码头，这里是指苹果操作系统中的停靠栏，这种停靠栏很有个性，而且很适用，Dock 栏如图 12-13 所示。

图 12-13　Dock 栏

它显示了当前视图的最上层对象，每个视图都至少有一个 View Controller 对象、一个 First Responder 对象、一个 Exit 项目。它也可能会有其他的最上层对象。使用 Dock 去连接 outlets 和 actions 变得非常容易，当想把某个对象连接到视图控制器时，只需简单地把它的图标拖拽到 Dock 上即可。

First Responder 是指任何物体在任何给定时间具有第一响应状态的代理对象。它原来在 nib 中使用。例如，按钮中有个 Touch Up Inside 事件，把它拖给 First Responder 的 cut:selector。如果在某一点的文本字段具有输入焦点，那么可以按下该按钮，使文字栏位，也就是现在的第一响应者，剪切其文本到剪贴板。

运行该 App，它看起来应该和在编辑器中设计的一样（此处使用了运行 iOS 7 的 4-inch Retina iPhone simulator）。运行效果如图 12-14 所示。

图 12-14　运行效果

12.3.3 程序代码分析

如果使用 MainWindow.xib 文件,这个文件有一个 UIWindow 对象指向 App Delegate,以及其他视图控制器。当想用 Storyboard 做这些工作的时候,就不需要 MainWindow.xib 了。那么 Storyboard 是如何加载到应用中的?

看看应用的 application delegate 类。打开 AppDelegate.h 文件,代码如下:

```
#import <UIKit/UIKit.h>
@interface AppDelegate:UIResponder <UIApplicationDelegate>
@property (strong,nonatomic) UIWindow * window;
@end
```

使用 Storyboard,应用代理必须继承自 UIResponder,并且有一个 UIWindows 属性。下面打开 AppDelegate.m 文件,这里什么都没有,所有的方法都是空的,甚至 application:didFinishLaunchingWithOptions 也仅仅是返回了一个"YES"。

秘密藏在 Ratings-Info.plist 文件里,在 Supporting Files group 配置里可以找到并打开它。

Storyboard 使用 UIMainStoryboardFile 或者"Main storyboard file base name"来指定应用启动的时候,哪一个 Storyboard 是必须被加载的。当设置生效,UIApplication 会自动加载这里所配置的 Storyboard 文件,并把它第一个视图控制器显示到 UIWindow 中。这些都不必写代码去实现,可以通过修改 Project Settings 下面的 Project Settings 来实现,如图 12-15 所示。

图 12-15 Project Settings 设置

打开 main.m,代码如下:

```
#import <UIKit/UIKit.h>
#import "AppDelegate.h"
int main(int argc, char * argv[])
{
```

```
    @autoreleasepool {
        return UIApplicationMain(argc, argv, nil,
            NSStringFromClass([AppDelegate class]));
    }
}
```

App delegate 并不是 Storyboard 的一部分,需要把它的名字指定给 UIApplicationMain(),否则,应用就找不到它。

12.3.4 main 函数及程序启动过程

main 函数是 Objective-C 程序的入口函数。Xcode 4.2 之前版本的 main 函数如下:

```
int main(int argc,char * argv[])
{
    NSAutoreleasePool * pool = [[NSAutoreleasePoolalloc] init];
    int retVal = UIApplicationMain(argc, argv,nil, nil);
    [pool release];
    return retVal;
}
```

Xcode 4.2 版本工程中的 main 函数如下:

```
int main(int argc,char * argv[])
{
    @autoreleasepool {
        return UIApplicationMain(argc, argv, nil,
NSStringFromClass([TCAppDelegate class]));
    }
}
```

可以看出一个重要的变化是在 Xcode 4.2 使用了 ARC 技术后,NSAutoreleasePool 被废弃,改用@autoreleasepool,这里请不要改回原先的方式。如果改变,在开启 ARC 选项后,程序将不能通过编译。

不论哪个版本,UIApplicationMain 函数都是程序的关键点,下面是对这个函数的分析:

UIApplicationMain() 函数是初始化程序的核心,它接收四个参数。其中 argc 和 argv 参数来自于 main() 接受的两个参数;另外两个 String 型参数分别表示程序的主要类(principal class)和代理类(delegate class)。如果主要类(principal class)为 nil,则默认为 UIApplication;如果代理类(delegate class)为 nil,则程序假设程序的代理来自 Main.nib 文件。如果这两个参数任意一个不为 nil,则 UIApplicationMain() 函数则会根据参数创建相应的功能类。因此,如果程序中使用自定义的 UIApplication 类的子类，需要将自定义类名作为第 3 个参数传进来。

iOS 程序的生命周期如图 12-16 所示。每一个 iOS 应用程序都包含一个 UIApplication 对象，iOS 系统通过该 UIApplication 对象监控应用程序生命周期全过程。一个 iOS 应用程序都要为其 UIApplication 对象指定一个代理对象，并由该代理对象处理 UIApplication 对象监测到的应用程序生命周期事件。

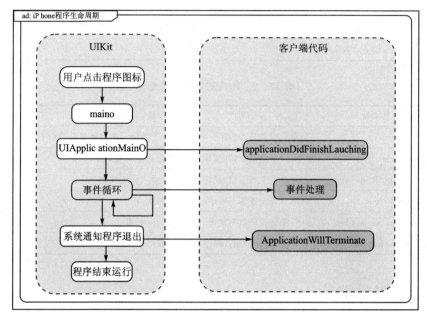

图 12-16　iOS 程序的生命周期

12.3.5　UIResponder 类

在 iOS 中，一个 UIResponder 对象表示一个可以接收触摸屏上的触摸事件的对象，也就是表示一个可以接收事件的对象。所有显示在界面上的对象都是从 UIResponder 直接或间接继承。

UIResponder 类定义的触摸事件相关信息如表 12-1 所列，定义的运动事件相关信息如表 12-2 所列，定义的响应对象链相关信息如表 12-3 所列。

表 12-1　触摸事件相关信息

方法名称	说　明
touchesBegan:withEvent	当用户触摸到屏幕时调用此方法
tochesMoved:withEvent	当用户触摸到屏幕并移动时调用此方法
tochesEnded:withEvent	当触摸离开屏幕时调用此方法
tochesCancelled:withEvent	当触摸被取消时调用此方法

运动事件是指当用户以特定方式移动设置,如摇摆设置时,设置会产生运动事件,由表 12-2 和表 12-3 所列的几个方法进行处理。

表 12-2 运动事件相关信息

方法名称	说　明
motionBegan:withEvent	运动开始时执行
motionEnded:withEvent	运动结束时执行
motionCancelled:withEvent	运动被取消时执行

表 12-3 响应对象链相关信息

方法名称	说　明
isFirstResponder	指示对象是否为第一响应者,这里的第一响应者就是当前有焦点的对象
nextResponder	下一个响应者,在实现中,一般会返回父级对象
canBecomeFirstResponder	获取一个布尔值,指定对象是否可以获取焦点
becomeFirstResponder	把对象设置为 FirstResponder 对象
canResignFirstResponder	对象是否可以取消 FirstResponder 对象
resignFirstResponder	取消对象为 FirstResponder 对象

12.4　Objective-c

12.4.1　Objective-c 介绍

1. Objective-c

Objective-c(也有人把它简称为"Object-C",或者"OC")语言由 Brad J. Cox 于 20 世纪 80 年代早期设计,以 SmallTalk 为基础,建立在 C 语言之上。1988 年,NeXT 获得 Objective-c 的授权,开发出了 Objective-c 的语言库和一个名为 NEXTSTEP 的开发环境。1994 年,NeXT 公司与 Sun 公司联合发布了一个针对 NEXTSTEP 系统的标准规范,并命名为 OPENSTEP。OPENSTEP 在自由软件基金会的实现名称为 GNUStep,有 Linux 下的版本。1996 年,苹果公司收购了 NeXT 公司,将 NEXTSTEP/OPENSTEP 定为苹果操作系统下一个主要发行版本的基础。并发布了一个相关开发环境,名为 Cocoa,内置了对 Objective-c 的支持。2007 年,苹果公司发布了 Objective-c 2.0,并在 iPhone 上使用 Objective-c 进行开发。

2. Objective-c 和 C++不同的风格

Objective-c 和 C++不同的风格,其实也就是 Simula67 和 Smalltalk 两大学派的不同主张。

(1) 单一继承

Objective-c 不支持多重继承(与 Java 和 Smalltalk 相同),而 C++语言支持多重继承。

(2) 动　态

Objective-c 是动态类型(dynamicaly typed),所以它的类库比 C++要容易操作。Objective-c 在运行时可以允许根据字符串名字来访问方法和类,还可以动态连接和添加类。

C++跟随面向对象编程里的 Simula 67(一种早期的面向对象语言)学派,而 Objecive-C 属于 Smalltalk 学派。

在 C++里,对象的静态类型决定是否可以发送消息给它,而对于 Objective-c 来说,由动态类型来决定。Simula 67 学派更安全,因为大部分错误可以在编译时查出。而 Smalltalk 学派更灵活,比如一些 Smalltalk 看来无误的程序拿到 Simula 67 那里就无法通过。

从很多方面来看,C++和 Objective-c 的差别,与其说是技术上的,不如说是思维方式上的。是否想更安全而舍弃灵活性? Simula 67 学派的支持者声称既然程序设计出色何必再要灵活性,而 Smalltalk 学派则辩称为了灵活可以容忍运行时多出错。

12.4.2　Objective-c 特点

1. 兼容性

Objective-c 可以说是一种面向对象的 C 语言,在 Objective-c 的代码中可以有 C 和 C++语句,它可以调用 C 的函数,也可以通过 C++对象访问方法。

2. 字符串

Objective-c 通常不使用 C 语言风格的字符串。大多数情况下是使用 Foundation 框架的 NSString 类型字符串。NSString 类提供了字符串的类包装,支持 Unicode、printf 风格的格式化工具。它是在普通的双引号字符串前放置一个@符号,例如:

NSString * myString = @"My String\n";
NSString * anotherString = [NSString stringWithFormat:@"%d %s", 1, @"String"];

3. 类

Objective-c 是一种面向对象的语言,定义类是它的基本能力。Objective-c 的类声明和实现包括两个部分:接口部分和实现部分。

4. 方　法

Objective-c 中方法不是在"."运算符,而是采用"[ ]"运算符。有时候方法调用也称为"消息发送"。

5. 属 性

属性是 Objective-c 2.0 提出的概念,它是替代对成员变量访问的"读取方法(getter)"和"设定方法(setter)"的手段,为了对类进行封装,一般情况下不直接访问成员变量,而是通过属性访问。

6. 协 议

Objective-c 中的协议类似于 Java 中的接口或 C++的纯虚类,只有接口部分定义,没有实现部分,即只有 h 文件,没有 m 文件。

7. 分 类

Objective-c 中的分类是类似于继承机制,通过分类能够扩展父类的功能。

12.4.3 Objective-c 和 C++/Java 比较

1. 函数的对比

不同语言之间的 Helloworld 方法如下:

```
Java 语言:
    public void helloWorld(bool ishelloworld) {
//其他
}
C++语言:
void helloWorld(bool ishelloworld) {
//其他
    }
Objective-c 语言:
    -(void)HelloWorld:(BOOL)ishelloworld{
//其他
}
```

前面带有减号"-"的方法为实例方法,必须使用类的实例才可以调用的。对应的有"+"号,代表是类的静态方法,不需要实例化即可调用。

2. 消 息

Objective-c 消息的定义:向对象发送信息。

消息是 iOS 的运行时环境特有的机制。和 C++、Java 下的类,或实例调用类或实例的方法类似。只能说是类似,实际上他们的机制有很大的差别。例如:

```
[object message:param1 withParameter:param2]
NSString *string;
string = [[NSString alloc] initWithString:@"Hello"];
```

上面的代码类似于:

```
java/c++:object.message()
java/c++:object.message(param1,param2)
java/c++:string * str;
str = new string("Hello");
```

3. Import

与C++类似,Objective-c 同样建议将声明和实现区分开。Objective-c 的头文件后缀名是 .h,源代码后缀名是 .m。Objective-c 使用 #import 引入其它头文件。与 #include 不同的是,#import 保证头文件只被引入一次。另外,#import 不仅仅针对 Objective-c 的头文件,即便是标准 C 的头文件,比如 stdlib.h,同样可以使用 #import 引入。

Objective-c 的 Import 使用方法如下:

```
import "Class.h"
import <Class.h>
import <director/Class.h>
```

4. Property 和 Synthesize

Property 定义:@property 声明用于自动创建 Property 属性变量的 getter 和 setter。
Synthesize 定义:@Synthesize 声明实现了 Property 属性变量的 getter 和 setter。
Property 和 Synthesize 用法:
interface:@property dataType variableName
implementation:@synthesiz variableName

5. 方法的声明

头文件中的方法如下:

```
-(returnType)method
-(returnType)method:(dataType)param1
-(returnType)method:(dataType)param1 withParam:(dataType)param2
```

withParam:(dataType)param2 类似于 C/C++/Java:

```
returnType  method()
returnType  method(param1)
returnType  method(param1,param2)
```

6. self

Objective C 提供了两个保留字 self 和 super,用于在方法定义中引用执行该方法

的对象。与C++比较,self 相当于 this,super 相当于调用父类的方法。写法如下:
　　[self method]
　　类似于:C++/Java
　　this.method();

7. 继承关系和接口实现

Objective-c 继承关系和接口实现如下:

```
ClassA:ParentA
ClassA:ParentA <Protocol>
ClassA <Protocol>
```

类似于 Java:

```
ClassA extends ParentA
ClassA extends ParentA implements interface
ClassA implements interface
```

Objective-c 的协议(Protocol)和 C++、Java 的接口类似。

8. 空指针

Objective-c 的空指针是没有存储任何内存地址的指针(NULL 指针);或者是被赋值为 0 的指针,在没有被具体初始化之前,其值为 0。
　　id obj = nil;
　　NSString * hello = nil;
　　nil 相当与 Java 中的 Null;

9. id

Objective-c 的 id 和 C++里的(Void *)类似。在 Objective-c 中,id 类型是一个独特的数据类型。在概念上,类似 Java 的 Object 类,可以转换为任何数据类型。换句话说,id 类型的变量可以存放任何数据类型的对象。在内部处理上,这种类型被定义为指向对象的指针,实际上是一个指向这种对象的实例变量的指针。例如,下面定义了一个 id 类型的变量和返回一个 id 类型的方法:
　　id anObject;
　　-(id)newObject:(int) type;
　　id 和 void * 并非完全一样。下面是 id 在 objc.h 中的定义:
　　typedef struct objc_object {
　　　　Class isa;
　　} * id;

从上面看出,id 是指向 struct objc_object 的一个指针。也就是说,id 是一个指向任何一个继承了 Object(或者 NSObject)类的对象。需要注意的是 id 是一个指针,所以在使用 id 的时候不需要加星号。

10. 垃圾回收

Objective-c 和 Java 一样，都有运行的环境，有内省的能力。Objective-c 和 Java 也有很多不同的地方，在 iOS 系统里，Objective-c 的内存需要自己管理，添加了 ARC 机制后编译器帮助了 Objective-c 添加 release 释放的代码。而 Java 是通过垃圾回收器管理内存。

12.5 iOS 基本控件

iOS 中提供了 UIButton、UILable、UITextField、UIImageView 等基础 UI 控件，继承自 UIView。大部分控件都继承自 UIControl，UIControl 派生自 UIView 类，每个控件都有很多视图的特性，包括附着于其他视图的能力，所有控件都拥有一套共同的属性和方法，包含显示内容、点击事件等，UIControl 的子类都有事件处理能力。

1. UILabel 控件

UILabel 控件用于显示多行文本，例如：

```
self.mylabel = [[UILabel alloc]initWithFrame:CGRectMake(20, 5, 200, 40)];
self.mylabel.backgroundColor = [UIColor redColor];
self.mylabel.text = @"你好,这是 label 测试.现在在测试换行";
self.mylabel.numberOfLines = 0;//以下两句就是实现换行的,不过要 frame 高度足够大。
self.mylabel.lineBreakMode = UILineBreakModeWordWrap;
// [self.view addSubview:self.mylabel];
```

备注：该 label 设置的 frame 高度足够显示两行，像这里高度 40 可以了，如果是高度 20 的话，还是只会显示一行文字。如果要 label 显示多行的话，还是建议使用 UItextview 控件。

2. UITextView 控件

UITextView 控件主要是实现多行文本的显示和编辑。如果需要设置为只读，则只需要在可视化编辑窗口中将它的 Behavior Editable 属性勾选框去掉即可。或者使用如下代码实现：

```
UITextView * txt = [[UITextView alloc]initWithFrame:CGRectMake(20, 100, 100, 50)];
txt.text = @"sd";
txt.editable = NO;//不可编辑
[self.view addSubview:txt];
```

3. UIButton 控件

UIButton 控件可以是个按钮点击，也可以由图片填充，例如：

```
UIButton * backbtn = [UIButton buttonWithType:UIButtonTypeRoundedRect];
backbtn.frame = CGRectMake(30,50,70,40);
[backbtn setTitle:@"按钮" forState:UIControlStateNormal];
//[backbtn setImage:[UIImage imageNamed:@"icon_top_enable.png"] forState:UIControl-
StateNormal];
//这个是图片填充按钮,如果是图片,则 Button 必须是 UIButtonTypeCustom 。当然也可以不用图片
[backbtn addTarget:self action:@selector(onclick) forControlEvents:
UIControlEventTouchUpInside];
//添加 button 点击事件 onclick
[self.view addSubview:backbtn];
```

同时也要为 onclick 事件做一些处理,代码如下:

(void)onclick{ NSLog(@"你点击了按钮");}

4. UITextField 控件

UITextField 可用于简单文本输入,可以作为密码输入框,输入时键盘控制、return 后隐藏键盘等,UITextField 控件常用属性:

（1）Text:要显示的文本。

（2）Placeholder:指定将要在文本字段中以灰色显示的占位符文本。

（3）Clear When Editing Begins:用户触摸此字段时是否删除字段中的值,即 text 框最右边有一个小叉清空按钮。

（4）Text Input Traits:文本输入特征。

5. UIImageView 控件

UIImageView 控件代表一个图片显示视图,它直接继承自 UIView 类,没有继承 UIControl,因此,UIImage 只能作为图片的显示控件,不能接收用户输入,也不能与用户交互,它只是一个静态控件。当程序需要使用 UIImageView 来显示图片时,即可直接在 Interface Builder 中把 UIImageView 拖入程序界面中,也可在程序中创建 UIImageView 对象。UIImageView 控件常用属性:

（1）image:指定图像文件。

（2）Mode:图像在视图内部的对齐方式以及是否缩放图像以适应视图。选择任何图像缩放的选项都会潜在地增加处理开销,因此,最好避开这些选项,并在导入图像之前调整好图像大小。通常 Mode 属性为 Center。

（3）Alpha:图像透明度。一般设置为 1.0。

（4）Background:该属性继承自 UIView,但它不会影响图像视图的外观,请忽略此属性。

（5）Drawing 复选框:选中 Opaque 表示视图后面的任何内容都不应该绘制,并且允许 iPhone 的绘图方法通过一些优化来加速绘图。

（6）Clear Context Before Drawing:选中它之后,iPhone 将使用透明黑色绘制控件

覆盖所有区域,然后才实际绘制控件,考虑到性能问题,并且适用情况很少,通常很少需要选中 ClearContext Before Drawing。

(7) Interaction 复选框:
- User Interaction Enabled:指定用户能否对此对象进行操作;
- Multiple Touch:是否能够接收多点触摸事件。

6. UISlider 控件

UISlider 控件就像其名字一样,是一个像滑动变阻器的控件。它处在不同的位置,这个 UISlider 会有不同的值。UISlider 控件常用属性:Value Changed。示例如下:

```
// 将 silder 的值反映到 sliderLabel
- (IBAction) sliderValueChanged: (id)sender
{
    UISlider * slider = (UISlider * )sender;
    int progressAsInt = (int)(slider.value + 0.5f);
    NSString * newText = [[NSStringalloc] initWithFormat:@"%d",
progressAsInt];
    sliderLabel.text = newText;
    [newText release];
}
```

7. UISwitch 控件(开关)

UISwitch 控件就是很像开关的那种控件,它只有两个状态:on 和 off。UISwitch 实现一个开关功能,属性 on:获取开关的状态是否为 on。方法 setOn 设置开关的状态,例如:

```
- (IBAction) switchChanged: (id)sender
{
UISwitch * whichSwitch = (UISwitch * )sender;
BOOL setting = whichSwitch.on;
[leftSwitch setOn:setting animated:YES];
[rightSwitch setOn:setting animated:YES];
}
```

8. UISegmentedControl 控件

UISegmentedControl 分段控件代替了桌面 iOS 上的单选按钮。不过它的选项个数非常有限,因为你的 iOS 设备屏幕有限。当需要使用选项非常少的单选按钮时它很合适。这个控件的可定制性比较强。UISegmentedControl 使用示例如下:

```
#definek SegmentIndex_Switches 0
#definek SegmentIndex_Button 1
- (IBAction)segmentChanged:(id)sender
{
switch([sender selectedSegmentIndex])
{
casek SegmentIndex_Switches:
leftSwitch.hidden = NO;
rightSwitch.hidden = NO;
doSomethingButton.hidden = YES;
break;
casek SegmentIndex_Button:
leftSwitch.hidden = YES;
rightSwitch.hidden = YES;
doSomethingButton.hidden = NO;
break;
}
}
```

9. UIActionSheet 控件和 UIAlertView 控件

UIActionSheet 控件用于迫使用户在两个或更多选项之间进行模式选择。操作表从屏幕底部弹出,显示一系列按钮供用户选择,用户只有单击了一个按钮后才能继续使用应用程序。

UIAlertView 控件以蓝色圆角矩形的形式出现在屏幕的中部,警报可显示一个或多个按钮。

为了让控制器类充当操作表的委托,控制器类需要遵从 UIActionSheetDelegate 协议。

例如,视图有一个 UISegmentedControl,"Switches"下有两个 UISwitch。"Button"下有一个"Do Something"的 UIButton,触摸"Do Something"Button 时弹出 UIActionSheet,触摸选择"Yes,I'm sure."时弹出 UIAlertView。

第 13 章

跨平台移动软件设计

13.1 跨平台移动软件设计目标

本章的设计目标是根据前面几章设计完成的物联网智能系统项目界面外观,采用 jQuery Mobile 重新进行界面的设计,设计的界面目标如图 13-1 所示,左边是第 11 章中"拍照更换界面图片设计目标"的界面,右边是本章设计目标的界面。

图 13-1 跨平台移动软件设计目标效果图

为了实现良好的界面效果,以及实现跨平台的目标,本章的设计底层采用 PhoneGap 库的 DroidGap 类作为活动类,界面采用 jQuery Mobile 库。相关 PhoneGap 和 jQuery Mobile 知识,本章后继部分将进行详细介绍。

13.2 物联网 App 跨平台程序基础

13.2.1 物联网 App 跨平台程序简介

最近几年,作者在为一些企业做了一些移动设备(嵌入式系统产品)的开发与设计时,发现与前几年比较具有特别大的变化,其中感受最深刻的是平台的变化。以前大部分项目是 Linux 或 WinCE 两个平台,并且只需要支持单一平台(一个产品,只选择

Linux 或 WinCE 其中之一即可)。而现在的移动设备,平台基本是 Android、iOS、Windows Phone、Linux,并且要求跨平台。一般要求至少支持 Android、iOS 这两个系统,有些要求支持 Android、iOS、Windows Phone 三个系统,还有些要求同时支持 Android、iOS、Windows Phone、Linux(例如 Ubuntu)等四个或四个以上的系统。

多平台开发,给开发人员带来了极大的困难,主要因为:

(1) 开发语言的差别:Linux 使用的是 C,Android 使用的是 Java,iOS 使用的是 Objective,Windows Phone 使用的是 C♯,Ubuntu 使用的是 C++。就算一个很简单的应用程序,如果要为这些平台都开发一个应用程序,那工作量是原来单一平台的五倍,甚至更加多。不管是单人开发几个平台的应用,或者是多人开发,都是具有很大的挑战性,光是掌握那么多个平台的开发语言和开发环境已经是具有很大的难度。

(2) 难以保持不同平台下界面和功能的一致性:由于各平台的 UI 机制不同,提供的 API 也不同,设计出来的界面,很难保证完全一致。在功能上也如此,有一些功能,只有部分平台支持,这样也很难保持功能上的一致。

基于上述原因,目前很多人把目光转移到跨平台移动开发上,争取做到开发一次,到处(各种平台上)可以使用,这样既可以减少开发工作量,又能保持不同平台下界面与功能的一致性。因此,跨平台移动开发,已经成为目前嵌入式软件设计所追求的目标之一。而移动 Web 开发,是跨平台移动开发最佳的解决方案,移动 Web 具有良好的发展机遇和技术优势。

虽然 Web 应用程序具有跨平台的优点,但用户体验不如原生态的 App(Native App)好。为此,人们提出了介于 Web-app、Native-app 这两者之间的 App,兼具"Native App 良好用户交互体验的优势"和"Web App 跨平台开发的优势"的新型 App——Hybrid App。

Hybrid(混合) App 是指混合模式移动应用。Hybrid App 主要以 JavaScript+Native 两者相互调用为主,从开发层面实现"一次开发,多处运行"的机制,成为真正适合跨平台的开发。目前已经有众多 Hybrid App 开发成功应用,比如百度、网易、街旁等知名移动应用,都是采用 Hybrid App 开发模式。

经过众多开发者与成功案例证明 Hybrid App 兼具了 Native App 的良好用户体验的优势,也兼具了 Web App 使用 HTML5 跨平台开发低成本的优势。

常用 Hybrid App 开发平台包括 PhoneGap、AppCan、appMobi、Titanium 等,它们基于 webkit 开源内核,使用 HTML5 标准开发,适配机型简单,支持开发者自定义插件,并能很好的应用于商业、教育、娱乐等行业,成为移动开发者的首选开发平台。

13.2.2 常见移动 Web 开发框架

移动 Web 开发框架,帮助开发者更加高效地开发移动 Web 应用。

1. Sencha Touch

Sencha Touch 是世界上第一个基于 HTML5 的移动 Web 开发框架,支持最新的

HTML5 和 CSS3 标准，全面兼容 Android 和 Apple iOS 设备，提供了丰富的 Web UI 组件，可以快速地开发出运行于移动终端的应用程序。

2. jQuery Mobile

jQuery Mobile 框架把"write less，do more"精神提升到更高的层次。jQuery 移动框架可以帮助设计一个可运行于所有流行智能手机和平板平台的应用程序，而不需要为每种移动终端都开发一个特别的版本。

3. jQTouch

jQTouch 是一款 jQuery 的插件，用于手机上实现动画、列表导航、默认应用样式等各种常见 UI 效果。支持 iPhone、Android 等手机。

4. The-M-Project

The-M-Project 是一个包含各种 UI 组件，基于 jQuery 开发 HTML5 应用程序的移动 Web 应用框架，支持 iOS、Android、Palm WebOS 和 BlackBerry 等平台。

5. DHTMLX Touch

DHTMLX Touch 是一个基于 HTML5 的免费 JavaScript 库，用于构建跨平台的移动 Web 应用程序。这不只是一组 UI 部件，而是一个完整的框架，它允许为手机等触摸设备创建强大的 Web 应用程序。

6. Web App.Net

Web App.Net 提供了很多的 API，因此可以帮助节省很多工作了。不需要花时间去进行 Ajax 调用的编码，因为已经内置了，另外还有很多其它内置功能，提供了详细的文档和应用演示。

7. Wijmo

Wijmo 是一个 jQuery UI Widgets 框架，混合了 JavaScript、CSS3、SVG 和 HTML5，拥有 30 多个组件，是 jQuery UI 的一个扩展。

8. jquery-mobile-960

jquery-mobile-960 是一个用于移动 Web 开发的网格框架，综合了 960.gs 的灵活性和 jQuery Mobile 的方便性。它的目的是让 jQuery Mobile 布局更加灵活，使得应用在移动终端更加易用。

9. SproutCore

SproutCore 是一个 HTML5 移动 Web 开发框架，它的目标是在无需浏览器插件的情况下，在浏览器中为应用程序提供极佳的桌面效果。

10. NimbleKit

NimbleKit 是为 iOS 设备构建应用程序最快速的方式，不需要知道 Objective-c

或者iOS SDK,只需结合JavaScript代码编写HTML页面就可以了。

13.2.3 常见Hybrid App平台

Hybrid App开发,现阶段主流的平台包括PhoneGap、AppCan、appMobi、Titanium等,它们基于webkit开源内核,使用HTML5标准开发,适配机型简单,支持开发者自定义插件,并能很好地应用于商业、教育、娱乐等行业,成为移动开发者的首选开发平台。

1. PhoneGap(Cordova)

PhoneGap是一款国外的开源移动开发平台。目前已经将核心代码贡献给Apache cordova,它是基于HTML、CSS和JavaScript,可以使用一些开源的框架比如jQuery Mobile、Dojo Mobile、Sencha Touch等来提高用户体验,也提供了比较丰富的原生插件调用。

PhoneGap的特性:

(1) 可以使用DreamWeaver 5.5编码,现在使用appMobi提供的xdk进行模拟器开发;

(2) 代码开源,开发者可以放心使用;

(3) 兼容性,一次开发,多处运行;

(4) 使用JS+HTML5,成本低。

PhoneGap的优点:

(1) Native接口比较丰富,通过封装的API可以直接访问硬件,比如说加速、相机、指南针、GPS、文件访问等;

(2) 接口文档描述非常详细;

(3) 支持平台多,包括iOS、Android、Blackberry、Symbian、Windows Phone 7、Windows Phone 8等。

PhoneGap的缺点:

(1) 需要针对相应的平台环境配置,进行编译,打包测试,发布等。由于使用Hybrid开发的用户群,大部分是Web开发者,对原生开发基本不了解,这无疑给每一个开发者增加了沉重的负担,需要对各个平台的开发都需要了解,对硬件等都要配置,加大开发成本;

(2) 使用效果启动慢,页面切换响应慢,数据请求慢;

(3) 文档虽比较详细但是基本是英文,对于国内大部分用户英文水平较差的是比较大的挑战;

(4) 因为是国外的框架,技术支持不够到位,出现问题,无法排解,成为技术攻关的难点。

2. AppCan

AppCan是本土移动开发中使用最广的移动平台,网络舆论而言,AppCan是

PhoneGap 的中国化，但是从对 AppCan 实际使用，以及转向移动开发的朋友们互相交流反馈，他们是截然不同的两个移动平台，AppCan 不仅封装了类似于 PhoneGap 的本地调用功能，而且封装了 uexWindow 多窗口机制，实现了移动端的 iframe 效果，虽然不是开源项目，但一直都有面向开发者的免费版，并且也有定位于企业用户的企业版套装，目前最新版本为 2.2.X。

AppCan 的特性：

(1) 提供的集成开发环境的 IDE 进行模拟器开发；

(2) 兼容性，一次开发，多处运行；

(3) 使用 JS＋HTML5，成本低；

(4) 在线打包；

(5) 代码加密保护机制。

AppCan 的优点：

(1) 支持在线上传证书打包，对于不了解苹果，以及 Android 环境开发的人是福音；

(2) 支持更多的原生调用，比如 UI 控件的封装，通讯类(Socket)，地图，支付宝等更多的原生控件支持；

(3) 拥有统一数据统计平台，便于运营管理开发的应用；

(4) 完善的技术支持，官方论坛以及 Q 群建设较为完善，使开发者更好地进行交流沟通。

AppCan 的缺点：

(1) 虽然有中文的开发文档，但描述比较简单，希望他们丰富他们的 API 文档；

(2) 免费版本不支持自定义插件(据说企业版可以自定义插件)；

(3) 暂时只支持 iOS、Android 两大平台；

(4) 许多功能需要企业版才能实现，不过是收费的。

3. Titanium

Titanium 移动平台是所有移动开发平台中比较另类的，它将 JavaScript 和本地库链接在一起，编译成字节码，针对 iOS 以及 Android 两个平台分别构建一个软件包。应用程序使用 HTML，JavaScript 和 CSS 进行开发，并支持 PHP，Ruby 和 Python。应用程序可以使用 Appcelerator API 访问本地特性。并提供 Appcelerator Studio 开发环境，由于编译成本地代码，所以用户体验是最好的。

Titanium 的特性：

(1) 针对不同平台生成对应的原生包；

(2) 供 Appcelerator Studio 开发。

Titanium 的优点：

(1) 针对 JS 解析生成原生控件，基本达到纯原生的用户体验；

(2) 支持自定义插件。

Titanium 的缺点：

(1) API 文档为英文，并且比较简单，对国内用户使用有一定挑战；

(2) 跟 PhoneGap 同样，国外框架，技术支持困难；

(3) 支持 Android、iOS、黑莓平台；

(4) 环境需要用户自己搭建，比较复杂。

4. appMobi

appMobi 推出了全新开发工具 XDK，这个工具使得开发者可以使用 HTML5 构建网络和移动平台的应用程序，可以进行屏幕仿真调试、设备实际调试和遥控调试等。

appMobi 的特性：

(1) 使用 XDK 进行开发；

(2) HTML5+CSS+JS；

(3) 一次开发，多处运行。

appMobi 的优点：

本地接口较为丰富，并且推出有游戏加速引擎，主要包括物理引擎、离线和动态缓存、媒体播放器、验证和加密、增强现实、二维码和 QR 扫描、更好的显示支持。

appMobi 的缺点：国外框架，技术支持差。

13.3　HTML5

1. 什么是 HTML5

HTML5 将成为 HTML、XHTML 以及 HTML DOM 的新标准。

HTML 的上一个版本诞生于 1999 年。从那以后，Web 世界已经经历了巨变。

HTML5 仍处于完善之中。然而，大部分现代浏览器已经具备了某些 HTML5 支持。

2. HTML5 是如何起步的

HTML5 是 W3C[①] 与 WHATWG[②] 合作的结果。

WHATWG 致力于 Web 表单和应用程序，而 W3C 专注于 XHTML 2.0。在 2006 年，双方决定进行合作，来创建一个新版本的 HTML。

HTML5 建立的一些规则：

(1) 新特性应该基于 HTML、CSS、DOM 以及 JavaScript；

(2) 减少对外部插件的需求（比如 Flash）；

(3) 更优秀的错误处理；

---

① W3C 指 World Wide Web Consortium，万维网联盟。
② WHATWG 指 Web Hypertext Application Technology Working Group。

(4) 更多取代脚本的标记；
(5) HTML5 应该独立于设备；
(6) 开发进程应对公众透明。

3. HTML5 新特性

HTML5 中的一些有趣的新特性：

(1) 用于绘画的 canvas 元素；
(2) 用于媒介回放的 video 和 audio 元素；
(3) 对本地离线存储的更好的支持；
(4) 新的特殊内容元素，比如 article、footer、header、nav、section；
(5) 新的表单控件，比如 calendar、date、time、email、url、search。

TIPS：最新版本的 Safari、Chrome、Firefox 以及 Opera 支持 HTML5 特性。Internet Explorer 9 以上支持某些 HTML5 特性。

13.4 PhoneGap 概述

13.4.1 PhoneGap 介绍

PhoneGap 是一个用基于 HTML、CSS 和 JavaScript 的创建移动跨平台移动应用程序的快速开发平台。它使开发者能够利用 iPhone、Android、Palm、Symbian、Windows phone、Bada 和 Blackberry 智能手机的核心功能——包括地理定位、加速器、联系人、声音和振动等，此外，PhoneGap 拥有丰富的插件，可以灵活和方便地扩展其功能。

PhoneGap 的优势如下：

(1) 跨平台。一次开发，多个平台共用。现主要包括了 Android、iOS、Apple iOS、Palm、Symbian、BlackBerry、Windows phone 等。

(2) 降低开发门槛。对于很多 Web 开发人员来说，熟悉 Objective－c 语言和 Java 语言都是比较痛苦的事情。有了 PhoneGap 就不用担心这些了。用熟悉的 Web 前端技术就可以开发出很专业的手机应用程序。

(3) 提供强大的硬件访问控制。比起传统的 Web 程序，PhoneGap 提供了一系列的 JS 的类，可以直接访问硬件，比如加速、相机、指南针、GPS、文件访问等，可以使用 JS 方便地调用系统的硬件，以弥补传统 Web 程序的一块错误。

(4) 方便的安装和使用。PhoneGap 的架构很复杂，但对于大多数开发者来说，只用很简单的配置就可以搭好环境。只要专注写好自己的 Web 页面，拷贝进去就可以了。

PhoneGap 的劣势如下：

(1) 运行速度慢。程序的载入和 UI 界面的反应都比原生的程序慢，因为它实际上还是在展示 Web 页面，所以载入、页面刷新等肯定是需要一定时间的。

（2）不适合部分程序。如果程序需要 3D 功能，或者对界面刷新有较高的要求，这样的程序现在来说还是用原生的语言会比较好。

PhoneGap API 主要包括以下 13 个模块：

- Accelerometer：点击设备屏幕的手势感应器。
- Camera：调用设备摄像头采集照片。
- Capture：使用设备的媒体应用程序调用媒体文件。
- Compass：获取设备移动的方向。
- Connection：快速检查网络状况以及蜂窝网络的信息。
- Contacts：设备联系人相关操作。
- Device：获取设备的相关信息。
- Events：通过 JavaScript 获取本地活动。
- File：通过 JavaScript 调用本地文件系统。
- Geolocation：使应用程序可以访问地理位置信息。
- Media：录制和播放音频文件。
- Notification：设备视觉、声音和触觉反馈。
- Storage：截获设备的本地存储选项。

13.4.2　PhoneGap 实例

1. 创建一个新项目

创建一个普通的 Android 新项目，项目名称为"myPhoneGap"。例如在 Obtain_Studio 下创建一个 Anroid 项目，模板选择 Android 4.3 模板，如图 13-2 所示。

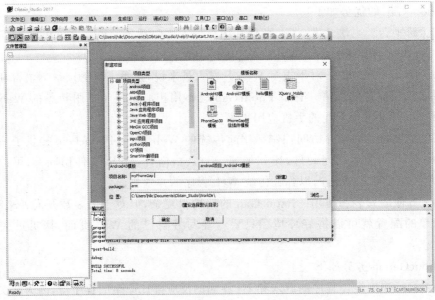

图 13-2　Android 4.3 模板

修改 Hello 类的代码,从 DroidGap 类派,代码如下:

```
package com.android.hello;
import android.os.Bundle;
import org.apache.cordova.*;
public class hello extends DroidGap
{
    @Override
    public void onCreate(Bundle savedInstanceState)
    {
        super.onCreate(savedInstanceState);
        //设置 <content src = "index.html" /> 在 config.xml
        //super.loadUrl(Config.getStartUrl());
        super.loadUrl("file:///android_asset/www/index.html");
    }
}
```

从自带例子中的 \ framework \ bin \ 目录下拷贝库文件 "classes.jar" 到 myPhoneGap 项目 libs 目录下。

从 C:\Users\Administrator\.cordova\lib\android\cordova\3.0.0\bin\templates\project\ 目录下,把 assets 所有文件拷贝到 myPhoneGap 项目 libs 目录下。

13.4.3 用 PhoneGap 开发 iOS 应用程序

1. 生成 iOS PhoneGap 项目

对于使用 PhoneGap 3.0 及以上版本,则按本章 "13.4.1 PhoneGap 介绍" 小节进行安装,完成之后,可以在 Windows 下使用以下命令生成一个 iOS PhoneGap 项目:

$ cordova create hello com.example.hello "HelloWorld"

$ cd hello

$ cordova platform add ios

$ cordova prepare   # or "cordova build"

命令运行完成之后,生成一个 HelloWorld 目录,该目录下就是一个完整的 iOS PhoneGap 项目了,COPY 到 MAC 操作系统下,采用 XCode 打开并编译运行即可。可以参考以下官方参考手册:http://docs.phonegap.com/en/edge/ guide_platforms_ios_index.md.html#iOS%20Platform%20Guide

2. 编译和运行 iOS PhoneGap 项目

创建后,可以使用 Xcode 打开它。双击打开 hello/platforms/ios/hello.xcodeproj 文件,如图 13-3 所示。

然后部署到模拟器,在 iOS 的模拟器中预览应用程序,步骤如下:

(1) 请确认在左侧面板中选择 xcodeproj 文件;

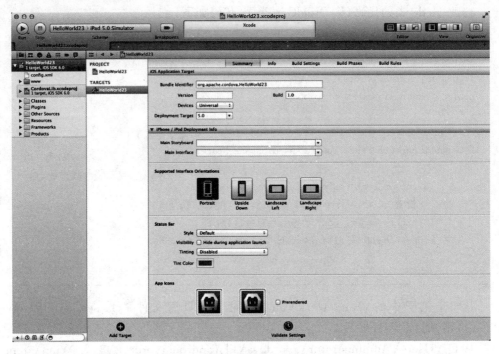

图 13-3 Xcode 打开 iOS PhoneGap 项目

(2) 选择右侧面板中的 Hello 应用程序；

(3) 从工具栏的计划菜单中选择所使用的设备，如 iPhone 13.0 或者 iPhone 7.0 模拟器，如图 13-4 所示。

(4) 按工具栏左侧的运行按钮，开始编译、部署并在模拟器中运行应用。一个单独的模拟器应用程序将被打开，显示应用程序运行的效果如图 13-5 所示。

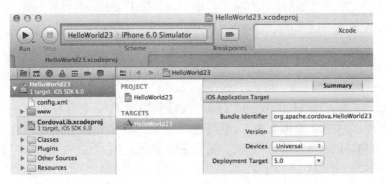

图 13-4 选择所使用的设备

图 13-5 项目运行效果

13.5　jQuery Mobile 概要

13.5.1　jQuery Mobile 介绍

jQuery Mobile 是 jQuery 在手机上和平板设备上的版本。jQuery Mobile（或简称"JQM"，或者"JM"）不仅会给主流移动平台带来 jQuery 核心库，而且会发布一个完整统一的 jQuery 移动 UI 框架。支持全球主流的移动平台。

目前 jQuery 驱动着 Internet 上的大量网站，在浏览器中提供动态用户体验，促使传统桌面应用程序越来越少。现在，主流移动平台上的浏览器功能都赶上了桌面浏览器，因此，jQuery 团队引入了 jQuery Mobile。JQM 的使命是向所有主流移动浏览器提供一种统一体验，不管使用哪种查看设备，使整个 Internet 上的内容更加丰富。

JQM 的目标是在一个统一的 UI 中交付超级 JavaScript 功能，兼容最流行的智能手机和平板电脑设备工作。与 jQuery 一样，JQM 是一个在 Internet 上直接托管、免费可用的开源代码基础。事实上，当 JQM 致力于统一和优化这个代码时，jQuery 核心库受到了极大关注。这种关注充分说明，移动浏览器技术在极短的时间内取得了多么大的发展。

与 jQuery 核心库一样，开发计算机上的应用不需要安装任何东西，只需将各种 *.js 和 *.css 文件直接包含到的 Web 页面中即可。jQuery Mobile 特点如下：

(1) 简单易用

jQuery Mobile 框架简单易用。页面开发主要使用标记，无需或仅需很少 JavaScript。

(2) 持续增强和优雅降级

尽管 jQuery Mobile 利用最新的 HTML5、CSS3 和 JavaScript，但并非所有移动设备都提供这样的支持。jQuery Mobile 的哲学是同时支持高端和低端设备，比如那些没有 JavaScript 支持的设备，尽量提供最好的体验。

(3) Accessibility

jQuery Mobile 在设计时考虑了访问能力，它拥有 Accessible Rich Internet Applications（WAI-ARIA）支持，以帮助使用辅助技术的残障人士访问 Web 页面。

(4) 小规模

jQuery Mobile 框架的整体比较小，JavaScript 库 12 KB，CSS 6 KB，还包括一些图标。

(5) 主题设置

jQuery Mobile 框架还提供一个主题系统，允许提供自己的应用程序样式。

(6) 浏览器支持

在移动设备浏览器支持方面取得了长足的进步，但并非所有移动设备都支持 HTML5、CSS 3 和 JavaScript。这个领域是 jQuery Mobile 的持续增强和优雅降级支持发挥作用的地方。如前面所述，jQuery Mobile 同时支持高端和低端设备，比如那些没有 JavaScript 支持的设备。

jQuery Mobile 目前支持以下移动平台：
- Apple iOS：iPhone、iPod Touch、iPad（所有版本）；
- Android：所有设备（所有版本）；
- Blackberry Torch（版本 6）；
- Palm WebOS Pre、Pixi；
- Nokia N900（进程中）。

13.5.2　jQuery Mobile 应用

jQuery Mobile 包含自动通过 AJAX 装载带有返回按钮的外部页面，以及一组页面切换动画效果和用来把页面显示为对话框的简单工具。

jQuery Mobile 的"page"模型被优化为可以支持单个页面或者页面内嵌的"page"。这里的 page 和传统意义上的页面有所不同，在 jQuery Mobile 里指的是 page 模型或者结构，data-role="page"的一个 div 就是一个 page。

这个模型的目标是允许开发者在创建站点时利用最佳实践——传统的链接就是起作用了，不需要任何特别配置——在创建一个富客户端，类似于本地应用程序可不能简单地依靠标准的 HTTP 请求来达到。

一个 jQuery Mobile 的站点必须采用 HTML5 的"doctype"标签才能充分利用框架的特性。在 head 标签里，需要包括 jQuery Mobile 库、jQuery 库和主题 CSS 文件等，例如：

```
<!DOCTYPE html>
<html>
    <head>
    <title>Page Title</title>
        <link rel="stylesheet" href="http://ajax.googleapis.com/ajax/libs/jquerymobile/1.4.2/jquery.mobile.min.css"/>
        <script src="http://ajax.googleapis.com/ajax/libs/jquery/1.13.2/jquery.min.js"></script>
        <script src="http://ajax.googleapis.com/ajax/libs/jquerymobile/1.4.2/jquery.mobile.min.js"></script>
    </head>
    <body>
    ...
    </body></html>
```

在标签里，每一个视图或者"page"被一个元素（通常是 div）设置 data-role="page"属性后所唯一标识：

第13章　跨平台移动软件设计

```
<div data-role="page">
    ...
</div>
```

在"page"容器内部,任何有效的 HTML 标记都可以使用,但是对于典型的 jQuery Mobile 页面而言,一个"page"的直接子元素是带有 data-role 的 header、content、footer 等,例如:

```
<div data-role="page">
    <div data-role="header">...</div>
    <div data-role="content">...</div>
    <div data-role="footer">...</div>
</div>
```

上面是 header、content 和 footer 的 div 代码。总的说来,这是标准的样板,实现应用如下:

```
<div data-role="page">
    <div data-role="header">
        <h1>Page Title</h1>
    </div><!-- /header -->
    <div data-role="content">
        <p>Page content goes here.</p>
    </div><!-- /content -->
    <div data-role="footer" data-position="fixed">
        <h4>Page Footer</h4>
    </div><!-- /footer -->
</div><!-- /page -->
```

13.5.3　jQuery Mobile 页面链接

1. AJAX 导航

为了实现在移动设备上的无缝客户体验,jQuery Mobile 默认采用 AJAX 的方式载入一个目的链接页面。因此,当在浏览器中点击一个链接来打开一个新的页面时,jQuery Mobile 接收这个链接,通过 AJAX 的方式请求链接页面,并把请求得到的内容注入到当前页面的 DOM 里。

这样的结果就是用户交互始终保存在同一个页面中。新页面中的内容也会轻松地显示到这个页面里。这种平滑的客户体验相比于传统打开一个新的页面并等待数秒的方式要好很多。当一个新的页面做为新的 data-role="page",div 插入到主页面时,主页面会有效地缓存取到的内容。使得当要访问一个页面时能够尽快地显示出来。这个工作过程听起来难以置信的复杂,但是做为开发人员的大部份不需要了解其中工作的具体细节，只要能看到效果就行。

注意：如果不想采用 AJAX 的方式加载页面，而想以原生的页面加载方式打开一个链接页面，只需要在打开的链接上添加属性 rel="external"即可。

2. 内部页面链接

jQuery Mobile 简化了创建 AJAX 站点和程序的过程。默认情况下，当点击一个链接时会指向一个外部页面（如：products.html），但是框架会解析该链接的 href 属性然后发出一个 AJAX 请求（Hijax，即渐进增强的 AJAX），并显示正在加载的提示。

如果 AJAX 请求成功，新页面内容会添加到 DOM 当中，所有 mobile widget 都是自动初始化的，然后新页面会通过动画过渡再显示出来。

如果 AJAX 请求失败，框架会显示一个小小的错误消息提示("e"主题的样式)，并会在一小段时间内消失，并且不会破坏当前的导航流，即页面不会刷新也不会对前进后退按钮有影响。

单个 HTML 文档可以包含多个"page"，只需要在一个页面包含多个 data-role="page"的 div 即可，每个 page div 必须由一个唯一的 ID（例如 id="foo"）链接到相应页面，链接时使用锚记（例如 href="#foo"）即可。当单击一个链接时，框架会寻找 id 为锚记 href 的内部"page"并显示到当前界面中。

要注意，如果正在通过 AJAX 从一个 mobile 页面链接到一个含有多个内部页面的页面，需要为该链接添加一个 rel="external" 或者 data-ajax="false"。该属性告知框架对页面进行重新加载，url hash 也将清零。这点十分关键，因为 AJAX 页面使用 hash(#)来追踪 AJAX 历史，当含有多个内部"page"的页面使用 hash 来指示内部"page"时会发生冲突。

举例来说，一个指向含有多个内部"page"的页面的链接会像这样：

```
<a href="multipage.html" rel="external">Multi-page link</a>
```

13.5.4　jQuery Mobile 内容格式

1. 基本的 HTML 样式

在移动设备上使用多列布局并不是推荐的，但是有时可能会需要把一些小的元素比如按钮导航 tab 等排成一行。

jQuery Mobile 框架提供了一个简单的方法来构建基于 CSS 的栅格布局，约定为 ui-grid。

有两个预设的配置布局：两列布局（class 含有 ui-grid-a）和三列布局（class 含有 ui-grid-b），几乎可满足在任何情况下使用的要求。网格宽度是100%的，且不可见（没有背景或边框），也没有 padding 和 margin，所以它们不应该影响它们内嵌元素的样式。

2. 两列网格

要创建一个两列(50/50%)布局,首先需要一个容器(class="ui-grid-a"),然后添加两个子容器(分别添加 ui-block-a 和 ui-block-b 的 class):

```
<div class="ui-grid-a">
    <div class="ui-block-a"><strong>I'm Block A</strong>
        and text inside will wrap </div>
    <div class="ui-block-b"><strong>I'm Block B</strong>
        and text inside will wrap </div>
</div><!-- /grid-a -->
```

网格 class 可以应用于任何容器。在下面的例子中为 fieldset 添加了 ui-grid-a 并为两个 Button 容器应用了 ui-block:

```
<fieldset class="ui-grid-a">
    <div class="ui-block-a"><button type="submit" data-theme="c">Cancel </button></div>
    <div class="ui-block-b"><button type="submit" data-theme="b">Submit </button></div>
</fieldset>
```

此外,网格块可以采用主题化系统中的样式——通过增加一个高度和颜色主题,就可以实现这种风格的外观:

3. 三列网格

另一种网格布局配置在父级容器中使用 class=ui-grid-b,而三个子级容器使用 ui-block-a/b/c,以创建三列的布局(33/33/33%):

```
<div class="ui-grid-b">
    <div class="ui-block-a">Block A </div>
    <div class="ui-block-b">Block B </div>
    <div class="ui-block-c">Block C </div>
</div><!-- /grid-a -->
```

4. 多列网格

(1) 四列网格

四列网格使用 class=ui—grid—c 来创建(25/25/25/25%)。

(2) 五列网格

五列网格使用 class=ui—grid—d 来创建(20/20/20/20/20%)。

(3) 多行网格

网格被设计用来折断多行的内容。举例来说,如果指定一个三列网格中包含九个子块,他们会折断成三行三列的布局。该布局需要为 class=ui-block-子块使用一个重复的序列(a,b,c,a,b,c 等)来创建。

5. 可折叠块

创建一个可折叠的内容块,创建一个容器,并添加 data-role="collapsible" 属性。

直接在容器里面添加任何标题元素(H1-H6)。框架会把标题自动转换为一个可点击的按钮并且添加一个"+"图标用来指明它是可以展开的。

在标题后面可以添加任何 HTML 标记。框架会自动把这些标记包裹在一个容器里用以折叠或显示(当点击标题时):

```
<div data-role="collapsible">
    <h3>I'm a header</h3>
    <p>我是可折叠的内容。默认我是显示的,但是可以点击标题来隐藏我!
    </p>
</div>
```

正如这个例子说明,默认情况下内容是展开的,要在页面加载时折叠内容,添加 data-collapsed="true" 属性:

```
<div data-role="collapsible" data-collapsed="true">
```

此代码将创建一个可折叠的块。

6. 主题化内容

容器的主要内容区(data-role="content"容器),应通过增加属性为 data-role="page"的父容器添加 data-theme 属性来主题化,以确保背景色能够在整个页面都应用,而不管内容的长度是多少。如果仅仅为内容容器添加了 data-theme,则背景色会在内容结束部分截断,可能会在固定 footer 和内容之间产生留白:

```
<div data-role="page" data-theme="a">
主题化可折叠块
要为折叠块的标题设置颜色请为容器添加 data-theme 属性。该图标和主体目前没有 data 属性来主题化,但可以直接使用自定义的 CSS 样式。
</p> <div data-role="collapsible" data-collapsed="true" data-theme="a">
```

13.5.5 jQuery Mobile 导航

jQuery Mobile 的导航模型是 jQuery Mobile 的核心所在,由于 jQuery Mobile 中区分了页面和"page",所以在下文中要注意页面和"page"出现时所代表的不同意思,另下文中的"页面更改"或者"页面变化"大多指的是从当前页面链接到 JQM 中的另一个"page"。

在 jQuery Mobile 里一个"page"由一个设置了 data-role="page"属性的元素构

成（通常是 div），通常里面包含"header"，"content"和"footer"，每个部分都可以包含普通的标签，表单和 jQuery Mobile 的自定义 widget。

页面载入的基本工作流程如下：首先，用户对页面发起一个正常的 HTTP 请求，随后的"page"会被插入到当前页面的 DOM 当中。正因为如此，DOM 每次可能会有"page"的一个数字，每个都可以通过连接到它的 data-url 属性来重新访问。

当一个 URL 在初始化请求时，可能有一个或多个"page"在响应，但只有第一个将被显示。存储多个"page"的优势是，它可以预读有可能被访问的静态页面。

13.5.6 jQuery Mobile 工具栏

工具栏（Toolbar）一般用于 header，footer 和 utility bar，他们遍及一个移动页面和程序，所以 jQuery Mobile 提供了一系列标准的工具栏和导航工具来涵盖大部分常见情况。

1. Toolbar 类型

在 jQuery Mobile 中，有两种标准类型的工具栏：Headers 和 Footers。

Header bar 充当页面标题的作用，通常是 mobile page 中的第一个元素，一般包含有一个页面标题和两个按钮。

Footer bar 通常是最后一个元素，相比于 header 在内容和功能上面更加自由，但是一般包含一些文字和按钮。

在 header/footer 中，一个水平的导航栏或者 tab 页是非常常见的；jQuery Mobile 包含了导航条部件，该部件能将一个无序列表（ul）链接变成一个水平分布的按钮栏。

2. Toolbar 位置选项

Header 和 footer 可以用不同方式调整它们在页面中的位置。默认情况下，toolbar 会使用"inline"模式，该模式中，header 和 footer 会位于自然文档流中（即默认的 HTML 行为），这样可以确保它们在所有的设备中可见（不管设备是否支持 JavaScript 和 CSS）。

"fixed"模式提供了一个无视人为滚动页面保持固定位置的 toolbar，toolbar 会像"inline"模式一样出现在它们在页面中原始的位置，但是拖动滚动条使 toolbar 离开视线时，框架会通过动态重新调整 toolbar 的位置，好让 toolbar 以动画的效果重新出现在当前浏览器视口的顶部或者底部。

在任何时候，轻按屏幕会切换固定工具栏的可视性：轻按一次显示，再按一次隐藏。这使用户可以选择隐藏工具栏，使得屏幕显示的内容更多。

要为 header 或 footer 设置此行为，为 toolbar 的容器添加 data-position="fixed" 属性。

"fullscreen"模式为全屏显示模式。

3. 顶部结构

header 通常是页面顶部包含页面标题文字和可选按钮以及定位到左侧和/或右导

航的工具条。

标题文本通常是一个 H1 标题元素,但它可以使用任何级别的标题(H1－H6)的,以体现语义的灵活性。例如,一个页面包含多个"page"时,可以使用 H1 表示的"首页"的标题,H2 元素表示二级"页面"的标题。默认情况下所有标题级别是相同的风格,以保持视觉上的一致性:

```
<div data-role="header">
<h1>页面标题</h1>
</div>
```

4. 默认 header 的特性

header 工具栏默认被设置为"a"主题(黑色),但是可以方便地设置主题。

5. Back 页面标题

框架会自动在每个页面生成该按钮,以简化创建通用导航条的过程,要阻止 header 中自动添加该按钮,可以自行在左边添加按钮或者为 header 容器添加 data-back-btn="false"属性。

6. 添加按钮

在标准 header 配置中,文本旁边有很多位置可供添加按钮。每个按钮通常都是一个"a"标签,但是任何可用的按钮标签都可以添加。为了节省空间,在工具栏中按钮被设置为 inline styling,所以按钮的宽度会和它所包含了文本、按钮所匹配。

7. 创建自定义后退按钮

如果对"a"标记使用 data-rel="back"属性,任何在此"a"上的点击都会模拟后退按钮,和浏览器的历史按钮一样,并会忽略"a"标记本身的 href。当链接到一个已有页面,比如"主页",或者生成后退按钮时或者一个按钮来关闭一个对话框时,该属性十分有用。当在源文件中使用此特性时,请确保提供一个有意义的 href 来指向引用页的 URL(这样才能使得用户在 C 级浏览器中也能使用该特性)。同样的,请记住如果只是想要一个反向过渡而并不实际回到上一页,应该使用 data-direction="reverse"属性来替代。

8. 默认按钮定位

header 插件会寻找 header 容器的直接子元素,并自动设置第一个链接在左边的位置,第二个链接在右边。在以下的例子中,"取消"按钮会出现在左边,而"保存"按钮会出现在右边,例如:

```
<div data-role="header" data-position="inline">
<a href="index.html" data-icon="delete">取消</a>
<h1>Edit Contact</h1>
<a href="index.html" data-icon="check">保存</a>
</div>
```

按钮会自动适应按钮所在工具栏的主题颜色,所以在一个主题为"a"的 header bar 里一个按钮也会被设置为"a",除非单独设置按钮的 data-theme 属性为其他值(例如"b"),例如:

```
<div data-role="header" data-position="inline">
<a href="index.html" data-icon="delete">Cancel</a>
<h1>Edit Contact</h1>
<a href="index.html" data-icon="check" data-theme="b">Save</a>
</div>
```

9. 用 class 控制按钮的位置

按钮位置同样可以用 class 而不是源代码的顺序来控制。当想按钮只在右边时就非常有用。为"a"标记添加 ui-btn-left 或者 ui-btn-right class 来指定按钮的位置。

data-icon 属性可以被用来创建如下图标:

- 左箭头 data-icon="arrow-l";
- 右箭头 data-icon="arrow-r";
- 上箭头 data-icon="arrow-u";
- 下箭头 data-icon="arrow-d";
- 删除 data-icon="delete";
- 添加 data-icon="Plus";
- 减少 data-icon="minus";
- 检查 data-icon="Check";
- 齿轮 data-icon="gear";
- 前进 data-icon="Forward";
- 后退 data-icon="Back";
- 网格 data-icon="Grid";
- 五角 data-icon="Star";
- 警告 data-icon="Alert";
- 信息 data-icon="info";
- 首页 data-icon="home";
- 搜索 data-icon="Search"。

在以下例子中,只在右边添加了一个按钮,所以必须要添加 data-backbtn="false"来防止出现后退按钮,而右边的按钮则需要添加 ui-btn-right class:

```
<div data-role="header" data-position="inline" data-backbtn="false">
<h1>页面标题</h1>
<a href="index.html" data-icon="gear" class="ui-btn-right">选项</a>
</div>
```

10. 自定义后退按钮的文本

如果想配置后退按钮的文本,可以使用 data-back-btn-text="previous" 属性,或者以编程方式设置插件的选项:

$.mobile.page.prototype.options.backBtnText = "previous";

如果采用编程方式,请在 mobileinit 事件的处理程序中设置该选项。

11. 自定义 header 配置

如果要创建一个自定义的 header,将自己的标记包裹在一个 div 容器中(在 header 容器中),插件不会应用自动按钮逻辑,所以可以编写自定义样式来布局 header 的内容。

12. footer 结构

footer 和 header 的基本结构相同,除了它使用 data-role="footer"来定义:

```
<div data-role="footer">
    <h4>Footer content</h4>
</div>
```

footer 工具栏默认会被设置为"a"主题(黑色),但是可以轻松地更改。在配置方面 footer 和 header 也非常相似,主要的区别是 footer 被设计为更轻便的结构来实现更灵活的配置,所以框架不会像在 header 中那样自动排列按钮。

任何链接或者合法的按钮标记都可以添加到 footer 中并自动转变成一个按钮。为了节省空间,工具栏中的按钮会自动被设置 inline styling,让按钮宽度能适应其内容。默认情况下,工具栏没有任何空白(padding)位置来填充导航条和其他小部件,要使工具栏包含空白,添加 class="ui-bar"属性:

```
<div data-role="footer" class="ui-bar">
    <a href="index.html" data-role="button" data-icon="delete">Remove</a>
    <a href="index.html" data-role="button" data-icon="plus">Add</a>
    <a href="index.html" data-role="button" data-icon="arrow-u">Up</a>
    <a href="index.html" data-role="button" data-icon="arrow-d">Down</a>
</div>
```

以下会创建一行包含按钮的工具栏。要为按钮分组则将链接包裹在一个 wrapper (通常是 div)中并为其添加 data-role="controlgroup"和 data-type="horizontal"

属性:

```
<div data-role="controlgroup" data-type="horizontal">
```

13. 固定 footer

在某些情况下,footer 会作为全局导航元素存在,可以需要 footer 在页面过渡和切换时也出现在固定的位置,而这一特性 jQuery Mobile 也提供了支持。

要实现 footer 的固定,可以为所有相关页面的 footer 添加一个 data-id 属性并使用相同的 code id 值。例如,为当前页和目标页添加一个 data-id="myfooter" 属性后,框架会在页面切换动画发生时自动保持 footer 元素的位置。请注意:这一效果只有在 footer/header 设置了 data-position="fixed" 时才会起作用。

14. navbar 示例

jQuery Mobile 有一个非常基本的 navbar 部件非常有用,它提供多达 5 个按钮和可选的图标,通常位于 header/footer 里面。

navbar 通常是一个包裹在一个容器里的无序链接列表,容器设置 data-role="navbar"属性。要设置其中一个链接是激活(被选择)的状态,为链接添加 class="ui-btn-active"属性。在下面这个例子中有一个位于 footer 中包含两个按钮的 navbar,其中一个被设置为激活状态:

```
<div data-role="footer">
    <div data-role="navbar">
        <ul>
            <li><a href="a.html" class="ui-btn-active">One </a></li>
            <li><a href="b.html">Two </a></li>
        </ul>
    </div><!-- /navbar -->
</div><!-- /footer -->
```

固定工具栏会在滚动后重新出现。这是一个 jQuery Mobile 框架里"fixed" header/footer 的演示。滚动会导致 header 和 footer 消失,停止滚动时会再度出现,任何时候轻触屏幕或者滑动屏幕都会使得工具栏重新出现或者消失。该行为使用 data-position="fixed"属性来设置:

```
<div data-role="header" data-position="fixed">
    <h1>Fixed toolbars </h1>
</div>
```

这个工具栏是用来处理在特殊情况下,想用内容来填充整个屏幕,并且希望 header 和 footer 工具栏在点击屏幕的时候出现或消失时所使用的模式,通常适用于照片、图像或者视频浏览的情况。

要开启该特性，需要将 data-fullscreen="true" 添加到的 page（data-role="page"的 div）中，并且为 header 和 footer 容器（div）同时添加 data-position="fixed"属性。

在这种模式下，如同 DEMO 展示的那样工具栏会覆盖页面内容，当工具栏可见时不是全部内容都能够看见的。

header 和 footer 都可以被主题化（默认为"a"主题），因为在视觉上工具栏通常是页面的主要层级。

主题化 headers 和 footers：要设置主题依然和其他一样，使用 data-theme 属性。例如（"b"主题，蓝色）：

```
<div data-role="header" data-theme="b">
    <h1>Page Title</h1>
</div>
```

设置工具栏中的 Button 主题：任何工具栏里的 Button 都会自动被设置为和工具栏一样的主题色，要额外设置也请使用 data-theme 属性。例如，为 header 设置的是"c"（亮灰）主题，则 Button 也会变成"c"主题，要设置其他主题，请为 Button 的"a"标签设置额外的 data-theme 属性：

```
<a href="add-user.php" data-theme="b">Save</a>
```

13.5.7　jQuery Mobile 按钮

Button 是 jQuery Mobile 中的核心部件，在其他插件中也被广泛使用。按钮在用作导航时应该被编码成"a"标记的链接，而提交表单的按钮则被编码成 Button 元素，它们会被框架提供相同的样式。

1. 为链接应用 Button 样式

在一个页面的的主要内容区域，可以把任何"a"标记都转变成 Button 样式，只需要添加 data-role="button"属性即可自动添加所有必须的样式 class 来把"a"标记都转变成 Button 样式。例如：

```
<a href="index.html" data-role="button">Link button</a>
```

2. 链接按钮表单按钮

为了简化编写样式代码，框架自动把任何带有 type="submit"/"reset"/"button"/"image" 的 Button 元素或者 input 元素转换为基于链接的 Button 样式，不需要添加 data-role="button" 属性。

基于表单（form-based）的按钮的原始按钮（input）是隐藏的，但是依然保留其标记。当一个按钮的点击事件触发时，也会在原始的表单按钮上触发点击事件。

默认情况下，在正文内容所有按钮都称为块级元素，所以他们会充满整个屏幕的宽度。

但是，如果想要一个更紧凑的按钮，让其只和它包含的文字和图标的宽度相适应，添加 data-inline="true" 即可。

Button 如果有多个按钮，并想使它们排成一排，那么包含按钮的容器也需要有一个 data-inline="true" 属性。这会使按钮排成一排并向左浮动：

```
<div data-inline="true">
    <a href="index.html" data-role="button">Cancel</a>
    <a href="index.html" data-role="button" data-theme="b">Save</a>
</div>
```

Cancel Save 如果想要按钮排成一排的同时也充满整个屏幕，可以使用内容网格系统来把普通按钮排列成 2 或 3 列。

有时候，可以把一组按钮设置在一起，形成一个单独的块，看起来像导航组件。要获得这种效果，把这组按钮包裹在一个含有 data-role="controlgroup" 属性的容器里，框架会创建一个垂直排列的按钮组，并去掉它们之间所有的填白和阴影，并只给第一个和最后一个按钮添加圆角：

```
<div data-role="controlgroup">
    <a href="index.html" data-role="button">Yes</a>
    <a href="index.html" data-role="button">No</a>
    <a href="index.html" data-role="button">Maybe</a>
</div>
```

通过为 controlgroup 容器添加 data-type="horizontal" 属性使得分组按钮能水平排布，如果对于屏幕来说它们的宽度太宽了，它们则会换行。

13.5.8　jQuery Mobile 列表视图

列表视图是 jQuery Mobile 中功能强大的一个特性。它会使标准的无序或有序列表应用更广泛。应用方法就是在 ul 或 ol 标签中添加 data-role="listview" 属性。

下面的一些情景将会用到创建列表视图：

（1）简单的文件列表，会有一个好看的盒环绕着每一个列表项。

（2）链接列表，框架会自动为每一个链接加一个箭头">"，显示在链接按钮的右侧。

（3）嵌套列表，如果在一个 li 中嵌套另一个 ul，jQuery Mobile 会为这个嵌套列表自动建立一个"page"，并为它的父 li 自动加一个链接，这样很容易实现树状菜单选项，设置功能等。

（4）分隔线的按钮列表，在一个 li 中存放两个链接，可以建立一个垂直分隔条，用户可点击左侧或右侧的列表选项，展现不同的内容。

（5）记数气泡，如果在列表选项中添加 class="ui-li-count" 属性，框架会在其中生成一个"小泡泡"图标显现于列表选项的右侧,并在"小泡泡"中显示一些内容。类似在收信箱中看到已经收到的信息条数。

(6) 查找过滤,在 ul 或 ol 中添加 data‑filter="true"属性,则这个列表项就具备查询的功能,"Filter result…"文本框将会显示在列表项的上面,允许用户根据条件来将一个大的列表项变小(过滤显示)。

(7) 列表分隔,将列表项分割,可以在任意列表项上添加属性 data‑role="list‑divider"。

(8) 列表缩略图和图标。将 img 元素放在在列表项的开始,jQuery Mobile 将会以缩略图的形式来展现,图片的大小为 80 像素×80 像素。如果添加 class="ui‑li‑icon"类样式 img 元素的大小将会以 16 像素×16 像素的图标展现。

1. 创建简单的列表

列表是在移动网站上能看到的一个常用元素。利用 jQuery Mobile 可以创建多种不同的列表格式,如基本链接列表、嵌套列表、编号列表、拆分按钮列表、带分隔符的列表、带图标的列表、缩略图或计数泡泡,以及包括搜索筛选栏的列表。常用的列表类型是基本链接列表。要创建一个基本链接列表,只需要创建一个标准的 HTML 无序列表,添加一个 data‑role 属性,并为它分配一个值 listview。

使用 jQuery Mobile 框架创建基本链接列表:

```
<ul data-role="listview">
  <li><a href="#">List item 1</a></li>
  <li><a href="#">List item 2</a></li>
</ul>
```

要创建一个编号列表,可以使用相同的代码,只需将 ul 修改为 ol,以将它转换为一个有序列表。

在 ul, ol 中再次嵌入 ul, ol 可以生成嵌套列表,例如下面代码所生成的即为嵌套列表:

```
<ul data-role="listview">
    <li>老师
        <ul>
            <li><a href="#home">老师 A</a></li>
            <li><a href="#home">老师 B</a></li>
        </ul>
    </li>
    <li>学生
        <ul>
            <li><a href="#home">学生 A</a></li>
            <li><a href="#home">学生 B</a></li>
        </ul>
    </li>
</ul>
```

jQuery Mobile 会以最高级的列表项内容生成列表,单击某列表项后会生成一个新的页面,该页面以被点击项的文字内容生成一个 header ,并显示子列表内容。

2. 增强列表

对列表提供更多功能的一个选项称为拆分按钮列表。拆分按钮列表能够在同一个列表项中提供两个可单击的选项。该功能对列表项很有用,例如,列表项包含有关该特定项的特定详细信息的一个链接,但可能还需要包含与该项有关的其他操作,例如,用于购买该项或将它分享到社交网络上的一个按钮。创建一个拆分按钮列表很简单:在使用 listview data-role 的一个列表项中添加两个彼此相邻的定位点标记。

使用 jQuery Mobile 框架创建拆分按钮列表:

```html
<ul data-role = "listview" data-split-icon = "gear">
<li>
    <a href = "item-detail.html">
    <h3>Item title </h3>
    <p>Item overview </p>
    </a>
    <a href = "item-purchase.html" data-rel = "dialog">Purchase item </a>
</li>
<li>
    <a href = "item-detail.html">
    <h3>Item title </h3>
    <p>Item overview </p>
    </a>
    <a href = "item-purchase.html" data-rel = "dialog">Purchase item </a>
</li>
</ul>
```

程序中的拆分按钮列表所提供的列表项之中包括一个标题和一个概述,用户可以单击它查看有关该项的更多详细信息。该列表项还包括一个用作在对话框中购买该列表项的一个超链接的图标。也可以使用 data-split-icon 属性,修改显示在列表项右侧的拆分按钮的默认图标。

另一个有用的基本列表增强是列表分隔符。列表分隔符提供一种对列表项进行分类的方式。例如,可以用字母标记的列表项,并使用列表分隔符按字母表上的每个字母来分隔它们,或者可能有一组与音乐相关的列表项,可以用列表分隔符将它们对应不同音乐流派进行分类。下面的程序显示了利用列表分隔符增强的一个简单列表示例。

将列表分隔符添加到 listview:

```html
<ul data-role = "listview">
    <li data-role = "list-divider">Alternative </li>
    <li><a href = "#">Nirvana </a></li>
```

```
    <li data-role="list-divider">Rock and Roll </li>
    <li> <a href="#">Jimi Hendrix </a> </li>
    <li> <a href="#">Led Zeppelin </a> </li>
</ul>
```

使用 data-role 属性值 list-divider，使这些列表项与其他列表项具有不同的视觉样式。

也可以通过使用图标、缩略图和计数泡泡来创建不同的视觉样式。可以通过使用 ul-li-count 类，将计数泡泡添加到一个列表项。

将计数泡泡添加到 jQuery Mobile 列表项：

```
<ul data-role="listview">
  <li>
    <a href="inbox.html"> Inbox
      <span class="ui-li-count">12 </span>
    </a>
  </li>
</ul>
```

就像将一个图片添加到 HTML 页面中一样，也可以添加一个缩略图。只需要将一个定位点元素添加到列表项，添加一个用作缩略图的图片，然后添加希望在它旁边显示的副本。jQuery Mobile 就会处理剩下的工作。

将缩略图添加到 jQuery Mobile 列表项：

```
<ul data-role="listview">
  <li>
    <a href="zeppelin.html#thank-you">
      <img src="images/album-cover.jpg" />
      <h3>Led Zeppelin </h3>
      <p>Thank You </p>
    </a>
  </li>
  <li>
    <a href="zeppelin.html#ten-years">
      <img src="images/album-cover.jpg" />
      <h3>Led Zeppelin </h3>
      <p>Ten Years </p>
    </a>
  </li>
</ul>
```

可以使用与添加缩略图一样的方法来添加图标；唯一的区别是要使用 ui-li-icon 类。将图标添加到 jQuery Mobile 列表项：

```
<ul data-role = "listview">
  <li>
    <a href = "zeppelin.html">
      <img src = "images/album-cover.jpg" class = "ui-li-icon" />
      Led Zeppelin Album
    </a>
  </li>
  <li>
    <a href = "hendrix.html">
      <img src = "images/album-cover.jpg" class = "ui-li-icon" />
      Jimi Hendrix Album
    </a>
  </li>
</ul>
```

ul-li-icon 类限制图片的大小,最大宽度和高度为 40 px,它还能够将图片放在列表项中的适当位置。

13.6 跨平台移动软件的实现

本小节将介绍如何实现一个简单的跨平台移动软件,采用 JQuery_Mobile 编程方式,实现前面所介绍的物联网智能软件系统界面,完成之后,再按本章 13.4.3 小节(用 PhoneGap 开发 IOS 应用程序),就可以移植到苹果手机上运行。

采用 Obtian_Studio 里的"android 项目\JQuery_Mobile 模板"模板,创建一个名为"jQuery_Mobile_001"的新项目,如图 13-6 所示。

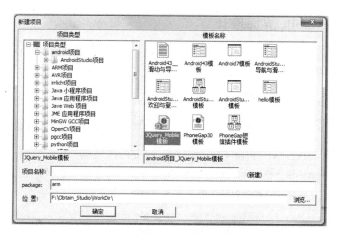

图 13-6 创建 jQuery_Mobile 模板项目

1. 实现类

实现类派生于 DroidGap 类。Cordova 提供了一个 Class(DroidGap extends CordovaActivity)和一个 interface(CordovaInterface)来让 Android 开发者开发 Cordova。一般情况下实现 DroidGap 即可,因为 DroidGap 类已经做了很多准备工作,可以说 DroidGap 类是 Cordova 框架的一个重要部分;如果在必要的情况下实现 CordovaInterface 接口,那么这个类中很多 DroidGap 的功能需要自己去实现。继承了 DroidGap 或者 CordovaInterface 的 Activity 就是一个独立的 Cordova 模块,独立的 Cordova 模块指的是每个实现了 DroidGap 或者 CordovaInterface 接口的 Activity 都对应一套独立的 WebView、Plugin、PluginManager,没有共享的。在初始化完 CordovaWebView 后调用 CordovaWebView.loadUrl()。此时完成 Cordova 的启动。程序代码如下:

```java
package com.android.hello;
import android.os.Bundle;
import org.apache.cordova.*;
public class hello extends DroidGap
{
    @Override
    public void onCreate(Bundle savedInstanceState)
    {
        super.onCreate(savedInstanceState);
        super.loadUrl("file:///android_asset/www/index.html");
    }
}
```

2. jQuery Mobile 界面设计

根据本章前面图 13-1 所示的要求进行设计,主要采用 jQuery Mobile 库。

(1) 导航栏的设计

```html
<div data-role="footer" data-theme="i" data-position="fixed">
    <div class="ui-grid-c" width="100%">
        <div class="ui-block-a">
            <div> <a href="javascript:home();" data-transition="slidedown"> <img src="images/message_selected.png" width="50%" height="5%" /> </a> </div>
                <div>  地点 </div>
        </div>
        <div class="ui-block-b">
            <div> <a href="javascript:Restaurant();" data-transition="slidedown"> <img src="images/contacts_unselected.png" width="50%" height="5%" /> </a> </div>
```

```
            <div>  控制</div>
        </div>
          <div class="ui-block-c">
            <div>   <a href="javascript:help();" data-transition="slidedown"><img src="images/news_unselected.png" width="50%" height="5%" /></a></div><div>  动态</div>
        </div>
          <div class="ui-block-d">
            <div>   <a href="javascript:syscolse();" data-transition="slidedown"><img src="images/setting_unselected.png" width="50%" height="5%" /></a></div><div>  设置</div>
        </div>
    </div>
</div><!-- /footer -->
```

(2) 主界面顶部的设计

```
    <div data-role="header" data-theme="i"  data-position="fixed">
      <table style="text-align: center; font-size: x-large; background-color: #9999ff; color: #FFFFFF; height: 17px; width: 100%;">
          <tr><td  style="width:300px"><strong>物联网智能系统</strong></td>
            <td></td>
            <td><input  onclick="myRegister();" id="Register" type="button" value="+" style="border-color: #FF9999; text-align: center; font-size:36px; background-color: #C09999; color: #FFFFFF;" />
          </td>
        </tr>
      </table>
    </div><!-- /header -->
```

(3) 地点列表的设计

```
    <div data-role="content"  data-position="fixed">
    <ul data-role="listview" class="listview">
    <li><a href="zeppelin.html#thank-you"><img src="images/alivingroom.png" />
      <h3>客厅</h3><p>103.204.177.76</p>
    </a></li>
    <li><a href="zeppelin.html#thank-you"><img src="images/bedroom.png" />
      <h3>主卧</h3><p>103.204.177.76</p>
    </a></li>
    <li><a href="zeppelin.html#thank-you"><img src="images/bedroom.png" />
      <h3>客房</h3>   <p>103.204.177.76</p>
    </a></li>
```

```
        <li> <a href = "zeppelin.html#thank-you"> <img src = "images/study.png" />
           <h3>书房</h3>   <p>103.204.177.76 </p>
        </a></li>
        <li> <a href = "zeppelin.html#thank-you"> <img src = "images/kitchen.png" />
           <h3>厨房</h3>  <p>103.204.177.76 </p>
        </a></li>
        <li> <a href = "zeppelin.html#thank-you"> <img src = "images/office.png" />
           <h3>办公室</h3> <p>103.204.177.76 </p>
        </a></li>
     </ul>
   </div>
```

从上面的程序代码可以看出,采用 jQuery Mobile 做的界面设计,程序代码要比安卓采用 XML 代码要简洁很多。本章最开始介绍的图 13-1 是运行于手机上的效果图,下面将比较运行于模拟器上以及运行于浏览器上的效果图,如图 13-7 所示。从图 13-7 可以看出,手机上 App 运行效果、模拟器上 App 运行效果以及电脑上浏览器直接打开 index.html 外观基本上相同,这有利于 jQuery Mobile 在底层对不同设备和不同浏览器兼容性的良好支持。

(a) 手机上App运行效果图　　(b) 模拟器上App运行效果图　　(c) 电脑上浏览器打开效果图

图 13-7　运行于手机上、模拟器上以及浏览器上的效果图

参考文献

[1] 王辰龙. 高级 Android 开发强化实战[M]. 北京:电子工业出版社,2018.

[2] 李明亮. Arduino 开发从入门到实战[M]. 北京:清华大学出版社,2018.

[3] (西班牙)恩里克·洛佩斯·马尼亚斯. Android 高性能编程[M]. 北京:电子工业出版社,2018.

[4] Roberto,(巴西)Ierusalimschy. Android 编程权威指南[M]. 2 版. 北京:电子工业出版社,2017.

[5] 朱元波. Android 传感器开发与智能设备案例实战[M]. 北京:人民邮电出版社,2017.

[6] 杨丰盛. Android 应用开发揭秘[M]. 北京:机械工业出版社,2010.

[7] 李宁. Android 开发权威指南[M]. 北京:人民邮电出版社,2011.

[8] 孙更新,邵长恒,宾晟. Android 从入门到精通[M]. 北京:电子工业出版社,2011.

[9] 余志龙. Android SDK 开发范例大全[M]. 北京:人民邮电出版社,2010.

[10] 郑昊,钟志峰,郭昊,许骏. 基于 Arduino/Android 的蓝牙通信系统设计[J]. 物联网技术,2012(05).

[11] 廖义奎. Cortex-A9 多核嵌入式系统设计[M]. 北京:中国电力出版社,2014.

[12] 廖义奎. Cortex-M3 之 STM32 嵌入式系统设计[M]. 北京:中国电力出版社,2012.

[13] 李刚. 疯狂 Android 讲义[M]. 北京:电子工业出版社,2010.

[14] (美)莱文,郑恩遥,房佩慈. 深入解析 Mac OS X & iOS 操作系统[M]. 北京:清华大学出版,2014.

[15] (美)霍尔茨纳. Objective——C2.0 编程快速上手[M]. 刘红伟,等译. 北京:机械工业出版,2010.

[16] (美) Gary Bennett,(美) Mitch Fisher,(美) Brad Lees. Objective——C 初学者指南[M]. 2 版. 王雷,译. 北京:人民邮电出版,2012.

[17] (美) Jesse Feiler. Objective——C 入门经典[M]. 陈昕昕,郭光伟,译. 北京:人民邮电出版,2013.

[18] 唐俊开. HTML5 移动 web 开发指南[M]. 北京:电子工业出版社,2012.

[19] 袁江. jquery 开发从入门到精通[M]. 北京:清华大学出版社,2013.

[20] 陶国荣.jquery mobile 权威指南[M].北京机械工业出版社,2012.

[21] APP 发展大势:这里也有"四化建设". http://www.weihom.com/index/detail/id/81.html.

[22] CES 2018 已发布新品大荟萃:一文感受尖端科技. https://www.ithome.com/html/it/342097.htm.

[23] Android 开发之旅:环境搭建及 HelloWorld. https://www.cnblogs.com/skynet/archive/2010/04/12/1709892.html.

[24] Android Studio 的安装与配置. https://blog.csdn.net/DAI_JICHEN/article/details/80782032.

[25] Android 研究院之应用程序界面五大布局. https://www.xuanyusong.com/archives/133.

[26] Android 样式之 shape 标签. https://blog.csdn.net/qq_36408196/article/details/82779247.

[27] Android xml 中 layout_weight 属性的工作原理. https://blog.csdn.net/gaugamela/article/details/56276516.

[28] Android XML shape 标签使用详解. https://www.cnblogs.com/popfisher/p/6238119.html.

[29] Android 之 Adapter 用法总结. https://blog.csdn.net/fznpcy/article/details/8658155/.

[30] Android 开发教程之 Adapter 用法总结. https://www.2cto.com/kf/201711/698782.html.

[31] 理解 Recyclerview 的使用. https://blog.csdn.net/u012124438/article/details/53495951.

[32] ViewPager 详解. https://blog.csdn.net/harvic880925/article/details/38453725.

[33] ViewPager 实现微信主界面. https://www.cnblogs.com/dadafeige/p/5370915.html.

[34] Fragment 简单使用. https://blog.csdn.net/bobo8945510/article/details/52790296.

[35] Android 使用 BottomNavigationView 实现底部导航栏. https://blog.csdn.net/afanbaby/article/details/79240620.

[36] 底部导航. https://blog.csdn.net/u011652925/article/details/78656397.

[37] BottomNavigationView + ViewPager + Fragment 实现左右滑动和下方导航栏. https://blog.csdn.net/htwhtw123/article/details/78441431?locationNum=4&fps=1.

[38] ESP8266 指令集汇总. http://blog.sina.com.cn/s/blog_b0c011190102w8wt.html.

[39] 蓝牙 4.0 和 3.0 的区别. https://www.kafan.cn/edu/80582686.html.

[40] Android4.3 蓝牙 BLE 初步. https://www.cnblogs.com/savagemorgan/p/3722657.html.

参考文献

[41] CC2540 AT 指令手册. https://wenku.baidu.com/view/da9a1d06e53a580217fcfe0a.html.

[42] Android 文件读写存储. https://www.cnblogs.com/CharlesGrant/p/8310507.html.

[43] Android 操作 SQLite 基本用法. https://www.cnblogs.com/foxy/p/7725010.html.

[44] Android SQLite 基本用法. https://blog.csdn.net/liu943367080/article/details/60873760.

[45] Android sqlite 操作. https://yq.aliyun.com/articles/12565.

[46] OkHttp3 的基本用法. https://www.jianshu.com/p/1873287eed87.

[47] Android WebView 基本使用. https://blog.csdn.net/lowprofile_coding/article/details/77928614.

[48] WebView 简书. https://www.jianshu.com/p/338bc827a74a.

[49] Android WebView 与 JS 交互全面详解. https://www.jb51.net/article/129222.htm.

[50] Android 传感器(OnSensorChanged)使用介绍. https://blog.csdn.net/crlqzg/article/details/51566136.

[51] 如何使用 TensorFlow API 构建视频物体识别系统. https://www.jiqizhixin.com/articles/2017-07-14-5.

[52] android Camera2 API 使用详解. https://www.cnblogs.com/lonelyxmas/p/9175746.html.

[53] Camera2 API 简介. https://www.cnblogs.com/lonelyxmas/p/9175768.html.

[54] Android 之各类传感器使用详解. https://blog.csdn.net/feishangbeijixing/article/details/44316387.

[55] MediaStore 类使用 Intent 录制音频和拍照. https://blog.csdn.net/tan313/article/details/46610417.